WITHDRAWN
UTSA LIBRARIES

RENEWALS 458-4574

Embedded Signal Processing with the Micro Signal Architecture

THE WILEY BICENTENNIAL–KNOWLEDGE FOR GENERATIONS

Each generation has its unique needs and aspirations. When Charles Wiley first opened his small printing shop in lower Manhattan in 1807, it was a generation of boundless potential searching for an identity. And we were there, helping to define a new American literary tradition. Over half a century later, in the midst of the Second Industrial Revolution, it was a generation focused on building the future. Once again, we were there, supplying the critical scientific, technical, and engineering knowledge that helped frame the world. Throughout the 20th Century, and into the new millennium, nations began to reach out beyond their own borders and a new international community was born. Wiley was there, expanding its operations around the world to enable a global exchange of ideas, opinions, and know-how.

For 200 years, Wiley has been an integral part of each generation's journey, enabling the flow of information and understanding necessary to meet their needs and fulfill their aspirations. Today, bold new technologies are changing the way we live and learn. Wiley will be there, providing you the must-have knowledge you need to imagine new worlds, new possibilities, and new opportunities.

Generations come and go, but you can always count on Wiley to provide you the knowledge you need, when and where you need it!

WILLIAM J. PESCE
PRESIDENT AND CHIEF EXECUTIVE OFFICER

PETER BOOTH WILEY
CHAIRMAN OF THE BOARD

Embedded Signal Processing with the Micro Signal Architecture

Woon-Seng Gan
Sen M. Kuo

IEEE PRESS

WILEY-INTERSCIENCE
A John Wiley & Sons, Inc., Publication

Copyright © 2007 by John Wiley & Sons, Inc. All rights reserved

Published by John Wiley & Sons, Inc., Hoboken, New Jersey
Published simultaneously in Canada

No part of this publication may be reproduced, stored in a retrieval system, or transmitted in any form or by any means, electronic, mechanical, photocopying, recording, scanning, or otherwise, except as permitted under Section 107 or 108 of the 1976 United States Copyright Act, without either the prior written permission of the Publisher, or authorization through payment of the appropriate per-copy fee to the Copyright Clearance Center, Inc., 222 Rosewood Drive, Danvers, MA 01923, (978) 750-8400, fax (978) 750-4470, or on the web at www.copyright.com. Requests to the Publisher for permission should be addressed to the Permissions Department, John Wiley & Sons, Inc., 111 River Street, Hoboken, NJ 07030, (201) 748-6011, fax (201) 748-6008, or online at http://www.wiley.com/go/permission.

Limit of Liability/Disclaimer of Warranty: While the publisher and author have used their best efforts in preparing this book, they make no representations or warranties with respect to the accuracy or completeness of the contents of this book and specifically disclaim any implied warranties of merchantability or fitness for a particular purpose. No warranty may be created or extended by sales representatives or written sales materials. The advice and strategies contained herein may not be suitable for your situation. You should consult with a professional where appropriate. Neither the publisher nor author shall be liable for any loss of profit or any other commercial damages, including but not limited to special, incidental, consequential, or other damages.

For general information on our other products and services or for technical support, please contact our Customer Care Department within the United States at (800) 762-2974, outside the United States at (317) 572-3993 or fax (317) 572-4002.

Wiley also publishes its books in a variety of electronic formats. Some content that appears in print may not be available in electronic formats. For more information about Wiley products, visit our web site at www.wiley.com.

Library of Congress Cataloging-in-Publication Data

Gan, Woon-Seng.
  Embedded signal processing with the Micro Signal Architecture / by Woon-Seng Gan and Sen M. Kuo.
      p. cm.
  Includes bibliographical references and index.
  ISBN: 978-0-471-73841-1
1. Signal processing—Digital techniques. 2. Embedded computer systems—Programming.
3. Computer architecture. I. Kuo, Sen M. (Sen-Maw) II. Title.
  TK5102.9.G364 2007
  621.382′2 – dc22

2006049693

Printed in the United States of America

10 9 8 7 6 5 4 3 2 1

Contents

Preface xi

Acknowledgments xvii

About the Authors xix

1. Introduction 1

    1.1   Embedded Processor: Micro Signal Architecture   1
    1.2   Real-Time Embedded Signal Processing   6
    1.3   Introduction to the Integrated Development Environment VisualDSP++   7

        1.3.1   Setting Up VisualDSP++   8
        1.3.2   Using a Simple Program to Illustrate the Basic Tools   9
        1.3.3   Advanced Setup: Using the Blackfin BF533 or BF537 EZ-KIT   12

    1.4   More Hands-On Experiments   15
    1.5   System-Level Design Using a Graphical Development Environment   18

        1.5.1   Setting up LabVIEW and LabVIEW Embedded Module for Blackfin Processors   19

    1.6   More Exercise Problems   21

Part A   Digital Signal Processing Concepts

2. Time-Domain Signals and Systems 25

    2.1   Introduction   25
    2.2   Time-Domain Digital Signals   26

        2.2.1   Sinusoidal Signals   26
        2.2.2   Random Signals   28

    2.3   Introduction to Digital Systems   33

        2.3.1   Moving-Average Filters: Structures and Equations   34
        2.3.2   Digital Filters   37
        2.3.3   Realization of FIR Filters   41

	2.4	Nonlinear Filters 45
	2.5	More Hands-On Experiments 47
	2.6	Implementation of Moving-Average Filters with Blackfin Simulator 50
	2.7	Implementation of Moving-Average Filters with BF533/BF537 EZ-KIT 52
	2.8	Moving-Average Filter in LabVIEW Embedded Module for Blackfin Processors 54
	2.9	More Exercise Problems 57

3. Frequency-Domain Analysis and Processing   59

- 3.1 Introduction   59
- 3.2 The z-Transform   60
  - 3.2.1 Definitions   60
  - 3.2.2 System Concepts   62
  - 3.2.3 Digital Filters   64
- 3.3 Frequency Analysis   70
  - 3.3.1 Frequency Response   70
  - 3.3.2 Discrete Fourier Transform   76
  - 3.3.3 Fast Fourier Transform   78
  - 3.3.4 Window Functions   83
- 3.4 More Hands-On Experiments   88
  - 3.4.1 Simple Low-Pass Filters   88
  - 3.4.2 Design and Applications of Notch Filters   91
  - 3.4.3 Design and Applications of Peak Filters   96
- 3.5 Frequency Analysis with Blackfin Simulator   98
- 3.6 Frequency Analysis with Blackfin BF533/BF537 EZ-KIT   102
- 3.7 Frequency Analysis with LabVIEW Embedded Module for Blackfin Processors   105
- 3.8 More Exercise Problems   110

4. Digital Filtering   112

- 4.1 Introduction   112
  - 4.1.1 Ideal Filters   113
  - 4.1.2 Practical Filter Specifications   115
- 4.2 Finite Impulse Response Filters   120
  - 4.2.1 Characteristics and Implementation of FIR Filters   121
  - 4.2.2 Design of FIR Filters   123
  - 4.2.3 Hands-On Experiments   126
- 4.3 Infinite Impulse Response Filters   129
  - 4.3.1 Design of IIR Filters   129
  - 4.3.2 Structures and Characteristics of IIR Filters   133
  - 4.3.3 Hands-On Experiments   136

Contents vii

4.4 Adaptive Filters   139

    4.4.1 Structures and Algorithms of Adaptive Filters   139
    4.4.2 Design and Applications of Adaptive Filters   142
    4.4.3 More Hands-On Experiments   148

4.5 Adaptive Line Enhancer Using Blackfin Simulator   151
4.6 Adaptive Line Enhancer Using Blackfin BF533/BF537 EZ-KIT   152
4.7 Adaptive Line Enhancer Using LabVIEW Embedded Module for Blackfin Processors   155
4.8 More Exercise Problems   158

Part B Embedded Signal Processing Systems and Concepts

5. Introduction to the Blackfin Processor   163

5.1 The Blackfin Processor: An Architecture for Embedded Media Processing   163

    5.1.1 Introduction to Micro Signal Architecture   163
    5.1.2 Overview of the Blackfin Processor   164
    5.1.3 Architecture: Hardware Processing Units and Register Files   165
    5.1.4 Bus Architecture and Memory   182
    5.1.5 Basic Peripherals   187

5.2 Software Tools for the Blackfin Processor   189

    5.2.1 Software Development Flow and Tools   189
    5.2.2 Assembly Programming in VisualDSP++   191
    5.2.3 More Explanation of Linker   195
    5.2.4 More Debugging Features   198

5.3 Introduction to the FIR Filter-Based Graphic Equalizer   200
5.4 Design of Graphic Equalizer Using Blackfin Simulator   202
5.5 Implementation of Graphic Equalizer Using BF533/BF537 EZ-KIT   206
5.6 Implementation of Graphic Equalizer Using LabVIEW Embedded Module for Blackfin Processors   211
5.7 More Exercise Problems   214

6. Real-Time DSP Fundamentals and Implementation Considerations   217

6.1 Number Formats Used in the Blackfin Processor   217

    6.1.1 Fixed-Point Formats   217
    6.1.2 Fixed-Point Extended Format   229
    6.1.3 Fixed-Point Data Types   231
    6.1.4 Emulation of Floating-Point Format   231
    6.1.5 Block Floating-Point Format   235

viii Contents

 6.2 Dynamic Range, Precision, and Quantization Errors 236

  6.2.1 Incoming Analog Signal and Quantization 236
  6.2.2 Dynamic Range, Signal-to-Quantization Noise Ratio, and Precision 238
  6.2.3 Sources of Quantization Errors in Digital Systems 240

 6.3 Overview of Real-Time Processing 250

  6.3.1 Real-Time Versus Offline Processing 250
  6.3.2 Sample-by-Sample Processing Mode and Its Real-Time Constraints 251
  6.3.3 Block Processing Mode and Its Real-Time Constraints 252
  6.3.4 Performance Parameters for Real-Time Implementation 255

 6.4 Introduction to the IIR Filter-Based Graphic Equalizer 260
 6.5 Design of IIR Filter-Based Graphic Equalizer Using Blackfin Simulator 261
 6.6 Design of IIR Filter-Based Graphic Equalizer with BF533/BF537 EZ-KIT 266
 6.7 Implementation of IIR Filter-Based Graphic Equalizer with LabVIEW Embedded Module for Blackfin Processors 266
 6.8 More Exercise Problems 270

7. Memory System and Data Transfer 274

 7.1 Overview of Signal Acquisition and Transfer to Memory 274

  7.1.1 Understanding the CODEC 274
  7.1.2 Connecting AD1836A to BF533 Processor 276
  7.1.3 Understanding the Serial Port 282

 7.2 DMA Operations and Programming 287

  7.2.1 DMA Transfer Configuration 289
  7.2.2 Setting Up the Autobuffer-Mode DMA 291
  7.2.3 Memory DMA Transfer 297
  7.2.4 Setting Up Memory DMA 298
  7.2.5 Examples of Using Memory DMA 298
  7.2.6 Advanced Features of DMA 302

 7.3 Using Cache in the Blackfin Processor 303

  7.3.1 Cache Memory Concepts 303
  7.3.2 Terminology in Cache Memory 305
  7.3.3 Instruction Cache 307
  7.3.4 Data Cache 310
  7.3.5 Memory Management Unit 313

 7.4 Comparing and Choosing Between Cache and Memory DMA 315
 7.5 Scratchpad Memory of Blackfin Processor 317

- 7.6 Signal Generator Using Blackfin Simulator 317
- 7.7 Signal Generator Using BF533/BF537 EZ-KIT 319
- 7.8 Signal Generation with LabVIEW Embedded Module for Blackfin Processors 321
- 7.9 More Exercise Problems 326

8. Code Optimization and Power Management 330

- 8.1 Code Optimization 330
- 8.2 C Optimization Techniques 331
  - 8.2.1 C Compiler in VisualDSP++ 332
  - 8.2.2 C Programming Considerations 333
  - 8.2.3 Using Intrinsics 339
  - 8.2.4 Inlining 342
  - 8.2.5 C/C++ Run Time Library 343
  - 8.2.6 DSP Run Time Library 343
  - 8.2.7 Profile-Guided Optimization 346
- 8.3 Using Assembly Code for Efficient Programming 349
  - 8.3.1 Using Hardware Loops 352
  - 8.3.2 Using Dual MACs 353
  - 8.3.3 Using Parallel Instructions 353
  - 8.3.4 Special Addressing Modes: Separate Data Sections 355
  - 8.3.5 Using Software Pipelining 356
  - 8.3.6 Summary of Execution Cycle Count and Code Size for FIR Filter Implementation 357
- 8.4 Power Consumption and Management in the Blackfin Processor 358
  - 8.4.1 Computing System Power in the Blackfin Processor 358
  - 8.4.2 Power Management in the Blackfin Processor 360
- 8.5 Sample Rate Conversion with Blackfin Simulator 365
- 8.6 Sample Rate Conversion with BF533/BF537 EZ-KIT 369
- 8.7 Sample Rate Conversion with LabVIEW Embedded Module for Blackfin Processors 371
- 8.8 More Exercise Problems 374

Part C   Real-World Applications

9. Practical DSP Applications: Audio Coding and Audio Effects 381

- 9.1 Overview of Audio Compression 381
- 9.2 MP3/Ogg Vorbis Audio Encoding 386
- 9.3 MP3/Ogg Vorbis Audio Decoding 390

x Contents

9.4 Implementation of Ogg Vorbis Decoder with BF537 EZ-KIT   391
9.5 Audio Effects   393

    9.5.1 3D Audio Effects   393
    9.5.2 Implementation of 3D Audio Effects with BF533/BF537 EZ-KIT   396
    9.5.3 Generating Reverberation Effects   398
    9.5.4 Implementation of Reverberation with BF533/BF537 EZ-KIT   399

9.6 Implementation of MDCT with LabVIEW Embedded Module for Blackfin Processors   400
9.7 More Application Projects   404

10. Practical DSP Applications: Digital Image Processing   406

10.1 Overview of Image Representation   406
10.2 Image Processing with BF533/BF537 EZ-KIT   409
10.3 Color Conversion   410
10.4 Color Conversion with BF533/BF537 EZ-KIT   412
10.5 Two-Dimensional Discrete Cosine Transform   413
10.6 Two-Dimensional DCT/IDCT with BF533/BF537 EZ-KIT   416
10.7 Two-Dimensional Filtering   417

    10.7.1 2D Filtering   418
    10.7.2 2D Filter Design   420

10.8 Two-Dimensional Filtering with BF533/BF537 EZ-KIT   422
10.9 Image Enhancement   422

    10.9.1 Gaussian White Noise and Linear Filtering   423
    10.9.2 Impulse Noise and Median Filtering   425
    10.9.3 Contrast Adjustment   428

10.10 Image Enhancement with BF533/BF537 EZ-KIT   432
10.11 Image Processing with LabVIEW Embedded Module for Blackfin Processors   433
10.12 More Application Projects   438

Appendix A: An Introduction to Graphical Programming with LabVIEW   441

Appendix B: Useful Websites   462

Appendix C: List of Files Used in Hands-On Experiments and Exercises   464

Appendix D: Updates of Experiments Using Visual DSP++ V4.5   473

References   475

Index   479

Preface

In this digital Internet age, information can be received, processed, stored, and transmitted in a fast, reliable, and efficient manner. This advancement is made possible by the latest fast, low-cost, and power-efficient embedded signal processors. Embedded signal processing is widely used in most digital devices and systems and has grown into a "must-have" category in embedded applications. There are many important topics related to embedded signal processing and control, and it is impossible to cover all of these subjects in a one- or two-semester course. However, the Internet is now becoming an effective platform in searching for new information, and this ubiquitous tool is enriching and speeding up the learning process in engineering education. Unfortunately, students have to cope with the problem of information overflow and be wise in extracting the right amount of material at the right time.

This book introduces just-in-time and just-enough information on embedded signal processing using the embedded processors based on the micro signal architecture (MSA). In particular, we examine the MSA-based processors called Blackfin processors from Analog Devices (ADI). We extract relevant and sufficient information from many resources, such as textbooks, electronic books, the ADI website, signal processing-related websites, and many journals and magazine articles related to these topics. The just-in-time organization of these selective topics provides a unique experience in learning digital signal processing (DSP). For example, students no longer need to learn advanced digital filter design theory before embarking on the actual design and implementation of filters for real-world applications. In this book, students learn just enough essential theory and start to use the latest tools to design, simulate, and implement the algorithms for a given application. If they need a more advanced algorithm to solve a more sophisticated problem, they are now more confident and ready to explore new techniques. This exploratory attitude is what we hope students will achieve through this book.

We use assembly programming to introduce the architecture of the embedded processor. This is because assembly code can give a more precise description of the processor's architecture and provide a better appreciation and control of the hardware. Without this understanding, it is difficult to program and optimize code using embedded signal processors for real-world applications. However, the use of C code as a main program that calls intrinsic and DSP library functions is still the preferred programming style for the Blackfin processor. It is important to think in low-level architecture but write in high-level code (C or graphical data flow). Therefore, we show how to balance high-level and low-level programming and introduce the techniques needed for optimization. In addition, we also introduce a very versatile graphical tool jointly developed by ADI and National Instruments (NI) that allows users to design, simulate, implement, and verify an embedded system with a high-level graphical data flow approach.

The progressive arrangement makes this book suitable for engineers. They may skip some topics they are already familiar with and focus on the sections they are interested in. The following subsections introduce the essential parts of this book and how these parts are linked together.

PART A: USING SOFTWARE TOOLS TO LEARN DSP—A JUST-IN-TIME AND PROJECT-ORIENTED APPROACH

In Chapters 2, 3, and 4, we explore fundamental DSP concepts using a set of software tools from the MathWorks, ADI, and NI. Rather than introducing all theoretical concepts at the beginning and doing exercises at the end of each chapter, we provide just enough information on the required concepts for solving the given problems and supplement with many quizzes, interactive examples, and hands-on exercises along the way in a just-in-time manner. Students learn the concepts by doing the assignments for better understanding. This approach is especially suitable for studying these subjects at different paces and times, thus making self-learning possible.

In addition to these hands-on exercises, the end of each chapter also provides challenging pen-and-paper and computer problems for homework assignments. These problem sets build upon the previous knowledge learned and extend the thinking to more advanced concepts. These exercises will motivate students in looking for different solutions for a given problem. The goal is to cultivate a learning habit after going through the book.

The theory portion of these chapters may be skipped for those who have taken a fundamental course on DSP. Nonetheless, these examples and hands-on exercises serve as a handy reference on learning important tools available in MATLAB, the integrated development environment VisualDSP++, and the LabVIEW Embedded Module for Blackfin Processors. These tools provide a platform to convert theoretical concepts into software code before learning the Blackfin processor in detail. The introduction to the latest LabVIEW Embedded Module for Blackfin Processors shows the advancement in rapid prototyping and testing of embedded system designs for real-world applications. This new tool provides exciting opportunities for new users to explore embedded signal processing before learning the programming details. Therefore, instructors can make use of these graphical experiments at the end of each chapter to teach embedded signal processing concepts in foundation engineering courses.

PART B: LEARNING REAL-TIME SIGNAL PROCESSING WITH THE BLACKFIN PROCESSOR—A BITE-SIZE APPROACH TO SAMPLING REAL-TIME EXAMPLES AND EXERCISES

Part B consists of Chapters 5, 6, 7, and 8, which concentrate on the design and implementation of embedded systems based on the Blackfin processor. Unlike a conventional user's manual that covers the processor's architecture, instruction set, and peripherals in detail, we introduce just enough relevant materials to get started on Blackfin-based projects. Many hands-on examples and exercises are designed in a step-by-step manner to guide users toward this goal. We take an integrated approach, starting from algorithm design using MATLAB with floating-point simulations to the fixed-point implementation on the Blackfin processor, and interfacing with external peripherals for building a stand-alone or portable device. Along this journey to final realization, many design and development tools are introduced to accomplish different tasks. In addition, we provide many hints and references and supplement with many challenging problems for students to explore more advanced topics and applications.

Part B is in fact bridging the gap from DSP concepts to real-time implementations on embedded processors, and providing a starting point for students to embark on real-time signal processing programming with a fixed-point embedded processor.

PART C: DESIGNING AND IMPLEMENTING REAL-TIME DSP ALGORITHMS AND APPLICATIONS—AN INTEGRATED APPROACH

The final part (Chapters 9 and 10) motivates users to take on more challenging real-time applications in audio signal processing and image processing. Students can use the knowledge and tools learned in the preceding chapters to complete the applications introduced in Chapters 9 and 10. Some guides in the form of basic concepts, block diagrams, sample code, and suggestions are provided to solve these application problems. We use a module approach in Part C to build the embedded system part by part, and also provide many opportunities for users to explore new algorithms and applications that are not covered in Parts A and B. These application examples also serve as good mini-projects for a hands-on design course on embedded signal processing. As in many engineering problems, there are many possible solutions. There are also many opportunities to make mistakes and learn valuable lessons. Users can explore the references and find a possible solution for solving these projects. In other words, we want the users to explore, learn, and have fun!

A summary of these three parts of the book is illustrated in Figure 1. It shows three components: (A) DSP concepts, (B) embedded processor architecture and real-time DSP considerations, and (C) real-life applications: a simple A-B-C

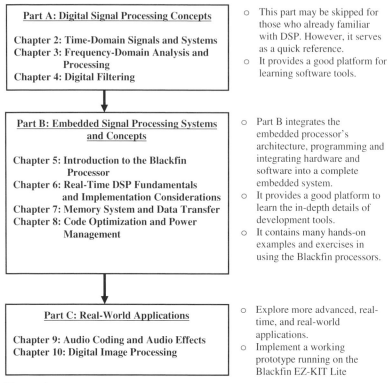

Figure 1 Summary of the book: contents and how to use them

approach to learning embedded signal processing with the micro signal architecture.

DESCRIPTION OF EXAMPLES, EXERCISES, EXPERIMENTS, PROBLEMS, AND APPLICATION PROJECTS

This book provides readers many opportunities to understand and explore the main contents of each chapter through examples, quizzes, exercises, hands-on experiments, exercise problems, and application projects. It also serves as a good hands-on workbook to learn different embedded software tools (MATLAB, VisualDSP++, and LabVIEW Embedded) and solve practical problems. These hands-on sections are classified under different types as follows.

1. **Examples** provide a just-in-time understanding of the concepts learned in the preceding sections. Examples contain working MATLAB problems to illustrate the concepts and how problems can be solved. The naming convention for software example files is

```
example{chapter number}_{example number}.m
```
They are normally found in the directory
```
c:\adsp\chap{x}\MATLAB_ex{x}\
```
where {x} is the chapter number.

2. **Quizzes** contain many short questions to challenge the readers for immediate feedback of understanding.

3. **Experiments** are mostly hands-on exercises to get familiar with the tools and solve more in-depth problems. These experiments usually use MATLAB, VisualDSP++, or LabVIEW Embedded. The naming convention for software experiment files is
```
exp{chapter number}_{example number}
```
These experiment files can be found in the directory
```
c:\adsp\chap{x}\exp{x}_{no.}_<option>
```
where {no.} indicates the experiment number and <option> indicates the BF533 or BF537 EZ-KIT.

4. **Exercises** further enhance the student's learning of the topics in the preceding sections, examples, and experiments. They also provide the opportunity to attempt more advanced problems to strengthen understanding.

5. **Exercise Problems** are located at the end of Chapters 1 to 8. These problem sets explore or extend more interesting and challenging problems and experiments.

6. **Application Projects** are provided at the end of Chapters 9 and 10 to serve as mini-projects. Students can work together as a team to solve these application-oriented projects and submit a report that indicates their approaches, algorithms, and simulations, and how they verify the projects with the Blackfin processor.

Most of the exercises and experiments require testing data. We provide two directories that contain audio and image data files. These files are located in the directories `c:\adsp\audio_files` and `c:\adsp\image_files`.

COMPANION WEBSITE

A website, www.ntu.edu.sg/home/ewsgan/esp_book.html, has been set up to support the book. This website contains many supplementary materials and useful reference links for each chapter. We also include a set of lecture slides with all the figures and tables in PowerPoint format. This website will also introduce new hands-on exercises and new design problems related to embedded signal processing. Because the fast-changing world of embedded processors, the software tools and the Blackfin processor will also undergo many changes as time evolves. The versions of software tools used in this book are:

- MATLAB Version 7.0
- VisualDSP++ Version 4.0
- LabVIEW 8.0
- LabVIEW Embedded Edition 7.1

This website keeps track of the latest changes and new features of these tools. It also reports on any compatibility problems when running existing experiments with the newer version of software.

All the programs mentioned in the exercises and experiments are available for download in the Wiley ftp site: ftp://ftp.wiley.com/public/sci_tech_med/embedded_signal/.

We also include a feedback section to hear your comments and suggestions. Alternatively, readers are encouraged to email us at ewsgan@ntu.edu.sg and kuo@ceet.niu.edu.

Learning Objective:

We learn by example and by direct experience because there are real limits to the adequacy of verbal instruction.

Malcolm Gladwell, *Blink: The Power of Thinking Without Thinking, 2005*

Acknowledgments

We are very grateful to many individuals for their assistance in developing this book. In particular, we are indebted to Todd Borkowski of Analog Devices, who got us started in this book project. Without his constant support and encouragement, we would not have come this far. We would like to thank Erik B. Luther, Jim Cahow, Mark R. Kaschner, and Matt Pollock of National Instruments for contributing to the experiments, examples, and writing in the LabVIEW Embedded Module for Blackfin Processors. Without their strong commitment and support, we would never have been able to complete many exciting demos within such a short time. The first author would like to thank Chandran Nair and Siew-Hoon Lim of National Instruments for providing their technical support and advice. We would also like to thank David Katz and Dan Ledger of Analog Devices for providing useful advice on Blackfin processors. Additionally, many thanks to Mike Eng, Dick Sweeney, Tonny Jiang, Li Chuan, Mimi Pichey, and Dianwei Sun of Analog Devices.

Several individuals at John Wiley also provided great help to make this book a reality. We wish to thank George J. Telecki, Associate Publisher, for his support of this book. Special thanks must go to Rachel Witmer, Editorial Program Coordinator, who promptly answered our questions. We would also like to thank Danielle Lacourciere and Dean Gonzalez at Wiley for the final preparation of this book.

The authors would also like to thank Dahyanto Harliono, who contributed to the majority of the Blackfin hands-on exercises in the entire book. Thanks also go to Furi Karnapi, Wei-Chee Ku, Ee-Leng Tan, Cyril Tan, and Tany Wijaya, who also contributed to some of the hands-on exercises and examples in the book. We would also like to thank Say-Cheng Ong for his help in testing the LabVIEW experiments.

This book is also dedicated to many of our past and present students who have taken our DSP courses and have written M.S. theses and Ph.D. dissertations and completed senior design projects under our guidance at both NTU and NIU. Both institutions have provided us with a stimulating environment for research and teaching, and we appreciate the strong encouragement and support we have received. Finally, we are greatly indebted to our parents and families for their understanding, patience, and encouragement throughout this period.

WOON-SENG GAN AND SEN M. KUO

About the Authors

Woon-Seng Gan is an Associate Professor in the Information Engineering Division in the School of Electrical and Electronic Engineering at the Nanyang Technological University in Singapore. He is the co-author of the textbook *Digital Signal Processors: Architectures, Implementations, and Applications* (Prentice Hall 2005). He has published more than 130 technical papers in peer-reviewed journals and international conferences. His research interests are embedded media processing, embedded systems, low-power-consumption algorithms, and real-time implementations.

Sen M. Kuo is a Professor and Chair at the Department of Electrical Engineering, Northern Illinois University, DeKalb, IL. In 1993, he was with Texas Instruments, Houston, TX. He is the leading author of four books: *Active Noise Control Systems* (Wiley, 1996), *Real-Time Digital Signal Processing* (Wiley, 2001, 2nd Ed. 2006), *Digital Signal Processors* (Prentice Hall, 2005), and *Design of Active Noise Control Systems with the TMS320 Family* (Texas Instruments, 1996). His research focuses on real-time digital signal processing applications, active noise and vibration control, adaptive echo and noise cancellation, digital audio applications, and digital communications.

Note

- A number of illustrations appearing in this book are reproduced from copyright material published by Analog Devices, Inc., with permission of the copyright owner.
- A number of illustrations and text appearing in this book are reprinted with the permission of National Instruments Corp. from copyright material.
- VisualDSP++® is the registered trademark of Analog Devices, Inc.
- LabVIEW™ and National Instruments™ are trademarks and trade names of National Instruments Corporation.
- MATLAB® is the registered trademark of The MathWorks, Inc.
- Thanks to the Hackles for providing audioclips from their singles "Leave It Up to Me" copyright ©2006 www.TheHackles.com

Chapter 1

Introduction

1.1 EMBEDDED PROCESSOR: MICRO SIGNAL ARCHITECTURE

Embedded systems are usually part of larger and complex systems and are usually implemented on dedicated hardware with associated software to form a computational engine that will efficiently perform a specific function. The dedicated hardware (or embedded processor) with the associated software is embedded into many applications. Unlike general-purpose computers, which are designed to perform many general tasks, an embedded system is a specialized computer system that is usually integrated as part of a larger system. For example, a digital still camera takes in an image, and the embedded processor inside the camera compresses the image and stores it in the compact flash. In some medical instrument applications, the embedded processor is programmed to record and process medical data such as pulse rate and blood pressure and uses this information to control a patient support system. In MP3 players, the embedded processor is used to process compressed audio data and decodes them for audio playback. Embedded processors are also used in many consumer appliances, including cell phones, personal digital assistants (PDA), portable gaming devices, digital versatile disc (DVD) players, digital camcorders, fax machines, scanners, and many more.

Among these embedded signal processing-based devices and applications, digital signal processing (DSP) is becoming a key component for handling signals such as speech, audio, image, and video in real time. Therefore, many of the latest hardware-processing units are equipped with embedded processors for real-time signal processing.

The embedded processor must interface with some external hardware such as memory, display, and input/output (I/O) devices such as coder/decoders to handle real-life signals including speech, music, image, and video from the analog world. It also has connections to a power supply (or battery) and interfacing chips for I/O data transfer and communicates or exchanges information with other embedded processors. A typical embedded system with some necessary supporting hardware is shown in Figure 1.1. A single (or multiple) embedded processing core is used to

---

Embedded Signal Processing with the Micro Signal Architecture. By Woon-Seng Gan and Sen M. Kuo
Copyright © 2007 John Wiley & Sons, Inc.

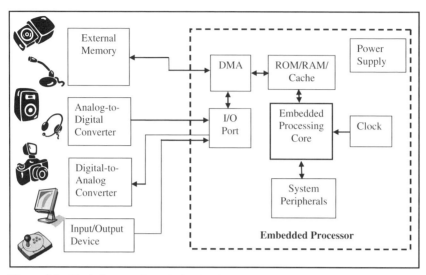

Figure 1.1 Block diagram of a typical embedded system and its peripherals

perform control and signal processing functions. Hardware interfaces to the processing core include (1) internal memories such as read-only memory (ROM) for bootloading code and random-access memory (RAM) and cache for storing code and data; (2) a direct memory access (DMA) controller that is commonly used to transfer data in and out of the internal memory without the intervention of the main processing core; (3) system peripherals that contain timers, clocks, and power management circuits for controlling the processor's operating conditions; and (4) I/O ports that allow the embedded core to monitor or control some external events and process incoming media streams from external devices. These supporting hardware units and the processing core are the typical building blocks that form an embedded system. The embedded processor is connected to the real-world analog devices as shown in Figure 1.1. In addition, the embedded processor can exchange data with another system or processor by digital I/O channels. In this book, we use hands-on experiments to illustrate how to program various blocks of the embedded system and how to integrate them with the core embedded processor.

In most embedded systems, the embedded processor and its interfaces must operate under real-time constraints, so that incoming signals are required to be processed within a certain time interval. Failure to meet these real-time constraints results in unacceptable outcomes like noisy response in audio and image applications, or even catastrophic consequences in some human-related applications like automobiles, airplanes, and health-monitoring systems. In this book, the terms "embedded processing" and "real-time processing" are often used interchangeably to include both concepts. In general, an embedded system gathers data, processes them, and makes a decision or responds in real time.

To further illustrate how different blocks are linked to the core embedded processor, we use the example of a portable audio player shown in Figure 1.2. In this

1.1 Embedded Processor: Micro Signal Architecture 3

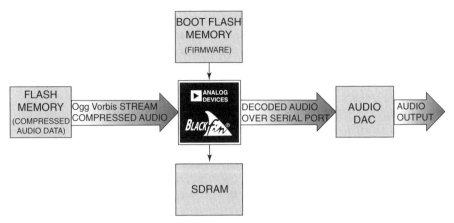

Figure 1.2 A block diagram of the audio media player (courtesy of Analog Devices, Inc.)

system, the compressed audio bit stream in Ogg Vorbis format (introduced in Chapter 9) is stored in the flash memory external to the embedded processor, a Blackfin processor. A decoding program for decoding the audio bit stream is loaded from the boot flash memory into the processor's memory. The compressed data are streamed into the Blackfin processor, which decodes the compressed bit stream into pulse code-modulated (PCM) data. The PCM data can in turn be enhanced by some postprocessing tasks like equalization, reverberation, and three-dimensional audio effects (presented in Chapter 9). The external audio digital-to-analog converter (DAC) converts the PCM data into analog signal for playback with the headphones or loudspeakers.

Using this audio media player as an example, we can identify several common characteristics in typical embedded systems. They are summarized as follows:

1. *Dedicated functions:* An embedded system usually executes a specific task repeatedly. In this example, the embedded processor performs the task of decoding the Ogg Vorbis bit stream and sends the decoded audio samples to the DAC for playback.
2. *Tight constraints:* There are many constraints in designing an embedded system, such as cost, processing speed, size, and power consumption. In this example, the digital media player must be low cost so that it is affordable to most consumers, it must be small enough to fit into the pocket, and the battery life must last for a long time.
3. *Reactive and real-time performance:* Many embedded systems must continuously react to changes of the system's input. For example, in the digital media player, the compressed data bit stream can be decoded in a number of cycles per audio frame (or operating frequency for real-time processing). In addition, the media player also must respond to the change of mode selected by the users during playback.

Therefore, the selection of a suitable embedded processor plays an important role in the embedded system design. A commonly used approach for realizing signal processing tasks is to use fixed-function and hardwired processors. These are implemented as application-specific integrated circuits (ASICs) with DSP capabilities. However, these hardwired processors are very expensive to design and produce, as the development costs become significant for new process lithography. In addition, proliferation and rapid change of new standards for telecommunication, audio, image, and video coding applications makes the hardwired approach no longer the best option.

An alternative is to use programmable processors. This type of processor allows users to write software for the specific applications. The software programming approach has the flexibility of writing different algorithms for different products using the same processor and upgrading the code to meet emerging standards in existing products. Therefore, a product can be pushed to the market in a shorter time frame, and this significantly reduces the development cost compared to the hardwired approach. A programmable digital signal processor is commonly used in many embedded applications. DSP architecture has evolved greatly over the last two decades to include many advanced features like higher clock speed, multiple multipliers and arithmetic units, incorporation of coprocessors for control and communication tasks, and advanced memory configuration. The complexity of today's signal processing applications and the need to upgrade often make a programmable embedded processor a very attractive option. In fact, we are witnessing a market shift toward software-based microarchitectures for many embedded media processing applications.

One of the latest classes of programmable embedded signal processors is the micro signal architecture (MSA). The MSA core [43] was jointly developed by Intel and Analog Devices Inc. (ADI) to meet the computational demands and power constraints of today's embedded audio, video, and communication applications. The MSA incorporates both DSP and microcontroller functionalities in a single core. Both Intel and ADI have further developed processors based on the MSA architecture for different applications. The MSA core combines a highly efficient computational architecture with features usually only available on microcontrollers. These features include optimizations for high-level language programming, memory protection, and byte addressing. Therefore, the MSA has the ability to execute highly complex DSP algorithms and basic control tasks in a single core. This combination avoids the need for a separate DSP processor and microcontroller and thus greatly simplifies both hardware and software design and implementation. In addition, the MSA has a very efficient and dynamic power management feature that is ideal for a variety of battery-powered communication and consumer applications that require high-intensity signal processing on a strict power budget. In fact, the MSA-based processor is a versatile platform for processing video, image, audio, voice, and data.

Inside the computational block of the MSA, there are two multiply-add units, two arithmetic-logic units, and a single shifter. These hardware engines allow the MSA-based processor to efficiently perform several multiply-add operations in parallel to support complex number crunching tasks. The availability of these hardware units coupled with high clock speed (starts from 300 MHz and rises steadily to the 1-GHz range) has created a substantial speed improvement over conventional microprocessors. The detailed architecture of the MSA core is explained with simple instructions and hands-on experiments in Chapter 5.

The MSA core uses a simple, reduce-instruction-set-computer (RISC)-like instruction set for both control and signal processing applications. The MSA also comes with a set of multifunction instructions that allows different sizes of op-codes to be combined into a single instruction. Therefore, the programmer has the flexibility of reducing the code size, and at the same time, maximizing the usage of available resources. In addition, some special instructions support video and wireless applications. This chapter introduces some basic features of programming and debugging the MSA core and uses examples and exercises to highlight the important features in the software tools. In subsequent chapters, we introduce more advanced features of the software tools.

The MSA core is a fixed-point processor. It operates on fixed-point fractional or integer numbers. In contrast to the floating-point processors, such as the Intel Pentium processors, fixed-point processors require special attention to manipulating numbers to avoid wrong results or extensive computation errors. The concepts of real-time implementation using a fixed-point processor are introduced and examined in Chapter 6.

As explained above, the embedded core must be combined with internal and external memories, serial ports, mixed signal circuits, external memory interfaces, and other peripherals and devices to form an embedded system. Chapter 7 illustrates how to program and configure some of these peripherals to work with the core processor. Chapter 8 explains and demonstrates several important techniques of optimizing the program written for the MSA core. This chapter also illustrates a unique feature of the MSA core to control the clock frequency and supply voltage to the MSA core via software, so as to reduce the power consumption of the core during active operation.

In this book, we examine the MSA core with the latest Blackfin processors (BF5xx series) from ADI. The first generation of Blackfin processors is the BF535 processor, which operates at 300 MHz and comes with L1 and L2 cache memories, system control blocks, basic peripheral blocks, and high-speed I/O. The next generation of Blackfin processors consists of BF531, BF532, BF533, and BF561 processors. These processors operate at a higher clock speed of up to 750 MHz and contain additional blocks like parallel peripheral interface, voltage regulator, external memory interface, and more data and instruction cache. The BF531 and BF532 are the least expensive and operate at 400 MHz, and the BF561 is a dual-core Blackfin processor that is targeted for very high-end applications. The BF533 processor operates at 750 MHz and provides a good platform for media processing applications. Recently released Blackfin processors include BF534, BF536, and BF537. These processors operate at around 400–500 MHz and feature a strong support for Ethernet connection and a wider range of peripherals.

Because all Blackfin processors are code compatible, the programs written for one processor can be easily ported to other processors. This book uses BF533 and BF537 processors to explain architecture, programming, peripheral interface, and implementation issues. The main differences between the BF533 and the BF537 are the additional peripheral supports and slightly less internal instruction memory of the BF537 processors. Therefore, explanation of the Blackfin processing core can be focused solely on the BF533, and additional sections are introduced for the extra peripheral supports in the BF537.

There are several low-cost development tools introduced by ADI. In this book, we use the VisualDSP++ simulator to verify the correctness of the algorithm and the EZ-KIT (development board that contains the MSA processor, memory, and other peripherals) for the Blackfin BF533 and BF537 processors for real-time signal processing and control applications. In addition, we also use a new tool (LabVIEW Embedded Module for Blackfin Processors) jointly developed by National Instruments (NI) and ADI to examine a new approach in programming the Blackfin processor with a blockset programming approach.

1.2 REAL-TIME EMBEDDED SIGNAL PROCESSING

DSP plays an important role in many real-time embedded systems. A real-time embedded system is a system that requires response to external inputs within a specific period. For example, a speech-recognition device sampling speech at 8 kHz (bandwidth of 4 kHz) must respond to the sampled signal within a period of 125 µs. Therefore, it is very important that we take special attention to define the real-time system and highlight some special design considerations that apply to real-time embedded signal processing systems.

Generally, a real-time system must maintain a timely response to both internal and external signal/data. There are two types of real-time system: hard and soft real-time systems. For the hard real-time system, an absolute deadline for the completion of the overall task is imposed. If the hard deadline is not met, the task has failed. For example, in the case of speech enhancement, the DSP software must be completed within 125 µs; otherwise, the device will fail to function correctly. In the case of a soft real-time system, the task can be completed in a more relaxed time range. For example, it is not critical how long it takes to complete a credit card transaction. There is no hard deadline by which the transaction must be completed, as long as it is within a reasonable period of time.

In this book, we only examine hard real-time systems because all embedded media processing systems are hard real-time systems. There are many important challenges when designing a hard real-time system. Some of the challenges include:

1. *Understanding DSP concepts and algorithms.* A solid understanding of the important DSP principles and algorithms is the key to building a successful real-time system. With this knowledge, designers can program and optimize the algorithm on the processor using the best parameters and set-

tings. Chapters 2 to 4 introduce the fundamentals of DSP and provide many hands-on experiments to implement signal processing algorithms.

2. **Resource availability.** The selection of processor core, peripherals, sensors and actuators, user interface, memory, development, and debugging tools is a complex task. There are many critical considerations that make the decision tough. Chapter 5 shows the architecture of the Blackfin processor and highlights its strength in developing the embedded system.

3. **Arithmetic precision.** In most embedded systems, a fixed-point arithmetic is commonly used because fixed-point processors are cheaper, consume less power, and have higher processing speed as compared to floating-point processors. However, fixed-point processors are more difficult to program and also introduce many design challenges that are discussed in Chapter 6.

4. **Response time requirements.** Designers must consider hardware and software issues. Hardware considerations include processor speed, memory size and its transfer rate, and I/O bandwidth. Software issues include programming language, software techniques, and programming the processor's resources. A good tool can greatly speed up the process of developing and debugging the software and ensure that real-time processing can be achieved. Chapter 7 explains the peripherals and I/O transfer mechanism of the Blackfin processor, whereas Chapter 8 describes the techniques used in optimizing the code in C and assembly programming.

5. **Integration of software and hardware in embedded system.** A final part of this book implements several algorithms for audio and image applications. The Blackfin BF533/BF537 EZ-KITs are used as the platform for integration of software and hardware.

To start the design of the embedded system, we can go through a series of exercises using the development tools. As we progress to subsequent chapters, more design and debugging tools and features of these tools are introduced. This progressive style in learning the tools will not overload the users at the beginning. We use only the right tool with just enough features to solve a given problem.

1.3 INTRODUCTION TO THE INTEGRATED DEVELOPMENT ENVIRONMENT VISUALDSP++

In this section, we examine the software development tool for embedded signal processors. The software tool for the Blackfin processor is the VisualDSP++ [33] supplied by ADI. VisualDSP++ is an integrated development and debugging environment (IDDE) that provides complete graphical control of the edit, build, and debug processes. In this section, we show the detailed steps of loading a project file into the IDDE, building it, and downloading the executable file into the simulator (or the EZ-KIT). We introduce some important features of the VisualDSP++ in this chapter, and more advanced features are introduced in subsequent chapters.

8   Chapter 1   Introduction

1.3.1 Setting Up VisualDSP++

The Blackfin version of VisualDSP++ IDDE can be downloaded and tested for a period of 90 days from the ADI website. Once it is downloaded and installed, a **VisualDSP++ Environment** icon will appear on the desktop. Double-click on this icon to activate the VisualDSP++. A **New Session** window will appear as shown in Figure 1.3. Select the **Debug target**, **Platform**, and **Processor** as shown in Figure 1.3, and click **OK** to start the VisualDSP++. Under the **Debug target** option, there are two types of Blackfin simulator, a cycle-accurate interpreted simulator and a functional compiled simulator. When **ADSP_BF5xx Blackfin Family Simulators** is selected, the cycle-accurate simulator is used. This simulator provides a more accurate performance, and thus is usually used in this book. The compiled simulator is used when the simulator needs to process a large data file. The **Processor** option allows users to select the type of processor. In this book, only the Blackfin BF533 and BF537 processors are covered. However, the code developed for any Blackfin processor is compatible with other Blackfin processors. In Figure 1.3, the ADSP-BF533 simulator is selected. Alternatively, users can select the ADSP-BF537 simulator.

A VisualDSP++ Version 4 window is shown in Figure 1.4. There are three subwindows and one toolbar menu in the VisualDSP++ window. A **Project Window** displays the files available in the project or project group. The **Disassembly** window displays the assembly code of the program after the project has been built. The **Output Window** consists of two pages, **Console** and **Build**. The **Console** page displays any message that is being programmed in the code, and the **Build** page shows any errors encountered during the build process. The toolbar menu contains all the tools, options, and modes available in the VisualDSP++ environment. We will illustrate these tools as we go through the hands-on examples and exercises in

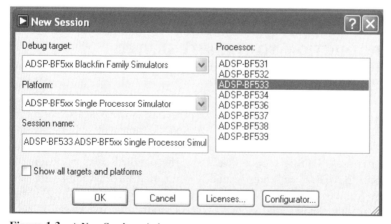

Figure 1.3   A **New Session** window

1.3 Introduction to the Integrated Development Environment VisualDSP++ 9

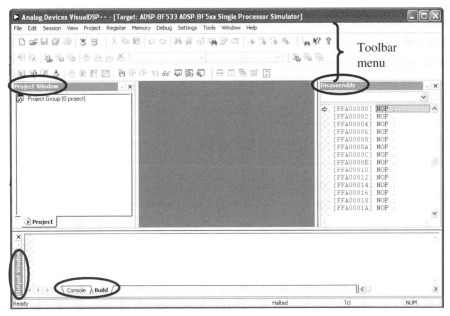

Figure 1.4 VisualDSP++ window

this and following chapters. Click on **Project → Project Options** and make sure that the right processor is selected for the targeted simulator.

1.3.2 Using a Simple Program to Illustrate the Basic Tools

In this section, we illustrate the basic tools and features in the VisualDSP++ IDDE through a series of simple exercises. We use a sample project that consists of two source files written in C for the Blackfin processor.

HANDS-ON EXPERIMENT 1.1

In this experiment, we first start the VisualDSP++ environment as explained above. Next, click on the **File** menu in the toolbar menu and select **Open → Project....** Look for the project file exp1_1.dpj under directory c:\adsp\chap1\exp1_1 and double-click on the project file. Once the project file is loaded into the VisualDSP++ environment, we can see a list of source files. Double-click on dotprod_main.c to see the source code in the editor window (right side) as shown in Figure 1.5.

Scroll through the source code in both dotprod_main.c and dotprod.c. This is a simple program to perform multiply-accumulate (or dot product) of two vectors, a and b. From the **Settings** menu, choose **Preferences** to open the dialog box as shown in Figure 1.6.

10 Chapter 1 Introduction

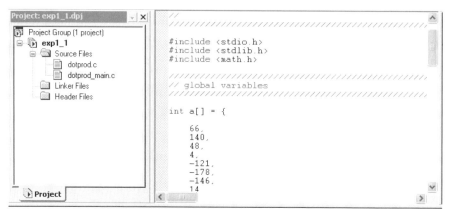

Figure 1.5 Snapshot of the C file dotprod_main.c displayed in the source window

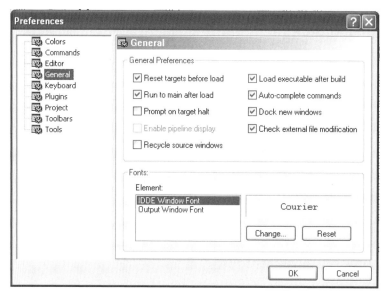

Figure 1.6 **Preferences** dialog box

Under the **General** preference, click on **Run to main after load** and **Load executable after build**. The first option starts executing from the void main() of the program after the program is loaded into the simulator. The second option enables the code to be loaded into the processor memory after the code is being built. The rest of the options can be left as default. Click **OK** to close the **Preferences** dialog box.

Now, we are ready to build the project. We can either click on **Project → Build Project** in the toolbar menu or press the function key **F7**. There is a build icon that

1.3 Introduction to the Integrated Development Environment VisualDSP++ 11

can be used to perform the build operation. The build operation basically combines the compile, assembler, and link processes to obtain an executable file (.dxe). Users will find the executable file exp1_1.dxe being created in directory c:\adsp\chap1\exp1_1\debug after the build operation. Besides the **Build Project** option, there is the **Rebuild All** option (or icon). The **Rebuild All** option builds the project regardless of whether the project build is up to date. The message Build completed successfully is shown in the **Build** page of the **Output Window** if the build process detects no error. However, users will notice that the build process detects an undefined identifier, as shown in Figure 1.7.

Users can double-click on the error message (in red), and the cursor will be placed on the line that contains the error. Correct the typo by changing results to result and save the source file by clicking on **File → Save → File dotprod_main.c**. The project is now built without any error, as indicated in the **Output** window.

Once the project has been built, the executable file exp1_1.dxe is automatically downloaded into the target (enabled in the **Preferences** dialog box), which is the BF533 (or BF537) simulator. Click on the **Console** page of the **Output Window**. A message appears stating that the executable file has been completely loaded into the target, and there is a Breakpoint Hit at <ffa006f8>. A solid red circle (indicates breakpoint) and a yellow arrow (indicates the current location of the program pointer) are positioned at the left-hand edges of the source code and disassembly code, as shown in Figure 1.8.

The VisualDSP++ automatically sets two breakpoints, one at the beginning and the other at the end of the code. The location of the breakpoint can be viewed by clicking on **Setting → Breakpoints** as shown in Figure 1.9. Users can click on the breakpoint under the **Breakpoint list** and click the **View** button to find the location of the breakpoint in the

```
.\dotprod_main.c
".\dotprod_main.c", line 81: cc0020:  error: identifier "results" is undefined
      results = a_dot_b( a, b );
      ^

1 error detected in the compilation of ".\dotprod_main.c".
cc3089: fatal error: Compilation failed
Tool failed with exit/exception code: 1.
Build was unsuccessful.
```

Figure 1.7 Error message appears after building the project

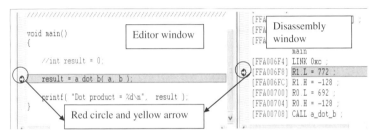

Figure 1.8 Breakpoint displayed in both editor and disassembly windows

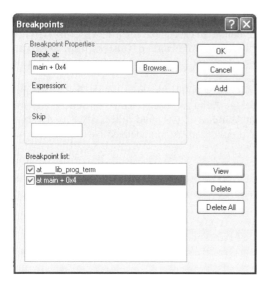

Figure 1.9 **Breakpoint** dialog box

Disassembly window. The breakpoint can be set or cleared by double-clicking on the gray gutter (Fig. 1.8) next to the target code in the editor and disassembly window.

The project is now ready to run. Click on the run button ▤ or **Debug → Run (F5)**. The simulator computes the dot product (sum of products or multiply-accumulate) and displays the result in the **Console** page of the **Output Window**. What is the answer for the dot product between arrays a and b?

Modify the source files to perform the following tasks:

1. Add a new 20-element array c; perform the dot product computation between arrays a and c and display the result.
2. Recompute the dot product of the first 10 elements in the arrays.
3. To obtain the cycle count of running the code from start to finish, we can use the cycle registers. Simply click on **Register → Core → Cycles**. Reload the program by clicking on **File → Reload Program**. The program pointer will reset to the beginning of the program. In the **Cycle** window, clear the CYCLE register value to 0 to initialize the cycle counter. Run the program and note the cycle count. Note: To display the cycle count in unsigned integer format, right-click on the **Cycle** window and select **unsigned integer**.

1.3.3 Advanced Setup: Using the Blackfin BF533 or BF537 EZ-KIT

In the previous hands-on experiments, we ran the program with the BF533 (or BF537) simulator. In this section, we perform the same experiment with the Blackfin

1.3 Introduction to the Integrated Development Environment VisualDSP++ 13

Figure 1.10 The BF533 EZ-KIT

BF533 (or BF537) EZ-KIT. The EZ-KIT board is a low-cost hardware platform that includes a Blackfin processor surrounded by other devices such as audio coder/decoder (CODEC), video encoders, video decoders, flash, synchronous dynamic RAM (SDRAM), and so on. We briefly introduce the hardware components in the EZ-KIT and show the differences between the BF533 and BF537 boards.

Figure 1.10 shows a picture of the BF533 EZ-KIT [29]. This board has four audio input and six audio output channels via the RCA jacks. In addition, it can also encode and decode three video inputs and three video outputs. Users can also program the four general-purpose push buttons (SW4, SW5, SW6, and SW7) and six general-purpose LEDs (LED4–LED9). The EZ-KIT board is interfaced with the VisualDSP++ (hosted on personal computer) via the USB interface cable.

Figure 1.11 shows a picture of the BF537 EZ-KIT [30]. This board consists of stereo input and output jack connectors. However, the BF537 EZ-KIT does not have any video I/O. Instead, it includes the IEEE 802.3 10/100 Ethernet media access control and controller area network (CAN) 2.0B controller. Similar to the BF533 EZ-KIT, the BF537 EZ-KIT has four general-purpose push buttons (SW10, SW11, SW12, and SW13) and six general-purpose LEDs (LED1–LED6). Other feature differences between BF533 and BF537 EZ-KITs are highlighted in subsequent chapters.

14 Chapter 1 Introduction

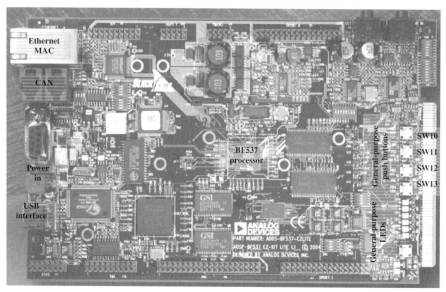

Figure 1.11 The BF537 EZ-KIT

This section describes the setup of the BF533 EZ-KIT [31]. The EZ-KIT's power connector is first connected to the power supply. Turn on the power supply and verify that the green LED is lit and LEDs 4–9 continue to roll (indicating that the board is not linked to the software). Next, the USB cable is connected from the computer to the EZ-KIT board. The window environment recognizes the new hardware and launches the **Add New Hardware Wizard**, which installs files located on the EZ-KIT CD-ROM. Once the USB driver is installed successfully, the yellow LED (USB monitor) should remain lit. A similar setup can also be carried out for the BF537 EZ-KIT [32]. Users can refer to the BF533 (or BF537) EZ-KIT Evaluation System Manual for more details on the settings.

The VisualDSP++ environment can be switched to the EZ-KIT target by the following steps. Click on **Session → New Session**. A **New Session** window will appear. Change the debug target and platform to that shown in Figure 1.12 (setup for the BF533 EZ-KIT). Click **OK** and note the change in the title bar of the VisualDSP++. We are now ready to run the same project on the BF533 EZ-KIT. Similarly, if the BF537 EZ-KIT is used, select the desired processor and its EZ-KIT.

Build the project and run the executable file on the EZ-KIT, using the same procedure as before. Do you notice any difference in the building process on the EZ-KIT platform compared to the simulator? Next, obtain the cycle count in running the same program on the EZ-KIT. Is there any change in the cycle count as compared to the simulator?

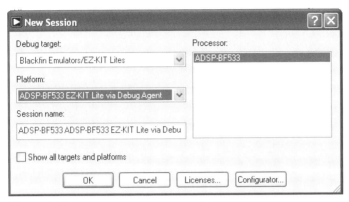

Figure 1.12 **New Session** window setup for BF533 EZ-KIT

1.4 MORE HANDS-ON EXPERIMENTS

We have learned how to load a project file into the Blackfin simulator and EZ-KIT. In this section, we create a new project file from scratch and use the graphic features in the VisualDSP++ environment. The following experiments apply to both BF533 and BF537 EZ-KITs.

HANDS-ON EXPERIMENT 1.2

1. Using the previous source files as templates, create two new source files vecadd_main.c and vecadd.c to perform vector addition of two arrays a and b. The result is saved in the third array c. Use **File** → **New** → **File** to create a blank page for editing in the VisualDSP++. Save these files in directory c:\adsp\chap1\exp1_2.

2. From the **File** menu, choose **New** and then **Project** to open the **Project Wizard**. Enter the directory and the name of the new project as shown in Figure 1.13. Click on **Finish** and **Yes** to create a new project.

3. An empty project is created in the **Project** window. Click on **Project** → **Project Options** to display the **Project Options** dialog box as shown in Figure 1.14. The default settings are used, and the project creates an executable file (.dxe). Because **Settings for configuration** is set to **Debug**, the executable file also contains debug information for debugging.

4. Click on **Compile** → **General (1)**, and click on **Enable optimization** to enable the optimization for speed as shown in Figure 1.15. Click **OK** to apply changes to the project options.

5. To add the source files to the new project, click on **Project** → **Add to Project** → **File(s)....** Select the two source files and click **Add**. The sources files are now added to the project file.

16 Chapter 1 Introduction

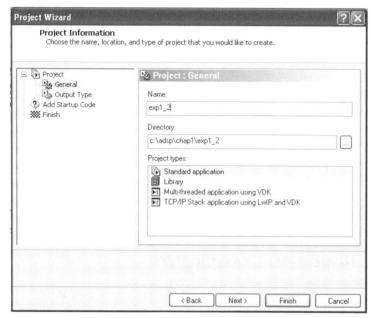

Figure 1.13 Project Wizard window

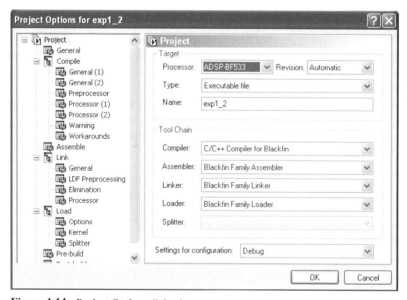

Figure 1.14 Project Options dialog box

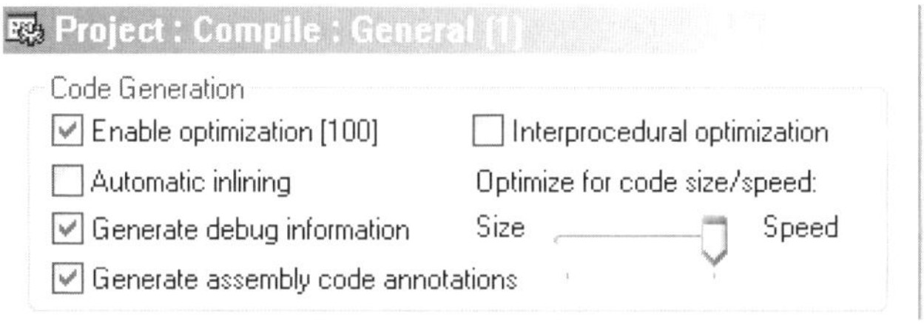

1.4 More Hands-On Experiments   17

Figure 1.15 Project wizard option for compile

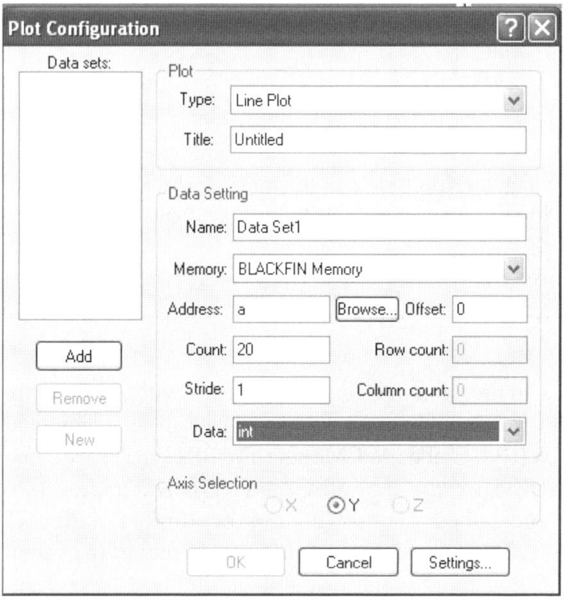

Figure 1.16 **Plot Configuration** dialog box

6. Build the project by following the steps given in Hands-On Experiment 1.1. Run the project and verify that the correct result in array c is displayed in the **Output Window**.

HANDS-ON EXPERIMENT 1.3

1. In this hands-on experiment, we introduce some useful graphic features in the VisualDSP++ environment. We plot the three arrays, a, b, and c, used in the previous experiment.

2. Make sure that the project is built and loaded into the simulator. Click on **View** → **Debug Windows** → **Plot** → **New**. A **Plot Configuration** dialog box appears as shown in Figure 1.16. We type in a and 20 and select int in **Address**, **Count**, and

18 Chapter 1 Introduction

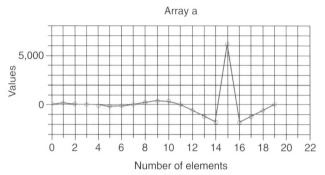

Figure 1.17 Display of array a

Data boxes, respectively. Click on **Add** to add in the array a. Use similar steps to plot the other two arrays. Modify the plot settings to make a graph as displayed in Figure 1.17.

3. Finally, add in the other two arrays in the same plot or create two new plots.

So far, we have learned how to program in C and run the code with the Blackfin simulator and EZ-KITs. In the following section, we introduce a new graphical development environment, LabVIEW Embedded Module for Blackfin Processors, jointly developed by NI and ADI. This new tool provides an efficient approach to prototyping embedded signal processing systems. This new rapid prototyping tool uses a graphical user interface (GUI) to control and select parameters of the signal processing algorithms and view updates of graphical plots on the fly.

1.5 SYSTEM-LEVEL DESIGN USING A GRAPHICAL DEVELOPMENT ENVIRONMENT

Graphical development environments, such as National Instruments LabVIEW, are effective means to rapidly prototype and deploy developments from individual algorithms to full system-level designs onto embedded processors. The graphical data flow paradigm that is used to create LabVIEW programs or virtual instruments (VIs) allows for rapid, intuitive development of embedded code. This is due to its flowchart-like syntax and inherent ease in implementing parallel tasks.

In this section and sections included at the end of each subsequent chapter, we present a common design cycle that engineers are using to reduce product development time by effectively integrating the software tools they use on the desktop for deployment and testing. This will primarily be done with the LabVIEW Embedded

1.5 System-Level Design Using a Graphical Development Environment

Development Module for the ADI Blackfin BF533 and BF537 processors which is an add-on module for LabVIEW to target and deploy to the Blackfin processor. Other LabVIEW add-ons, such as the Digital Filter Design Toolkit, may also be discussed.

Embedded system developers frequently use computer simulation and design tools such as LabVIEW and the Digital Filter Design Toolkit to quickly develop a system or algorithm for the needs of their project. Next, the developer can leverage his/her simulated work on the desktop by rapidly deploying that same design with the LabVIEW Embedded Module for Blackfin Processors and continue to iterate on that design until the design meets the design specifications. Once the design has been completed, many developers will then recode the design using VisualDSP++ for the most efficient implementation. Therefore, knowledge of the processor architecture and its C/assembly programming is still important for a successful implementation. This book provides balanced coverage of both high-level programming using the graphical development environment and conventional C/assembly programming using VisualDSP++.

In the first example using this graphical design cycle, we demonstrate the implementation and deployment of the dot product algorithm presented in Hands-On Experiment 1.1 using LabVIEW and the LabVIEW Embedded Module for Blackfin Processors.

1.5.1 Setting up LabVIEW and LabVIEW Embedded Module for Blackfin Processors

LabVIEW and the LabVIEW Embedded Module for Blackfin Processors (trial version) can be downloaded from the book companion website. A brief tutorial on using these software tools is included in Appendix A of this book. Once they are installed, double-click on **National Instruments LabVIEW 7.1 Embedded Edition** under the **All Programs** panel of the start menu. Hands-On Experiment 1.4 provides an introduction to the LabVIEW Embedded Module for Blackfin Processors to explore concepts from the previous hands-on experiments.

HANDS-ON EXPERIMENT 1.4

This exercise introduces the NI LabVIEW Embedded Module for Blackfin Processors and the process for deploying graphical code on the Blackfin processor for rapid prototyping and verification. The dot product application was created with the same vector values used previously in the VisualDSP++ project file, `exp1_1.dpj`. This experiment uses arrays, functions, and the **Inline C Node** within LabVIEW.

From the LabVIEW **Embedded Project Manager** window, open the project file `DotProd - BF5xx.lep` located in directory `c:\adsp\chap1\exp1_4`. Next, double-click on `DotProd_BF.vi` from within the project window to open the front panel. The front panel is the graphical user interface (GUI), which contains the inputs and outputs of the

20 Chapter 1 Introduction

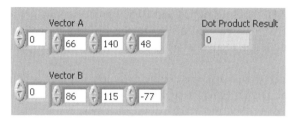

Figure 1.18 Front panel of DotProd.vi

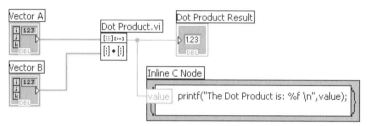

Figure 1.19 Block diagram of DotProd.vi

program as shown in Figure 1.18. Select **View → Block Diagram** to switch to the LabVIEW block diagram, shown in Figure 1.19, which contains the source code of the program. The dot product is implemented by passing two vectors (or one-dimensional arrays) to the **Dot Product** function. The result is then displayed in the **Dot Product Result** indicator. The value is also passed to the standard output buffer, because controls and indicators are only available in JTAG debug or instrumented debug modes. The graphical LabVIEW code executes based on the principle of data flow, in this case from left to right.

This graphical approach to programming makes this program simple to implement and self-documenting, which is especially helpful for larger-scale applications. Also note the use of the **Inline C Node**, which allows users to test existing C code within the graphical framework of LabVIEW.

Now run the program by clicking on the **Run** arrow to calculate the dot product of the two vectors. The application will be translated, linked, compiled, and deployed to the Blackfin processor. Open the **Processor Status** window and select **Output** to see the numeric result of the dot product operation.

Another feature available for use with the LabVIEW Embedded Module for Blackfin Processors is instrumented debug mode, which allows users to probe wires on the LabVIEW block diagram and interact with live-updating front panel controls and indicators while the code is actually running on the Blackfin. Instrumented debug can be accomplished through TCP (Ethernet) on the BF537 and through JTAG (USB) on both the BF537 and BF533 processors. To use instrumented debug, run the application in debug mode, using the **Debug** button. Try changing vector elements on the front panel and probing wires on the block diagram. For additional configuration, setup, and debugging information, refer to *Getting Started with the LabVIEW Embedded Module for Analog Devices Blackfin Processors* [52], found in the book companion website.

1.6 MORE EXERCISE PROBLEMS

1. List 10 embedded systems in Table 1.1 that are required to perform some forms of DSP. Explain the signal processing tasks.
2. Survey the key components in a typical iPod player.
3. Survey the key components in a typical digital camera.
4. Survey the differences between fixed-function processors and programmable processors. State the advantages and disadvantages of these processors.
5. In Hands-on Experiment 1.1, C language is used to program the Blackfin processor. A low-level assembly program can also be used to compute the dot product. The low-level assembly code uses a standard set of assembly syntaxes. These assembly syntaxes are introduced in Chapter 5 onward. Use the Blackfin simulator (either BF533 or BF537) to create a new project file that includes the source codes available in directory c:\adsp\chap1\problem1_5. Examine the source files and understand the differences between the C function code in Experiment 1.1 and the assembly function code listed in this exercise. Benchmark and compare the cycle count for performing the same dot product with assembly code with that obtained with C code. Also benchmark the C code with optimization enabled.
6. Implement the same project in the Blackfin (either BF533 or BF537) EZ-KIT. Any difference in the cycle counts compared to the Blackfin simulator?

Table 1.1 DSP Tasks in Embedded Systems

Embedded Systems	DSP Tasks
(1)	
(2)	
(3)	
(4)	
(5)	
(6)	
(7)	
(8)	
(9)	
(10)	

7. The Fibonacci series can be computed by adding the two successive numbers to form the next number in the series. Generate the Fibonacci series for the first 10 numbers of 1, 1, 2, 3, 5, 8, 13, 21, 34, 55 ..., using the Blackfin simulator, starting from the first two numbers. Verify your answer in the Blackfin memory window.

8. Refer to the ADI manual on getting started with VisualDSP++ [34] and go through all the experiments described in the manual. The manual can be found in the ADI website.

Part A

Digital Signal Processing Concepts

Chapter 2

Time-Domain Signals and Systems

This chapter uses several noise reduction examples and experiments to introduce important time-domain techniques for processing digital signals and analyzing simple DSP systems. The detailed analysis and more advanced methods are introduced in Chapter 3, using frequency-domain techniques.

2.1 INTRODUCTION

With the easy accessibility of increasingly powerful personal computers and the availability of powerful and easy-to-use MATLAB [48] software for computer simulations, we can now learn DSP concepts more effectively. This chapter uses hands-on methods to introduce fundamental time-domain DSP concepts because it is more interesting to examine real-world DSP applications with the help of interactive MATLAB tools.

In particular, this chapter uses the latest powerful graphical user interface (GUI) tool called Signal Processing Tool, which comes with the *Signal Processing Toolbox* [49]. Because each experiment requires a set of general problem-solving skills and related DSP principles, we provide multiple contexts including the necessary DSP theory, computer simulations, and hands-on experiments for achieving thorough understanding. Most of the DSP subjects are covered in the introduction to hands-on exercises and experiments. These experiments are organized to introduce DSP subjects from the simple introductory subjects in this chapter and gradually introduce more complicated experiments and applications in subsequent chapters. Each experiment introduces and applies just enough information at that time to complete the required tasks. This is similar to using a "spiral learning" technique to continually circle back and cover concepts in more and more depth throughout Chapters 2, 3, and 4.

Embedded Signal Processing with the Micro Signal Architecture. By Woon-Seng Gan and Sen M. Kuo
Copyright © 2007 John Wiley & Sons, Inc.

Projects introduced in this chapter are based on designing simple filters to remove broadband (white) noise that corrupts the desired narrowband (sinusoidal) signal. First, a MATLAB example shows how to generate a digital signal and use it to introduce a sampling theorem. A quiz is provided immediately afterward to ensure that we understand the relationship between analog and digital worlds. A hands-on experiment implements the moving-average filter with length $L = 5, 10,$ and 20, using the MATLAB code. We find that when the filter is working for $L = 5$, it reduces more noise when $L = 10$ with higher undesired signal attenuation, but are surprised to learn that the filter output approaches zero when $L = 20$. Finally, for more complicated problems of adding different noises (sinusoidal and white) to the speech, we have to enhance the desired speech. We use the simple moving-average filter but fail. Now our interest is piqued to learn more advanced DSP techniques in Chapters 3 and 4. In this fashion, we learn important DSP concepts repeatedly at each project by doing hands-on experiments and exercises. We continually circle back the DSP subjects and cover concepts in more and more depth throughout the book.

2.2 TIME-DOMAIN DIGITAL SIGNALS

A digital signal $x(n)$ is defined as a function of time index n, which corresponds to time at nT_s seconds if the signal is sampled from an analog signal $x(t)$ with the sampling period T_s seconds. The sampling period can be expressed as

$$T_s = \frac{1}{f_s}, \qquad (2.2.1)$$

where f_s is the sampling frequency (or sampling rate) in hertz (or cycles per second). For many real-world applications, the required sampling rates are defined by the given applications. For example, the sampling rate for telecommunications is 8,000 Hz (or 8 kHz), and for compact discs (CDs) it is 44.1 kHz.

2.2.1 Sinusoidal Signals

An analog sine wave can be expressed as

$$x(t) = A\sin(2\pi f t) = A\sin(\Omega t), \qquad (2.2.2)$$

where A is the amplitude, f is the frequency of the sine wave in hertz, and $\Omega = 2\pi f$ is the frequency in radians per second. If we sample this analog sine wave with sampling rate f_s, we obtain a digital sine wave $x(n)$ with samples at time $0, T_s, 2T_s, \ldots nT_s, \ldots$. This digital signal can be expressed by replacing t with nT_s in Equation 2.2.2 as

$$x(n) \equiv x(nT_s) = A\sin(2\pi f n T_s) = A\sin(\omega n), \qquad (2.2.3)$$

where the digital frequency ω in radians per sample (or simply radians) is defined as

2.2 Time-Domain Digital Signals

$$\omega = 2\pi f T_s = 2\pi f / f_s, \quad -\pi \le \omega \le \pi. \tag{2.2.4}$$

It is important to note that the sampling rate must satisfy the Nyquist sampling theorem expressed as

$$f_s \ge 2 f_N, \tag{2.2.5}$$

where f_N is the maximum frequency component (or bandwidth) of the signal, which is also called a Nyquist frequency. If $f_s < 2f_N$, frequency components higher than $f_s/2$ will fold back to the frequency range from 0 Hz to $f_s/2$ Hz, which results in a distortion called aliasing. The sampling theorem implies that digital signals can only have meaningful representation of signal components from 0 Hz to $f_s/2$ Hz.

EXAMPLE 2.1

We can generate a 200-Hz sine wave that is sampled at 4,000 Hz (or 4 kHz) using the MATLAB program example2_1.m. A partial code is listed as follows:

```
fs = 4000;                % sampling rate is 4 kHz
f  = 200;                 % frequency of sinewave is 200 Hz
n  = 0:1:fs/f;            % time index n that cover one cycle
xn = sin(2*pi*f*n/fs);    % generate sinewave
plot(n,xn,'-o');grid on;
```

The generated sine wave is shown in Figure 2.1, in which the digital samples are marked by open circles. Digital signal $x(n)$ consists of those discrete-time samples; however, we usually

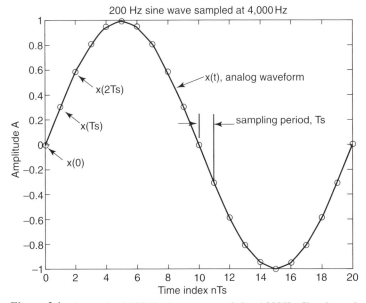

Figure 2.1 One cycle of 200-Hz sine wave sampled at 4,000 Hz. Signal samples are marked by open circles

plot a digital signal by connecting those samples with a line such as $x(t)$ shown in Figure 2.1.

Based on Equation 2.2.1, the sampling period $T_s = 1/4{,}000$ s. As shown in Figure 2.1, one cycle of sine wave consists of 20 samples. Therefore, the period of sine wave is $(1/4{,}000) \times 20 = 1/200$ s, which is equivalent to the frequency of 200 Hz.

QUIZ 2.1

1. If the sine wave shown in Figure 2.1 is obtained by sampling an analog sine wave with a sampling rate of 100 Hz, what is the frequency of the sine wave? Why?
2. If the frequency of the sine wave shown in Figure 2.1 is 20 Hz, what is the sampling period used for obtaining these digital samples? Why?
3. If we want to produce 3 s of analog sine wave by converting a digital sine wave with a digital-to-analog (D/A) converter (DAC) with a sampling rate of 4 kHz, how many digital samples are needed?

Quiz 2.1 shows that the frequency of a digital signal is dependent on the sampling rate f_s. Therefore, it is more convenient to use the normalized digital frequency defined as

$$F \equiv f/(f_s/2), \quad -1 \leq F \leq 1 \tag{2.2.6}$$

with unit cycles per sample. Comparing this definition with Equation 2.2.4, we show that $\omega = F\pi$. For example, the digital frequency of the sine wave shown in Figure 2.1 is $F = 0.1$ or $\omega = 0.1\pi$.

In many real-world applications, the operation of sampling analog signals is implemented by an analog-to-digital (A/D) converter (ADC). Similarly, the operation of converting digital signals to analog forms is realized by a D/A converter. These devices are introduced in Section 2.7 for real-time experiments.

2.2.2 Random Signals

The sine wave shown in Figure 2.1 is a deterministic signal because it can be defined exactly by a mathematical equation (Eq. 2.2.3). In practice, the signals encountered in the real world such as speech, music, and noise are random signals. In addition, the desired signals are often corrupted by noises such as thermal noise generated by thermal agitation of electrons in electrical devices. To enhance the signals, we must use different techniques based on the characteristics of signals and noise to reduce the undesired noise components.

In many practical applications, a complete statistical characterization of a random variable may not be available. The useful measures associated with a random signal are mean, variance, and autocorrelation functions. For stationary signals, the mean (or expected value) is independent of time and is defined as

2.2 Time-Domain Digital Signals

$$m_x = E[x(n)]$$
$$\cong \frac{1}{N}[x(0)+x(1)+\ldots+x(N-1)] = \frac{1}{N}\sum_{n=0}^{N-1} x(n), \qquad (2.2.7)$$

where the expectation operator $E[.]$ extracts an average value. The variance is defined as

$$\sigma_x^2 = E\left[(x(n)-m_x)^2\right] = E[x^2(n)] - m_x^2. \qquad (2.2.8)$$

Note that the expected value of the square of a random variable is equivalent to the average power. The MATLAB function `mean(x)` gives the average of the data in vector x. The function `y=var(x)` returns the variance of the values in the vector x, and the function `std(x)` computes the standard derivation σ_x.

The mean of a uniformly distributed random variable in the interval (X_1, X_2) is given as

$$m_x = \frac{X_2 + X_1}{2}. \qquad (2.2.9)$$

The variance is

$$\sigma_x^2 = \frac{(X_2 - X_1)^2}{12}. \qquad (2.2.10)$$

The MATLAB function `rand` generates arrays of random numbers whose elements are uniformly distributed in the interval (0, 1). The function `rand` with no arguments is a scalar whose value changes each time it is referenced. In addition, MATLAB provides the function `randn` for generating normally distributed random numbers with zero mean and unit variance.

QUIZ 2.2

1. Compute the mean and variance of random numbers generated by the MATLAB function `rand`.
2. How do we generate the zero mean ($m_x = 0$) and unit variance ($\sigma_x^2 = 1$) random numbers that are uniformly distributed with `rand`?

EXAMPLE 2.2

Similar to Example 2.1, we generate 60 samples of a sine wave that is corrupted by noise, using the MATLAB script `example2_2.m`. The generated noisy samples are saved in data file `sineNoise.dat`, and the original sine wave and the corrupted sine wave are shown in Figure 2.2.

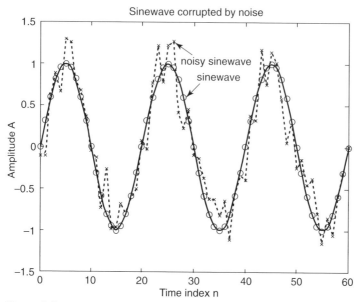

Figure 2.2 Original (open circles) and corrupted (x) signals

EXERCISE 2.1

1. Modify the MATLAB code example2_2.m to generate 3 s of sine wave and 3 s of sine wave corrupted by noise, and save the generated data in files sine3sec.dat and sineNoise3sec.dat, respectively, using ASCII format.

2. Use the function save filename x to save the generated signals to the binary files named sine3sec.mat and sineNoise3sec.mat. The MATLAB code for generating these data files is given in sineNoiseGen.m.

3. In sineNoiseGen.m, we use the following code to mix sine wave with noise

   yn = xn + 0.25*randn(size(n));

   The number 0.25 determines the signal-to-noise ratio (SNR), which can be defined as

   $$SNR = 10\log_{10}\left(\frac{P_x}{P_y}\right), \quad (2.2.11)$$

   where P_x and P_y denote the powers of signal and noise, respectively. Modify the MATLAB code by replacing 0.25 with values 0.1, 0.5, and 1, and save the generated signals in files sineNoise3sec_01.mat,

2.2 Time-Domain Digital Signals 31

`sineNoise3sec_05.mat`, and `sineNoise3sec_1.mat`, respectively. The modified MATLAB program is given as `sineNoiseGenSNR.m`.

HANDS-ON EXPERIMENT 2.1

In this experiment, we use a hands-on exercise to introduce the Signal Processing Tool (SPTool), which is an interactive GUI software for digital signal processing. The SPTool can be used to analyze signals, design and analyze filters, filter the signals, and analyze the spectrum of signals. We can open this tool by typing

`sptool`

in the MATLAB command window. The SPTool main window appears as shown in Figure 2.3.

As indicated by Figure 2.3, there are three main functions that can be accessed within the SPTool: The **Signal Browser** is for analyzing time-domain signals. We can also play portions of signals with the computer's audio hardware. The **Filter Designer** is for designing or editing digital filters. This GUI supports most of the *Signal Processing Toolbox* filter design methods. Additionally, we can design a filter by using the pole/zero editor (introduced in Chapter 3) to graphically place poles and zeros on the z-plane. The **Spectrum Viewer** uses the spectral estimation methods supported by the *Signal Processing Toolbox* to estimate the power spectral density of signals.

In this experiment, we use the **Signal Browser** for listening and viewing the digital signals. Signals from the workspace or file can be loaded into the SPTool by clicking on **File → Import**. An **Import to SPTool** window appears and allows the user to select the data

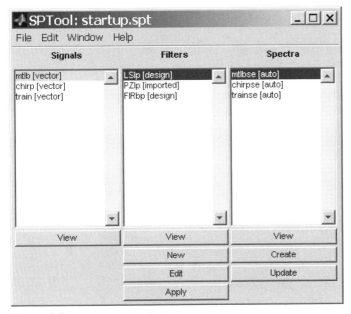

Figure 2.3 SPTool startup window

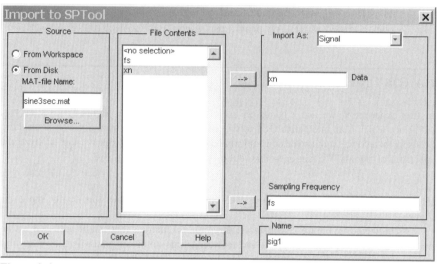

Figure 2.4 Import to SPTool window

from the file or workspace. For example, we can view the sine wave generated in Exercise 2.1 that was saved in the file sine3sec.mat. As shown in Figure 2.4, we click on the radio button **From Disk MAT-file Name** in the **Source** region, browse to the folder that contains the data file, and select the file sine3sec.mat. From the **File Contents** region, we highlight xn (signal vector defined in sineNoiseGen.m) and click on the top arrow to import it to the **Data** box. We then select fs (sampling rate) and click on the bottom arrow to import it to the **Sampling Frequency** box. Note that the default vector name used by SPTool is sig1, as shown in the **Name** box at the bottom right corner. We can change the name by entering a meaningful name into this box. Finally, we click on **OK** and the vector name sig1 is displayed in the **Signals** region.

To view the signal, we simply highlight the signal sig1, and click on **View** in the **Signals** region of Figure 2.3. The **Signal Browser** window, shown in Figure 2.5, allows the user to zoom-in and zoom-out the signal, read the data values via markers, display format, and even play the selected signal with the computer's sound card. For example, we can click on the **Play Selected Signal** button to play the 3-s tone in sig1 with the computer's sound card.

EXERCISE 2.2

1. Use different tools available on the **Signal Browser** to evaluate the imported signal sig1.
2. Import the saved file sineNoiseGen.mat from Exercise 2.1 to the SPTool, name it sig2, and use **Signal Browser** to evaluate the noisy signal.

2.3 Introduction to Digital Systems   33

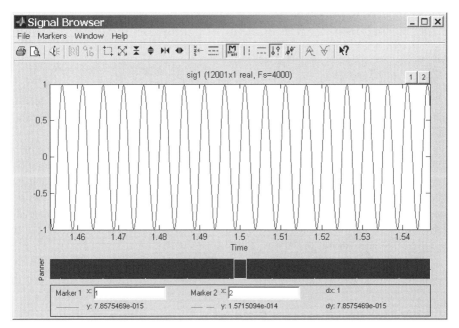

Figure 2.5 **Signal Browser** window

3. Play both sine wave `sig1` and noisy sine wave `sig2` and observe their differences.
4. Select both `sig1` and `sig2` in the SPTool window, and click on **View** to evaluate these two signals in the same **Signal Browser** window. Note that you may need to zoom in to see the differences between these two signals in details.
5. Repeat Exercise 2 for signals saved in files `sineNoise3sec_01.mat`, `sineNoise3sec_05.mat`, and `sineNoise3sec_1.mat`.

2.3 INTRODUCTION TO DIGITAL SYSTEMS

A DSP system (or algorithm) performs operations on digital signals to achieve predetermined objectives. In some applications, the single-input, single-output DSP system processes an input signal $x(n)$ to produce an output signal $y(n)$. The general relationship between $x(n)$ and $y(n)$ is described as

$$y(n) = F[x(n)], \qquad (2.3.1)$$

where F denotes the function of the digital system. A block diagram of the DSP system defined in Equation 2.3.1 is illustrated in Figure 2.6. The processing of digital signals can be described in terms of combinations of fundamental operations

Figure 2.6 General block diagram of digital system

including addition (or subtraction), multiplication, and time shift (or delay). Therefore, a DSP system consists of the interconnection of adders, multipliers, and delay units.

A digital filter alters the spectral content of input signals in a specified manner. Common filtering objectives include removing noises, improving signal quality, extracting information from signals, and separating signal components that have been previously combined. A digital filter is a mathematical algorithm that can be implemented in digital hardware and software and operates on a digital input signal for achieving filtering objectives. A digital filter can be classified as being linear or nonlinear, time invariant or time varying.

The objective of this section is to introduce several simple digital systems for reducing noises and thus enhancing the desired signals.

2.3.1 Moving-Average Filters: Structures and Equations

As shown in Figure 2.2, the effect of noise causes signal samples to fluctuate from the original values; thus the noise may be removed by averaging several adjacent samples. The moving (running)-average filter is a simple example of a digital filter. An L-point moving-average filter is defined by the following input/output (I/O) equation

$$y(n) = \frac{1}{L}[x(n) + x(n-1) + \ldots + x(n-L+1)] = \frac{1}{L}\sum_{l=0}^{L-1} x(n-l), \quad (2.3.2)$$

where each output signal $y(n)$ is the average of L consecutive input signal samples.

EXAMPLE 2.3

To remove the noise that corrupts the sine wave as shown in Figure 2.2, we implement the moving-average filter defined in Equation 2.3.2 with $L = 5$, 10, and 20, using the MATLAB code `example2_3.m`. The original sine wave, noisy sine wave, and filter output for $L = 5$ are shown in Figure 2.7. This figure shows that the moving-average filter with $L = 5$ is able to reduce the noise, but the output signal is different from the original sine wave in terms of amplitude and phase. In addition, the output waveform is not as smooth as the original sine wave, which indicates that the moving-average filter has failed to completely remove all the noise components.

In addition, we plot the filter outputs for $L = 5$, 10, and 20 in Figure 2.8 for comparison. This figure shows that the filter with $L = 10$ can remove more noise than the filter with

2.3 Introduction to Digital Systems

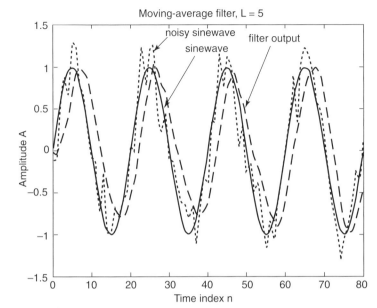

Figure 2.7 Performance of moving-average filter, $L = 5$

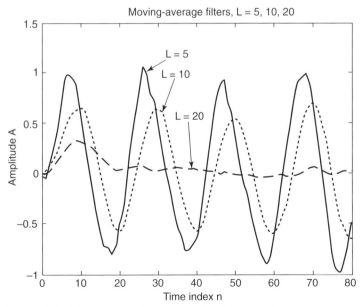

Figure 2.8 Moving-average filter outputs, $L = 5$, 10, and 20

$L = 5$ because the filter output is smoother for $L = 10$; however, the amplitude of the filter output is further attenuated. When the filter length L is increased to 20, the sine wave component is attenuated completely.

QUIZ 2.3

1. Figure 2.7 shows that the filtered ($L = 5$) output sine wave has an amplitude similar to that of the original sine wave, but is shifted to the right by 2 samples. Figure 2.8 also shows that the filtered output for $L = 10$ has further shifted to the right. Why?
2. Figure 2.8 shows that the filter with $L = 10$ produces smoother output than the filter with $L = 5$. Why? However, the amplitude of the output sine wave is further attenuated. Why?
3. Why does the filter output with $L = 20$ approach 0 after 20 samples? (Hint: Plot 20 samples of the waveform and you may find it exactly covers one period of sine wave. This question is examined further in Chapter 3 with frequency-domain techniques.)

We answer some questions in this chapter; however, the best answer is related to the frequency response of the filter, which is introduced in Chapter 3.

EXERCISE 2.3

1. Modify the MATLAB code `example2_3.m` by decreasing SNR as follows:

   ```
   xn = sn + randn(size(n));
   ```

   Run the code and compare the filter outputs with those given in Example 2.3. We observe that the moving-average filter has failed to reduce the noise to an acceptable level, even with $L = 10$. Why?
2. Modify `example2_3.m` by increasing the length of signal from 80 samples to 3s of samples. Run the code with different SNR settings, and save the results using .mat format.
3. Import the saved data files generated in Exercise 2 into SPTool, and evaluate and play these waveforms.

Implementation of Equation 2.3.2 requires $L - 1$ additions and L memory locations for storing signal sequence $x(n), x(n-1), \ldots, x(n-L+1)$ in a memory buffer. As illustrated in Figure 2.9, the signal samples used to compute the output signal $y(n)$ at time n are L samples included in the window at time n. These samples are almost the same as the samples used in the previous window at time $n-1$ to compute $y(n-1)$, except that the oldest sample $x(n-L)$ in the window at time $n-1$ is replaced by the newest sample $x(n)$ of the window at time n. Thus Equation 2.3.2 can be computed as

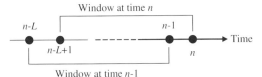

Figure 2.9 Concept of moving window in signal processing

$$y(n) = y(n-1) + \frac{1}{L}[x(n) - x(n-L)]. \quad (2.3.3)$$

Therefore, the averaged signal, $y(n)$, can be computed recursively based on the previous result $y(n-1)$. This recursive equation can be realized by using only two additions. However, we need $L+1$ memory locations for keeping $L+1$ signal samples $\{x(n), x(n-1) \ldots x(n-L)\}$ and another memory location for storing $y(n-1)$.

The recursive equation given in Equation 2.3.3 is often used in DSP algorithms, which involves the output feedback term $y(n-1)$ for computing current output signal $y(n)$. This type of filters is discussed further in Chapter 3.

2.3.2 Digital Filters

The I/O equation given in Equation 2.3.2 can be generalized as a difference equation with L parameters, expressed as

$$\begin{aligned} y(n) &= b_0 x(n) + b_1 x(n-1) + \ldots + b_{L-1} x(n-L+1) \\ &= \sum_{l=0}^{L-1} b_l x(n-l), \end{aligned} \quad (2.3.4)$$

where b_l are the filter coefficients. The moving-average filter coefficients are all equal as $b_l = 1/L$. We can use filter design techniques (introduced in Chapter 4) to determine different sets of coefficients for a given specification to achieve better performance.

Define a unit impulse function as

$$\delta(n) = \begin{cases} 1, & n = 0 \\ 0, & n \neq 0 \end{cases}. \quad (2.3.5)$$

Substituting $x(n) = \delta(n)$ into Equation 2.3.4, the output is called the impulse response of the filter, $h(n)$, and can be expressed as

$$h(n) = \sum_{l=0}^{L-1} b_l \delta(n-l) = b_0, b_1, \ldots, b_{L-1}, 0, 0, \ldots \quad (2.3.6)$$

Therefore, the length of the impulse response is L for the I/O equation defined in Equation 2.3.4. This type of filter is called a finite impulse response (FIR) filter.

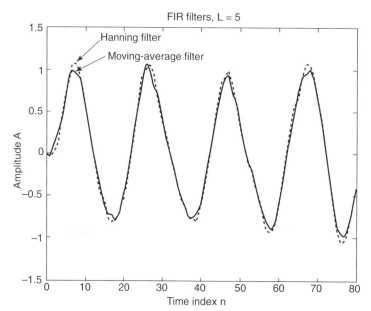

Figure 2.10 Comparison of moving-average and Hanning filters, $L = 5$

The impulse responses of the filter, $h(n)$, are the same as the FIR filter coefficients (weights or taps), b_l, $l = 0, 1, \ldots, L - 1$.

EXAMPLE 2.4

The moving-average filter defined by the I/O equation defined in Equation 2.3.2 is an FIR filter of length L (order $L - 1$) with the same coefficients $b_l = 1/L$. Consider an FIR Hanning filter of length $L = 5$ with the coefficient set {0.1 0.2 0.4 0.2 0.1}. Similar to Example 2.3, we implement this filter with MATLAB script example2_4.m and compare the performance with the moving-average filter of the same length. The outputs of both filters are shown in Figure 2.10. The results show that the five-point Hanning filter has less attenuation than the moving-average filter. Therefore, we show that better performance can be achieved by using different filter coefficients derived from filter design techniques.

HANDS-ON EXPERIMENT 2.2

Exercise 2.3 shows that the simple moving-average filter has failed to enhance signals with low SNR. In this experiment, we use hands-on exercises based on the SPTool for designing FIR filters for this purpose. Following procedures similar to those used in Experiment 2.1, we import four data files (generated in Exercise 2.1), sineNoise3sec_01.mat, sine-Noise3sec.mat, sineNoise3sec_05.mat, and sineNoise3sec_1.mat, into SPTool, and name them sig1, sig2, sig3, and sig4, respectively. We display the waveforms and play these signals.

2.3 Introduction to Digital Systems   39

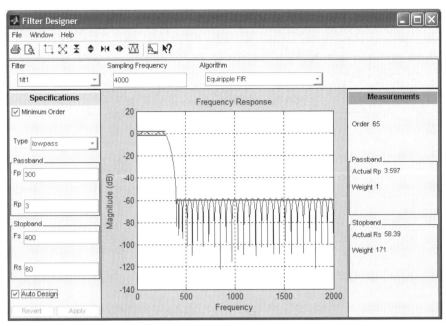

Figure 2.11 **Filter Designer** window

As shown in Figure 2.3, the **Filter Designer** (in middle column) can be used for designing digital filters. We can simply click on the **New** button to design a new filter or the **Edit** button for an existing filter under the **Filters** column in the SPTool to open the **Filter Designer** window as shown in Figure 2.11. We can design low-pass, high-pass, bandpass, and bandstop filters by using different filter design algorithms, which are introduced in Chapter 4. In this experiment, we learn how to design FIR filters for enhancing sinusoidal signals. As shown in Figure 2.11, we enter 4000 in the **Sampling Frequency** box. In the **Specifications** region, we use 300 as Passband Fp, 400 as Stopband Fs, and 60 as Rs. Note that the designed filter is called filt1, as shown in the **Filter** region, which will also appear in the **Filters** region of the SPTool window.

Once the filter has been designed, the frequency specification and other filter characteristics can be verified by using the **Filter Viewer**. Selecting the name (filt1) of the designed filter and clicking on the **View** button under the **Filters** column in the SPTool will open the **Filter Viewer** window as shown in Figure 2.12 for analyzing the designed filter.

When the filter characteristics have been confirmed, we can then select the input signal (sig1) and the designed filter (filt1). We click on the **Apply** button to perform filtering and generate the output signal. The **Apply Filter** window, shown in Figure 2.13, allows us to specify the file name of the output signal (sig1_out). The **Algorithm** list provides a choice of several filter structures, which are briefly introduced in the next section and discussed in detail in Chapter 4. We repeat this process for sig2, sig3, and sig4 and produce filter outputs sig2_out, sig3_out, and sig4_out, respectively. We can evaluate the filter performance by viewing and playing the waveforms.

40 Chapter 2 Time-Domain Signals and Systems

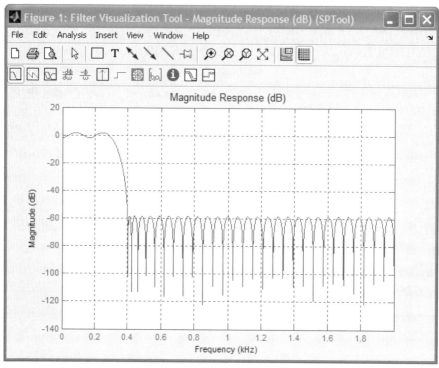

Figure 2.12 **Filter Viewer** window

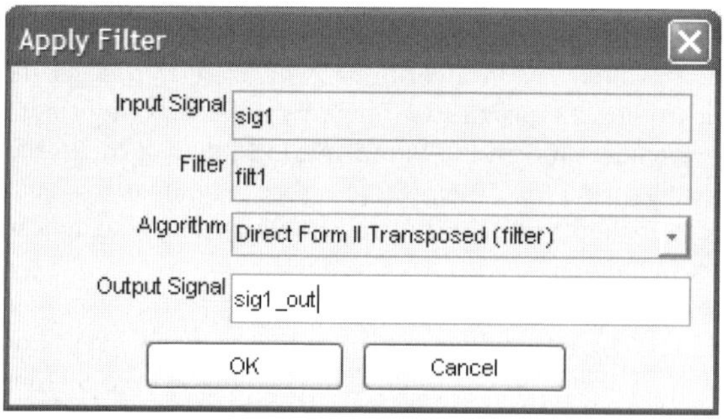

Figure 2.13 **Apply Filter** window

EXERCISE 2.4

1. View the input/output pairs, for example, sig1 with sig1_out, to compare the filter input and output on the same graph. Repeat this for other pairs.

2. Design a bandpass filter as shown in Figure 2.14. Use the designed filter to reduce noise in four files. Evaluate the filter performance by viewing the input/output pairs and also playing the tones.

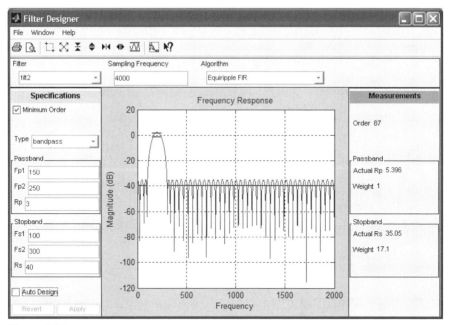

Figure 2.14 Bandpass filter design

3. Compare the performance of the low-pass filter shown in Figure 2.11 and the bandpass filter shown in Figure 2.14. Which filter has better performance? Why?
4. Compare the performance of the bandpass filter with the performance of the moving-average filter.

2.3.3 Realization of FIR Filters

As mentioned above, a DSP system consists of the interconnection of adders, multipliers, and delay units. In this section, we use the FIR filter given in Equation 2.3.4 as an example to show how to realize an FIR filter by using these basic building blocks.

A sample-by-sample addition of two sequences, $x_1(n)$ and $x_2(n)$, is illustrated in Figure 2.15(a), a multiplier is illustrated in Figure 2.15(b), and a delay unit is illustrated in Figure 2.15(c), where the box labeled z^{-1} represents a unit delay. A delay by M units can be implemented by cascading M delay units in a row, configured as a first-in first-out signal (or delay) buffer. This buffer is called the tapped-delay line, or simply the delay line. With these basic units, the FIR filter defined in Equation 2.3.4 can be realized in Figure 2.16.

The FIR filtering defined in Equation 2.3.4 is identical to the linear convolution of two sequences, $h(n)$ and $x(n)$, defined as

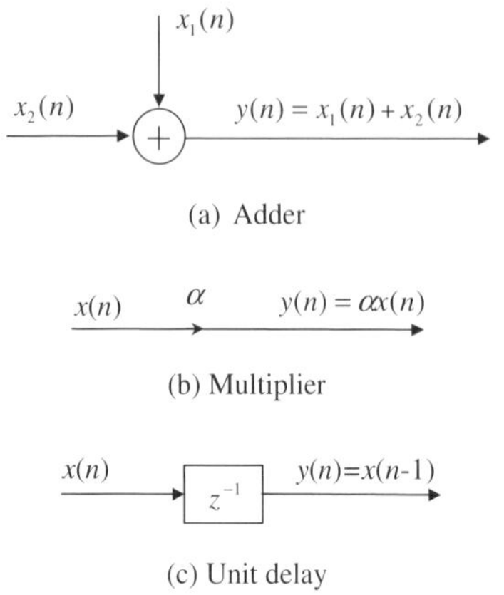

42 Chapter 2 Time-Domain Signals and Systems

(a) Adder: $y(n) = x_1(n) + x_2(n)$

(b) Multiplier: $y(n) = \alpha x(n)$

(c) Unit delay: $y(n) = x(n-1)$

Figure 2.15 Block diagram of basic units of digital systems: an adder (a), a multiplier (b), and a unit delay (c)

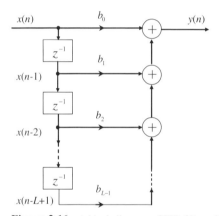

Figure 2.16 A block diagram of FIR filter with length L

$$y(n) = x(n) * h(n) = \sum_{i=-\infty}^{\infty} x(i)h(n-i) = \sum_{i=-\infty}^{\infty} h(i)x(n-i), \qquad (2.3.7)$$

where $*$ denotes linear convolution. For causal sequences (values equal to zero for $n < 0$), the summation in Equation 2.3.7 starts from $i = 0$.

EXAMPLE 2.5

Given two sequences $h(n) = \{1, 2, 2, 1\}$ and $x(n) = \{1, 2, 3, 4\}$ for $n = 0, 1, 2, 3$, the linear convolution operation defined by Equation 2.3.7 can be evaluated with the following graphical approach:

1. Reverse the time sequence $x(i)$ (reflected about the origin) to obtain $x(-i)$.
2. Shift $x(-i)$ to the right by n samples if $n > 0$ (or to the left n samples if $n < 0$) to form the sequence $x(n - i)$.
3. Compute the products of $h(i)x(n - i)$ for those i that have nonzero overlap between $h(i)$ and $x(n - i)$.
4. Summing all the products $h(i)x(n - i)$ yields $y(n)$.

Using this graphical method, we have

$y(0) = h(0)x(0) = 1 \times 1 = 1$
$y(1) = h(0)x(1) + h(1)x(0) = 1 \times 2 + 2 \times 1 = 4$
$y(2) = h(0)x(2) + h(1)x(1) + h(2)x(0) = 1 \times 3 + 2 \times 2 + 2 \times 1 = 9$
$y(3) = h(0)x(3) + h(1)x(2) + h(2)x(1) + h(3)x(0) = 1 \times 4 + 2 \times 3 + 2 \times 2 + 1 \times 1 = 15$
$y(4) = h(1)x(3) + h(2)x(2) + h(3)x(1) = 2 \times 4 + 2 \times 3 + 1 \times 2 = 16$
$y(5) = h(2)x(3) + h(3)x(2) = 2 \times 4 + 1 \times 3 = 11$
$y(6) = h(3)x(3) = 1 \times 4 = 4$
$y(n) = 0$ for $n = 7, 8, \ldots$

In general, if $h(n)$ and $x(n)$ are two sequences of length L and N, respectively, the resulting sequence $y(n)$ is of length $L + N - 1$. The MATLAB function y = conv(h, x) implements the linear convolution of two sequences in vectors h and x, with the output sequence stored in vector y.

EXAMPLE 2.6

Considering the sequences given in Example 2.5, the linear convolution of $h(n)$ with $x(n)$ can be implemented by using MATLAB (example2_6.m) as follows:

```
xn = [1 2 3 4];
hn = [1 2 2 1];
yn = conv(xn, hn)
```

Execution of this program results in

```
>> example2_6
yn =
     1     4     9    15    16    11     4
```

In many real-world applications, the designed FIR filters have symmetric coefficients as follows:

$$b_l = b_{L-1-l}, \qquad (2.3.8)$$

for $l = 0, 1, \ldots, L/2 - 1$ if L is an even number or $(L - 1)/2$ if L is an odd number. In addition, those coefficients may be antisymmetric (or negative symmetric), expressed as

$$b_l = -b_{L-1-l}. \qquad (2.3.9)$$

Therefore, there are four types of symmetric FIR filters depending on whether L is even or odd and whether coefficients b_l have positive or negative symmetry. We revisit this issue in Chapter 3.

EXERCISE 2.5

1. Realize the moving-average filter defined by both Equations 2.3.2 and 2.3.3 with $L = 5$.
2. Realize the Hanning filter defined in Example 2.4.
3. Draw a signal-flow diagram of a positive symmetric FIR filter with $L = 5$.
4. Draw a signal-flow diagram of a positive symmetric FIR filter with $L = 6$.
5. Draw a signal-flow diagram of a positive symmetric FIR filter with general length L.

For the symmetric FIR filter defined in Equation 2.3.8, the difference equation given in Equation 2.3.4 can be simplified as

$$y(n) = b_0 x(n) + b_1 x(n-1) + \ldots + b_{L-1} x(n-L+1)$$
$$= b_0 [x(n) + x(n-L+1)] + b_1 [x(n-1) + x(n-L+2)] + \ldots \quad (2.3.10)$$

This shows that the number of multiplications required to implement the symmetric FIR filtering can be reduced to half if L is even. In addition, we only have to store half the amount ($L/2$) of coefficients because they are symmetric.

QUIZ 2.4

1. Does a symmetric FIR filter save the required number of additions?
2. Rewrite Equation 2.3.10 for antisymmetric FIR filters.
3. Redraw a signal-flow diagram for antisymmetric FIR filters if L is an even number.
4. Redraw a signal-flow diagram for antisymmetric FIR filters if L is an odd number.

A linear system is a system that satisfies the superposition principle, which states that the response of the system to a weighted sum of signals is equal to the corresponding weighted sum of the responses of the system to each of the individual input signals. That is, a system is linear if and only if

$$F[\alpha_1 x_1(n) + \alpha_2 x_2(n)] = \alpha_1 F[x_1(n)] + \alpha_2 F[x_2(n)] \quad (2.3.11)$$

for any arbitrary input signals $x_1(n)$ and $x_2(n)$ and for any arbitrary constants α_1 and α_2. If the input is the sum (superposition) of two or more scaled sequences, we can find the output due to each sequence alone and then add the separate scaled outputs.

QUIZ 2.5

Identify whether the following systems are linear or nonlinear:

1. $y(n) = b_0 x(n) + b_1 x(n-1)$
2. $y(n) = b_0 x(n) + b_1 x^2(n)$
3. $y(n) = b_0 x(n) + b_1 x(n) y(n-1)$

We have introduced moving-average, Hanning, low-pass, and bandpass filters in this Section. These digital filters can be classified as linear, time-invariant (LTI) filters. A digital system is time invariant if its input-output characteristics do not change with time.

2.4 NONLINEAR FILTERS

In this section, we present a simple nonlinear filter, the median filter, which is very effective for reducing impulse noises.

EXAMPLE 2.7

Example 2.2 shows that the desired sine wave is corrupted by random noise, which can be reduced by linear FIR filters. For some real-world applications, the signal is corrupted by impulse noise as shown in Figure 2.17. Similar to Example 2.3, we use a moving-average filter with $L = 5$ for the noisy sine wave (see example2_5.m). Figure 2.17 clearly shows that the energy presented in the impulse noise will still influence the linear filter output; thus the linear filter may not be as effective for reducing impulse noises.

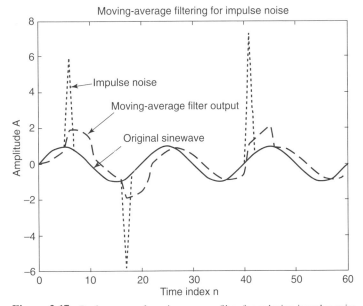

Figure 2.17 Performance of moving-average filter for reducing impulse noise

46 Chapter 2 Time-Domain Signals and Systems

An efficient technique for reducing impulse noise is the use of a nonlinear median filter. An L-point running median filter can be implemented as a first-in first-out buffer of length L to store input signals $x(n), x(n-1), \ldots, x(n-L+1)$. These samples are moved to a new sorting buffer, where the elements are ordered by magnitude. The output of the running median filter $y(n)$ is simply the median of the L numbers in the sorting buffer. Medians will not smear out discontinuities in the signal if the signal has no other discontinuities within $L/2$ samples and will approximately follow low-order trends in the signal. Note that running medians is a nonlinear algorithm that does not obey the superposition property described by Equation 2.3.11.

EXAMPLE 2.8

MATLAB provides a function `medfilt1` for implementing a one-dimensional median filter. The following command

    y = medfilt1(x,L);

returns the output of the order L (default is 3), one-dimensional median filtering of x. The output vector y is the same size as x; for the edge points, zeros are assumed to the left and right of x.

A five-point median filter is implemented in `example2_8.m` for reducing the same impulse noise shown in Figure 2.17. The original sine wave, the corrupted sine wave, and the median filter output are shown in Figure 2.18. It is clearly shown that the nonlinear median filter is very effective in removing impulse noise.

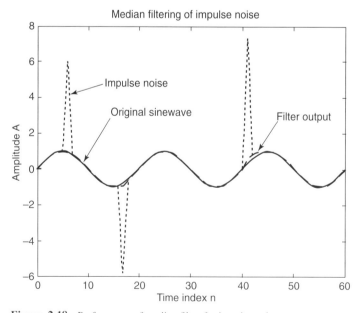

Figure 2.18 Performance of median filter for impulse noise

Figure 2.19   Cascade of median and moving-average filters

Although median filters generally preserve sharp discontinuities in a signal, they often fail to provide sufficient reduction of broadband noises. A good compromise is to use a combination of a linear filter such as a Hanning filter and a nonlinear median filter. See Exercise 2.6 for details.

EXERCISE 2.6

1. Generate a noisy sine wave that is corrupted by random noise and impulse noise.
2. Apply the 5-point moving-average filter or the Hanning filter to the noisy sine wave. Is it effective?
3. Apply the 5-point median filter to the noisy sine wave. Is it effective?
4. Apply both the 5-point moving-average filter and the median filter to the noisy sine wave. These two filters are connected in cascade form as illustrated in Figure 2.19. Compare the result with the results obtained in Exercises 2 and 3.
5. Change the order of filters shown in Figure 2.19, that is, use the moving-average filter first. Compare the results with Exercise 4. Which one is better? Why?

2.5 MORE HANDS-ON EXPERIMENTS

We used simple linear filters in Section 2.3 to enhance sinusoidal signals that were corrupted by random noise and introduced a nonlinear median filter in Section 2.4 for reducing impulse noise. In this section, we present different signals corrupted by noises and show that these filters have difficulty in solving the problems. This will give us the motivation to learn advanced filters in Chapters 3 and 4.

EXAMPLE 2.9

In this example, we use the MATLAB code `mulSineGen.m` for simulating the desired sine wave that is corrupted by another sine wave at a different frequency. This program generates the following digital signals and saves the files in both ASCII and .mat formats:

1. 200-Hz sine wave is corrupted by 400-Hz sine wave with sampling rate of 4,000 Hz (name the files `sine200plus400at4k.dat` and `.mat`).

2. 200-Hz sine wave is corrupted by 400-Hz sine wave with sampling rate of 1,000 Hz (sine200plus400at1k.dat and .mat).

3. 200-Hz sine wave is corrupted by 60-Hz power line hum with sampling rate of 4,000 Hz (sine200plus60at4k.dat and .mat).

EXERCISE 2.7

1. Import the three files in Example 2.9 into SPTool, view, and play the signals.

2. Design a moving-average filter to reduce the undesired tonal noise. Is the filter working? Why not?

3. In Exercise 2.4, we designed a bandpass filter for enhancing the sinusoidal signal corrupted by random noise, which achieved better performance than the moving-average filters. In this exercise, we use SPTool to design a bandstop filter to attenuate the undesired sinusoidal noises. For example, when a 200-Hz sine wave is corrupted by a 400-Hz sine wave, we can reduce the undesired 400-Hz sine wave by designing a bandstop filter at 400 Hz. Design different bandstop filters for other cases.

4. Design a low-pass filter to attenuate the 400-Hz sine wave and pass the desired 200-Hz sinewave.

5. Design a high-pass filter to attenuate 60-Hz hum, thus enhancing the desired 200-Hz sine wave.

6. Similar to Exercise 5, design a bandpass filter to attenuate 60-Hz hum.

7. Similar to Exercise 5, design a bandstop filter to attenuate 60-Hz hum.

8. Evaluate the performance of filters designed in Exercises 5, 6, and 7 and compare the required filter lengths.

EXERCISE 2.8

In MATLAB script exercise2_8.m, a speech file timit1.asc (with 8-kHz sampling rate) is corrupted by (1) 1,000-Hz tonal noise (the corrupted signal is saved in data file speech_tone.mat) and (2) random noise (the corrupted signal is saved in data file speech_random.mat).

1. Use SPTool to view and play the original speech and the speech corrupted by tonal noise and random noise.

2. Design and use moving-average, Hanning, and nonlinear median filters to reduce noise. Are these filters working?

3. Design a bandstop filter at 1,000 Hz for reducing the tonal noise in speech_tone.mat. View and play the result. Is this filter working? Why?

4. Are you able to design a digital filter with SPTool to enhance the speech signal that was corrupted by random noise? Why not?

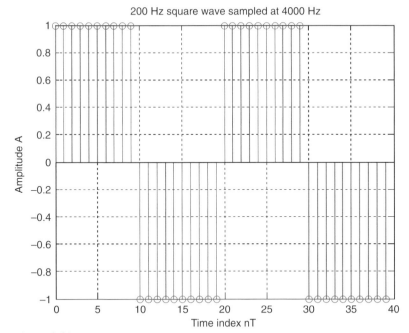

Figure 2.20 A square wave

EXAMPLE 2.10

Similar to Example 2.1, we use the MATLAB function `square` to generate a square wave with two periods (see `example2_10.m`). The generated waveform is displayed in Figure 2.20 with the MATLAB function `stem`.

Similar to Example 2.3, we use this square as input to the moving-average filter with $L = 5$. The output waveform is shown in Figure 2.21, where we observe that the corners of the square wave are smoothed.

EXERCISE 2.9

1. Modify the MATLAB code `example2_10.m` by increasing the length from 40 to 8,000, and use the MATLAB function `soundsc` to play the generated signal.
2. Modify the MATLAB code `example2_10.m` by generating a 200-Hz sine wave of length 8,000, and use the MATLAB function `soundsc` to play the generated signal. Observe the differences as compared with the square wave played in Exercise 1.
3. In Figure 2.20, we plot the square wave with `stem`. Display the waveform with the function `plot` and observe and explain the differences.
4. In Figure 2.21, we use the moving-average filter with $L = 5$. Try different filter lengths, display and play the outputs, and compare the differences.

50   Chapter 2   Time-Domain Signals and Systems

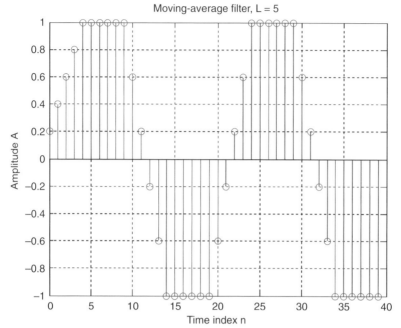

Figure 2.21   Output from the moving-average filter with $L = 5$

5. Use the Hanning filter defined in Example 2.4 to filter the square wave and compare the result with the output from the moving-average filter.

6. Add a white noise to the square wave and use the moving-average filter to reduce noise. Is this filter effective? Why?

2.6   IMPLEMENTATION OF MOVING-AVERAGE FILTERS WITH BLACKFIN SIMULATOR

In this section, we implement the moving-average filters with the Blackfin simulator (introduced in Chapter 1) to verify the correctness of programs written in C and assembly. We write a C program that implements the moving-average filter and test it on data files that contain the sine wave corrupted by noise. The processed signal from the simulator is saved in new data files and compared with the original one to evaluate filter performance.

HANDS-ON EXPERIMENT 2.3

In this experiment, we write a simple moving-average filter program in C based on Equation 2.3.2. This C program (ma.c) reads in the input noisy signal from the data file, sineNoise3sec_4k.dat, which consists of a sine wave and noise as described in

2.6 Implementation of Moving-Average Filters 51

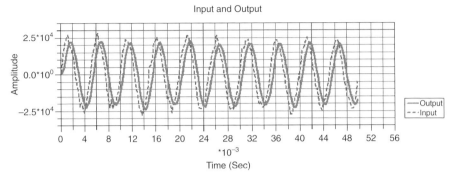

Figure 2.22 Graphical plot of the input and output signals

Hands-On Experiment 2.1. A moving-average filter of length $L = 5$ is implemented to remove the noise. The filter output is stored in the data file `filterout.dat`. Because the file is long, users can use the Blackfin compiled simulator for BF533 to speed up the simulation.

Open the project file `exp2_3.dpj` in directory `c:\adsp\chap2\exp2_3`. Study the program to understand its flow and statements. Note that the coefficients of the filter (`lpf`) are represented in integer format (`fract16`), instead of the double-precision floating-point (64 bits) format used in the MATLAB environment. Integer format is commonly used in programming fixed-point digital signal processors like the Blackfin processor. The DSP run time library for the Blackfin processor is written in integer format. A detailed explanation of the different number formats used in the Blackfin processor is provided in Chapter 4.

Set the options as stated in Hands-On Experiment 1.1. Build the project and run the loaded code by clicking on the run icon . Users can view both the input and processed data by using the graphical display of the VisualDSP++. With the steps described in Hands-On Experiment 1.3, a single graph is created to show both the input and output signals, as shown in Figure 2.22. Users are encouraged to explore different plotting features provided in the VisualDSP++.

The VisualDSP++ simulator can play back the displayed waveform by exporting the data to the sound card. First, display the input signal `in` and the output signal `out` in separate plots. Configure the sampling rate to 4,000 Hz. Right-click on the plot, and click on **Export → Sound Card → Export**. Extend the duration of play back.

EXERCISE 2.10

1. Modify the length of the moving-average filter to $L = 10$ and 20. Compare the results with those shown in Figure 2.8.

2. Instead of using the moving-average filter, implement the 5-tap FIR Hanning filter defined in Example 2.4 with the VisualDSP++ simulator and verify its result.

3. Replace the DSP library function `fir_fr16` from the main program by a simple C code that implements Equation 2.3.3.

2.7 IMPLEMENTATION OF MOVING-AVERAGE FILTERS WITH BF533/BF537 EZ-KIT

We used the Blackfin simulator to test the functionality of the moving-average filter in Section 2.6. In this section, we examine how to acquire and process signal in real time with the EZ-KIT. The Blackfin EZ-KIT provides a platform to take in a real-world signal and process it in real time. We can evaluate the effects of filtering by evaluating the input and output signals. In the following experiments, a noisy signal is sampled by the EZ-KIT via the audio I/O channel, and the Blackfin processor executes the moving-average filter to remove the tonal noise in the signal. The processed signal from the EZ-KIT is passed to a headphone (or loudspeaker) for real-time playback. Therefore, the EZ-KIT performs the entire DSP chain of A/D conversion, processing, and D/A conversion. More details on the process of CODEC (coder/decoder, including A/D and D/A converters and the associated low-pass filters) and how the digital sample is sampled and processed are provided in Chapters 6 and 7.

It is important to note that the EZ-KIT can only support a sampling rate of 48 or 96 kHz. In our experiments, the CODEC is set to 48 kHz by the initialization program. Detailed description and explanation of the CODEC settings is given in Chapter 7.

HANDS-ON EXPERIMENT 2.4

In this experiment, we connect devices such as the sound card and headphone to the ADSP-BF533 EZ-KIT board. The EZ-KIT has four input channels (or two stereo pairs) and six output channels (or three stereo pairs). In this experiment, only one pair of input and output channels are used, as indicated in Figure 2.23. Note that right connectors are needed for connecting the devices to the EZ-KIT. *Make sure that pins 5 and 6 of SW9 on the EZ-KIT*

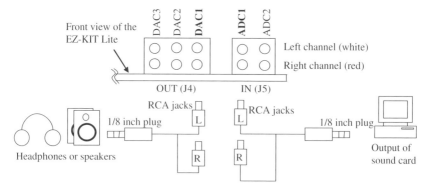

Figure 2.23 Connecting the input and output audio jacks of the BF533 EZ-KIT Lite

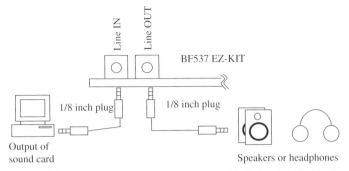

Figure 2.24 Connecting the input and output audio jacks of the BF537 EZ-KIT Lite

are set to ON (i.e., pins 5 and 6 are in line with the rest of the pins in SW9). Please refer to Figure 1.10 in Chapter 1 for the location of SW9 in the BF533 EZ-KIT.

If the BF537 EZ-KIT is used, there are only one pair of stereo input and one pair of stereo output channels available, as shown in Figure 2.24. All the connectors used in the BF537 EZ-KIT are $\frac{1}{8}$-in. stereo jack.

Activate the VisualDSP++ for the BF533 (or BF537) EZ-KIT target and open exp2_4_533.dpj (or exp2_4_537.dpj) in directory c:\adsp\chap2\exp2_4_533 (or c:\adsp\chap2\exp2_4_537). Build the project, and the executable file (exp2_4_533.dxe or exp2_4_537.dxe) is automatically loaded into the memories of the EZ-KIT. Select speech_tone_48k.wav in directory c:\adsp\audio_files and play this wave file continuously on the computer. Note that the file speech_tone_48k.wav is derived from the data file speech_tone.mat generated in Exercise 2.8. A MATLAB file, convert_ez_kit.m, is used to convert the data file into a wave file sampling at 48 kHz. Examine this MATLAB program and see how to convert the 8-kHz sampling rate to 48 kHz. Run the program and listen to the original wave file from the headphone. See the exact location of the switches in BF533 and BF537 EZ-KITs in Figure 1.10 and Figure 1.11, respectively. Press **SW5** on the BF533 EZ-KIT (or **SW11** on the BF537 EZ-KIT) to activate the moving-average filter; listen to the processed signal again, and compare it with the original wave file by pressing **SW4** on the BF533 EZ-KIT (or **SW10** on the BF537 EZ-KIT). Explain why the moving-average filter of length $L = 48$ is required to attenuate the 1,000 Hz sine wave from the speech signal sampling at 48 kHz.

EXERCISE 2.11

1. Modify the source code in the project, exp2_4_533.dpj (or exp2_4_537.dpj), to implement moving-average filters of $L = 24$ and 36. Listen to the differences and observe the output signals, using the VisualDSP++ plots. Are we able to remove the sine wave by using the moving-average filters of $L = 24$ and 36? Why?

2. Benchmark the cycle count needed to complete processing of one sample, using $L = 24$, 36, and 48.

3. Instead of using the I/O equation (Eq. 2.3.2) to implement the moving-average filter, a more computational efficient equation (Eq. 2.3.3) can be used. Implement the moving-average filter with Equation 2.3.3 in the EZ-KIT. Benchmark its cycle count, and comment on the differences.

4. Design a bandstop filter centered at 1,000 Hz with a sampling frequency of 48 kHz using SPTool to reduce the tonal noise in `speech_tone_48k.wav`.

2.8 MOVING-AVERAGE FILTER IN LABVIEW EMBEDDED MODULE FOR BLACKFIN PROCESSORS

Digital filters can be easily simulated, prototyped, deployed, and tested in LabVIEW. The following two experiments illustrate the process for simulating and prototyping a moving-average filter for the Blackfin BF533/BF537 processor. First, the Digital Filter Design toolkit is used to develop and simulate the filter on the computer and test its performance with simulated input signals. Next, the LabVIEW Embedded Module for Blackfin Processors is used to program the Blackfin EZ-KIT to perform real-time audio processing that filters an audio signal corrupted by a sine wave.

HANDS-ON EXPERIMENT 2.5

In this experiment, we design a moving-average filter to remove a 1,000-Hz tone that muffles the speech signal. An interactive LabVIEW application has been created to simulate and test filtering results with the actual audio signal.

To begin designing and simulating the filter, open the executable file `MA_Filter_Sim.exe`, located in directory `c:\adsp\chap2\exp2_5`. This LabVIEW application was created with LabVIEW for Windows and the LabVIEW Digital Filter Design toolkit. The simulator has been preloaded with the 3s of audio file, `speech_tone_48k.wav`, used in previous VisualDSP++ experiments. If there are loudspeakers attached to the computer, click on **Play** to hear the audio. Do you hear the sine wave obscuring the speech?

Use the **Taps** control to experiment with different numbers of FIR filter taps. Change the **Apply Filter?** control to **Filtered Signal** to see the output signal after the filter is applied. Click on **Play** again to hear the filtered signal. Note that the 48-tap moving-average filter places a zero precisely on the 1,000-Hz tone that distorts the speech, as shown in Figure 2.25. Use the **Zoom** tool to zoom in on the graph or the **Autoscale** feature to see the entire signal. We can also load a new audio file (.wav), and use the same moving-average filter to filter the wave file and listen to the results. Compare the filtered output with the original signal and comment on the results.

2.8 Moving-Average Filter in Labview Embedded Module   55

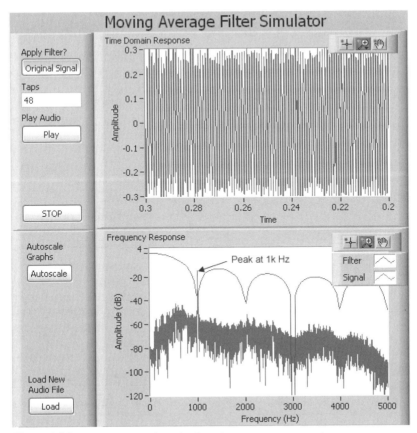

Figure 2.25 Moving-average filter simulator (MA_Filter_Sim.exe)

HANDS-ON EXPERIMENT 2.6

In this experiment, we use the LabVIEW Embedded Module for Blackfin Processors with the Blackfin EZ-KIT to test the designed moving-average filter. Based on the results from Hands-On Experiment 2.5, we implement a 48-tap moving-average filter on the BF533/BF537 EZ-KIT for real-time experiments.

When implementing audio experiments on the Blackfin processor, a good starting point is the Audio Talkthrough example shown in Figure 2.26. The project file Audio Talkthrough-BF533.lep for BF533 EZ-KIT (or Audio Talkthrough-BF537.lep for BF537) can be found in directory c:\adsp\chap2\exp2_6, which is useful for testing the audio setup consisting of computer, cables, and loudspeakers (or headphone). On the block diagram outside of the **While Loop**, the **Initialize Audio** function creates a circular buffer between the audio CODEC and the Blackfin processor. Inside the **While Loop**, the **Audio Write–Read** function writes right and left channels of audio data to the output **Audio Buffer**, all zeros on the first iteration of the loop, and then reads in newly acquired audio samples to the right and left channel arrays for processing. This step allows data to be read

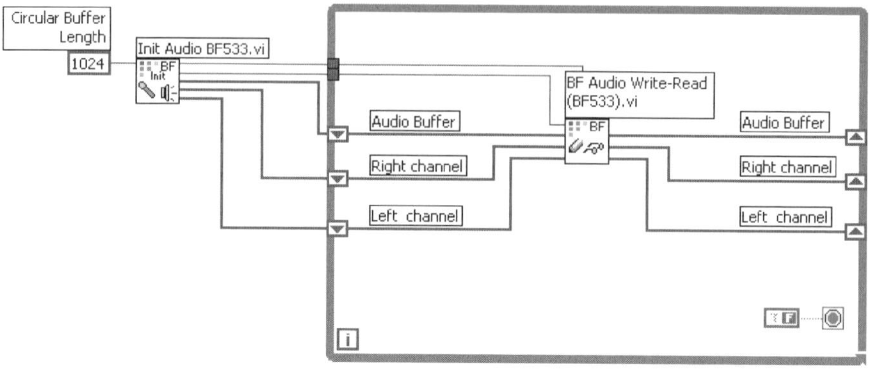

56 Chapter 2 Time-Domain Signals and Systems

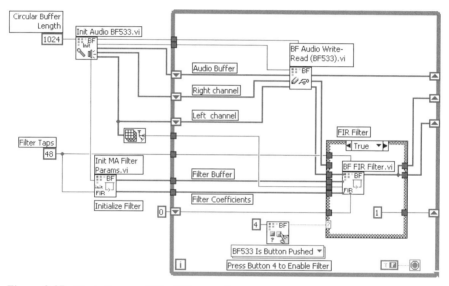

Figure 2.26 Audio talkthrough example

Figure 2.27 Block diagram of MA Filter.vi

and processed in one-loop iteration and passed out to the loudspeaker (or headphone) in the next iteration. The process of reading values in the first iteration and then writing them out in the next is called pipelining.

Once the audio talkthrough has been established, we still have to make two modifications. First, we must wire in an **FIR** function to filter the input signal; second, we must initialize that filter with proper parameters. Now open the complete moving-average filter program from the LabVIEW **Embedded Project Manager** window. The project is called MA Filter-BF533.lep for the BF533 EZ-KIT (or MA Filter-BF537.lep for BF537) and is located in directory c:\adsp\chap2\exp2_6. Figure 2.27 shows a block diagram of the top level VI in this project, MA Filter-BF533.vi (or MA Filter-BF537.vi).

In this experiment, we add two functions to complete the filtering exercise. Outside of the **While Loop**, the **Init MA Filter Params VI** sets up the moving-average filter designed

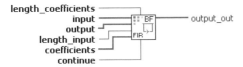

Figure 2.28 Context help information for the **BF FIR Filter** function

in Section 2.7 with the specified length (in this case $L = 48$), and outputs the filter coefficients, which have been converted from the decimal 1/48 (or 0.020833) to the fixed-point integer fract16 representation, a commonly used number format on embedded signal processors.

Inside the **While Loop**, the **BF Audio Write/Read** function writes data to the DAC and then reads values from the ADC through the circular buffer of 1,024 samples, 512 for each channel (left and right). The **BF FIR Filter** function (see Fig. 2.28) filters the input signal with the FIR filter defined by the filter coefficients. This filtering function is placed inside a **Case Structure** so that the signal passes through unchanged unless **SW4** is pressed on the BF533 EZ-KIT (or **SW10/PB4** is pressed on the BF537), allowing us to easily compare the effects of the filter.

When this project is compiled and run on the Blackfin target, play the speech_tone_48k.wav, using MA_Filter_Sim.exe from Hands-on Experiment 2.5. Note how the 1,000-Hz tone is filtered out when the filtered switch is pressed. Experiment with different filter lengths and see how they affect the results.

2.9 MORE EXERCISE PROBLEMS

1. The sampling rates for some applications are (a) 8 kHz for telecommunication systems, (b) 16 kHz for broadband systems, (c) 44.1 kHz for CDs, and (d) 48 kHz for digital audio tapes. What are the sampling periods for those systems? If 2 s of samples are needed and each sample is represented by 2 bytes, how many bytes of memory are needed for those systems?

2. If a 2-kHz signal is used in systems defined in Problem 1, what are the digital frequencies in terms of (a) radians per samples and (b) cycles per sample in those four digital systems?

3. Digital signals can be nicely displayed by the MATLAB function stem, where stem(x) plots the data sequence x as stems from the x-axis terminated with circles for the data value. Similar to Example 2.1, plot one cycle of 2-kHz sine wave with four different sampling rates defined in Problem 1.

4. Generate 256 random numbers, using the MATLAB functions rand and randn. Compute the mean and variance, using Equations 2.2.9 and 2.2.10, and verify the results by using the MATLAB functions mean and var.

5. How can we use rand for generating zero-mean random numbers? How can we make this zero-mean random variable have unit variance?

6. The power of the sine wave given in Equation 2.2.3 is $A^2/2$, and the power of random numbers is defined in Equation 2.2.10. Calculate the SNR for signals generated in Exercise 2.1(3).

7. Generate 256 samples of a sine wave that is corrupted by zero-mean white noise with SNR equal to 10, 0, and −10 dB. The frequency of the sine wave is 1 kHz, and the sampling rate is 8 kHz.

8. Apply the moving-average filter described in Equation 2.3.2 for noisy signals generated in Problem 7. What is the best filter length L for reducing noise? Why? Does this filter work for SNR = −10 dB?

9. Compute the impulse responses of filters described by Equations 2.3.2 and 2.3.3. Also, implement the filter defined in Equation 2.3.3 and compare the filter outputs from both Equations 2.3.2 and 2.3.3. Are those two outputs identical? Why or why not?

10. Compute the linear convolution of sequence $x(n) = \{1, 2, 3, 4\}$ with the impulse response of the Hanning filter defined in Example 2.4, using the graphical method. Also, use MATLAB function `conv` to realize this operation and verify both results.

11. Modify the C code given in this chapter to implement a median filter in the VisualDSP++ simulator or the EZ-KIT, as described in Examples 2.7 and 2.8.

12. Convert the MATLAB data file `speech_random.mat` into a wavefile and try to reduce the random noise by using the EZ-KIT, based on a bandpass filter that retains most of the speech energy.

13. Instead of acquiring signal directly from a sound card as in Hands-On Experiment 2.4, we can capture live sound by using a microphone connected to the EZ-KIT. The noisy signal can be played back with a pair of loudspeakers, and the volume can be adjusted to a suitable level. The noisy signal can then be captured by using a microphone that is fitted to the line in of the EZ-KIT. However, the microphone must be preamplified before connecting to the audio input of the EZ-KIT. Run the program in Hands-On Experiment 2.4 and listen to the output with a pair of headphones. Is the moving-average filter working? Compare the performance with the direct input connection from the sound card to that with the EZ-KIT.

14. Modify the source files in Hands-On Experiment 2.4 to implement a direct pass through without filtering. Further modify this program to perform the following operations on the left and right input signals:

$$\text{left output} = \frac{\text{left input} + \text{right input}}{2}$$

$$\text{right output} = \frac{\text{left input} - \text{right input}}{2}$$

Is there any difference between the left and right outputs? Why? Make sure that you are playing a stereo sound at the input.

15. Modify the project file in Hands-On Experiment 2.6 to implement the Hanning filter in Example 2.4.

Chapter 3

Frequency-Domain Analysis and Processing

This chapter introduces frequency-domain techniques for analyzing digital signals and systems. In particular, we focus on z-transform, system concepts, and discrete Fourier transform with their applications. We use these frequency-analysis methods to analyze and explain the noise reduction examples introduced in Chapter 2 and use frequency-domain techniques to design more advanced filters for reducing noises. In addition, we use several examples and hands-on experiments to introduce some useful MATLAB and Blackfin tools for analysis and design of DSP algorithms.

3.1 INTRODUCTION

In Chapter 2, we introduced simple time-domain techniques such as moving-average filters, Hanning filters, and nonlinear median filters for removing noises that corrupted the desired signals. In particular, we described those filters with time-domain methods such as I/O equations and signal-flow diagrams and used them to enhance sine waves embedded in white noise. We learned that they worked for some conditions, but failed for others. In this chapter, we introduce frequency-domain techniques to analyze those signals and systems and thus to understand their characteristics and explain the results obtained in Chapter 2. In addition, we use frequency-domain concepts and techniques to design effective filters for reducing noises.

In Example 2.3, we used a moving-average filter with length $L = 5$, 10, and 20 to enhance a sine wave corrupted by white noise. We found that this filter worked for $L = 5$ and 10, but failed for $L = 20$. We also found that the filter caused undesired effects such as attenuation of the desired sine wave amplitude and shifting of the phase of signal. In Example 2.9, we encountered cases in which the desired sine wave was corrupted by other sine waves at different frequencies, and we were not

---

Embedded Signal Processing with the Micro Signal Architecture. By Woon-Seng Gan and Sen M. Kuo
Copyright © 2007 John Wiley & Sons, Inc.

able to remove those narrowband noises. Again, in Exercise 2.8, we failed to attenuate a tonal noise in desired broadband signals such as speech. In this chapter, we use frequency-domain techniques to analyze those cases in order to understand the problems associated with the signals and the filters. We also use frequency-domain pole/zero concepts to develop notch and peak filters for reducing noises in those cases.

3.2 THE z-TRANSFORM

The z-transform is a powerful technique for analyzing digital signals and systems. This transform uses polynomials and rational functions to represent block diagrams and I/O equations of digital systems introduced in Section 2.3.

3.2.1 Definitions

The z-transform of a digital signal $x(n)$ is defined as

$$X(z) = \sum_{n=-\infty}^{\infty} x(n) z^{-n}, \tag{3.2.1}$$

where z is a complex variable. The set of z values for which $X(z)$ exists is called the region of convergence. We can recover $x(n)$ from $X(z)$ by using the inverse z-transform. For a causal sequence (i.e., $x(n) = 0$ for $n < 0$), the summation starts from $n = 0$. In addition, for a finite-length causal sequence $x(n)$, $n = 0, 1, \ldots, N - 1$, the summation starts from $n = 0$ and ends at $n = N - 1$.

EXAMPLE 3.1

Consider the causal signal defined as

$$x(n) = a^n u(n),$$

where $u(n) = 1$ for $n \geq 0$ and $u(n) = 0$ for $n < 0$ is the unit step function. From Equation 3.2.1 and using the formulas of the infinite-geometric series

$$\sum_{n=0}^{\infty} x^n = \frac{1}{1-x}, \quad |x| < 1, \tag{3.2.2}$$

the z-transform of the signal is calculated as

$$X(z) = \sum_{n=-\infty}^{\infty} a^n u(n) z^{-n} = \sum_{n=0}^{\infty} a^n z^{-n} = \sum_{n=0}^{\infty} (az^{-1})^n = \frac{1}{1-az^{-1}}, \quad |az^{-1}| < 1.$$

The region of convergence is defined as $|az^{-1}| < 1$ or

$$|z| > |a|,$$

which is the exterior of a circle with radius $|a|$.

EXERCISE 3.1

1. Find the z-transform of the signal $x(n) = Ae^{-j\omega n}u(n)$.
2. Find the z-transform of the unit impulse sequence $\delta(n)$ defined in Equation 2.3.5.
3. Find the z-transform of the unit step sequence $u(n)$ defined in Example 3.1.
4. Find the z-transform of the impulse response of a 5-point Hanning filter defined in Example 2.4 as $h(n) = \{0.1\ 0.2\ 0.4\ 0.2\ 0.1\}$.
5. Find the z-transform of the impulse response of a 4-point moving-average filter defined as $h(n) = \{0.25\ 0.25\ 0.25\ 0.25\}$. We will revisit this exercise later.

The complex variable z can be expressed in both Cartesian and polar forms as

$$z = \text{Re}[z] + j\,\text{Im}[z] = re^{j\theta}, \qquad (3.2.3)$$

where $\text{Re}[z]$ and $\text{Im}[z]$ represent the real part and the imaginary part of z, respectively. In the polar form, $r = |z|$ is the magnitude of z and θ is an angle with the real axis. The geometric representation in the complex z-plane is illustrated in Figure 3.1. In the figure, the circle labeled as $|z| = 1$ is called the unit circle. This representation is used to evaluate the frequency response and stability of a digital system.

If the signal $y(n)$ is a delayed version of $x(n)$ by k samples, that is, $y(n) = x(n - k)$, the z-transform of $y(n)$ is given as

$$Y(z) = z^{-k}X(z). \qquad (3.2.4)$$

Thus the delay of k samples in the time domain corresponds to the multiplication of z^{-k} in the z-domain. A unit delay, z^{-1}, was shown in Figure 2.15. This element has the effect of delaying the sampled signal by one sampling period (T_s seconds in time) and is one of the three basic operators for implementing DSP systems.

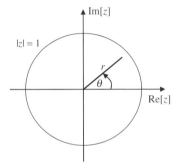

Figure 3.1 Polar representation of complex variable z in the z-plane

EXAMPLE 3.2

The I/O equation of an FIR filter is expressed in Equation 2.3.4. Taking the z-transform of both sides using the property given in Equation 3.2.4, we obtain

$$Y(z) = b_0 X(z) + b_1 z^{-1} X(z) + \ldots + b_{L-1} z^{-(L-1)} X(z)$$
$$= \left(\sum_{l=0}^{L-1} b_l z^{-l} \right) X(z). \qquad (3.2.5)$$

EXERCISE 3.2

1. Compute the z-transform of the I/O equation for the symmetric FIR filter defined in Equation 2.3.10.
2. Compute the z-transform of the I/O equation for the moving-average filter defined in Equation 2.3.2.
3. The z-transform of an N-periodic sequence is defined as

$$x(n) = \{x(0)\ x(1) \ldots x(N-1)\ x(0)\ x(1) \ldots x(N-1)\ x(0) \ldots\}.$$

The first period of the sequence is $x_1(n) = \{x(0)\ x(1) \ldots x(N-1)\}$, and its z-transform is $X_1(z)$. Show that the z-transform of $x(n)$ is

$$X(z) = \frac{X_1(z)}{1 - z^{-N}}.$$

3.2.2 System Concepts

If $y(n)$ is the result of linear convolution of two sequences $x(n)$ and $h(n)$ as expressed in Equation 2.3.7, the z-transform of $y(n)$ is given as

$$Y(z) = X(z) H(z) = H(z) X(z). \qquad (3.2.6)$$

Therefore, the convolution in the time domain is equivalent to the multiplication in the z-domain. The transfer (or system) function of a given system is defined as the z-transform of the system's impulse response, or the ratio of the system's output and input in the z-domain, expressed as $H(z) = Y(z)/X(z)$. From Equation 3.2.5, the transfer function of the FIR filter is expressed as

$$H(z) = \frac{Y(z)}{X(z)} = \sum_{l=0}^{L-1} b_l z^{-l}, \qquad (3.2.7)$$

which also represents the z-transform of the impulse responses $h(l) = b_l$, $l = 0, 1, \ldots, L - 1$.

Because the impulse-response samples of an FIR filter are the same as the filter coefficients, an FIR filtering can be described by the linear convolution defined in Equation 2.3.7 or the I/O equation given in Equation 2.3.4, which express the output in terms of input samples weighted by filter coefficients. Connecting delay units,

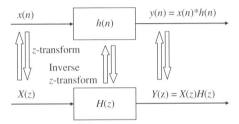

Figure 3.2 Representation of LTI systems in both time- and z-domains

multipliers, and adders pictorially represent I/O equations as the signal-flow diagram given in Figure 2.16. The transfer function $H(z)$ describes how the system operates on the input signal $x(n)$ to produce the output signal $y(n)$. A system that is both linear and time invariant is called a linear time-invariant (LTI) system. This system can be represented in both the time domain and the z-domain given in Equations 2.3.4 and 3.2.6 as illustrated in Figure 3.2, where $h(n)$ is the impulse response of the system.

EXAMPLE 3.3

The moving-average filter is defined in Equation 2.3.2. Taking the z-transform of both sides, we have

$$Y(z) = \frac{1}{L}[1 + z^{-1} + \ldots + z^{-(L-1)}]X(z) = \left(\frac{1}{L}\sum_{l=0}^{L-1} z^{-l}\right)X(z).$$

By arranging the terms, we obtain the system function as

$$H(z) = \frac{1}{L}\sum_{l=0}^{L-1} z^{-l}. \qquad (3.2.8)$$

Using the finite-geometric series identity,

$$\sum_{l=0}^{L-1} x^l = \frac{1-x^L}{1-x}, \quad x \neq 1. \qquad (3.2.9)$$

Equation 3.2.8 can be rewritten as

$$H(z) = \frac{Y(z)}{X(z)} = \frac{1}{L}\left(\frac{1-z^{-L}}{1-z^{-1}}\right). \qquad (3.2.10)$$

This equation can be rearranged as

$$\frac{1}{L}(1-z^{-L})X(z) = (1-z^{-1})Y(z).$$

Taking the inverse z-transform of both sides and rearranging terms, we obtain

$$y(n) = y(n-1) + \frac{1}{L}[x(n) - x(n-L)]. \qquad (3.2.11)$$

This is identical to the recursive equation described in Equation 2.3.3.

EXERCISE 3.3

Derive the transfer functions of digital systems described by the following difference equations:

1. $y(n) - 2y(n-1) + y(n-2) = 0.5[x(n) + x(n-1)]$
2. $y(n) = y(n-1) + [x(n) - x(n-10)]$
3. $y(n) = 0.1x(n) + 0.2x(n-1) + 0.1x(n-2)$

3.2.3 Digital Filters

Filtering is one of the most commonly used signal processing techniques to remove or attenuate undesired signal components while enhancing the desired portions of the signals. There are two classes of digital filters based on the length of the impulse response: FIR and IIR (infinite impulse response) filters.

A filter is called an FIR filter if its response to an impulse input becomes zero after a finite number of L samples. The system described by Equation 2.3.4 or Equation 3.2.7 is an FIR filter with a finite impulse response $\{h(i) = b_i, i = 0, 1, \ldots, L-1\}$ of length L. The FIR filter coefficients (taps or weights) are the same as the impulse response samples of the filter. By setting $H(z) = 0$ in Equation 3.2.7, we obtain $L-1$ zeros. Therefore, the FIR filter of length L has order $L-1$.

In practical application, an FIR filter is implemented by using the direct-form structure (or realization) illustrated in Figure 2.16. The FIR filter requires $2L$ memory locations for storing L input samples and L filter coefficients. The signal buffer $\{x(n), x(n-1), x(n-2), \ldots, x(n-L+1)\}$ is also called a delay buffer or a tapped-delay line, which is implemented as a first-in first-out buffer in memory.

The finite length of the impulse response guarantees that the FIR filters are stable. In addition, a perfect linear-phase response can be easily designed with an FIR filter, allowing a signal to be processed without phase distortion. The disadvantage of FIR filters is the computational complexity, because it may require a higher-order filter to fulfill a given frequency specification. There are a number of techniques for designing FIR filters for given specifications. FIR filter design methods and tools are introduced in Chapter 4.

If the impulse response of a filter is not a finite-length sequence, the filter is called an IIR filter. Equation 3.2.11 can be generalized to the transfer function of the IIR filter as

$$y(n) = b_0 x(n) + b_1 x(n-1) + b_2 x(n-2) + \ldots + b_{L-1} x(n-L+1)$$
$$- a_1 y(n-1) - a_2 y(n-2) - \ldots - a_M y(n-M)$$
$$= \sum_{l=0}^{L-1} b_l x(n-l) - \sum_{m=1}^{M} a_m y(n-m), \qquad (3.2.12)$$

where the coefficient sets $\{b_l\}$ and $\{a_m\}$ are constants that determine the filter's characteristics. Taking the z-transform of both sides in Equation 3.2.12 and arranging terms, we obtain

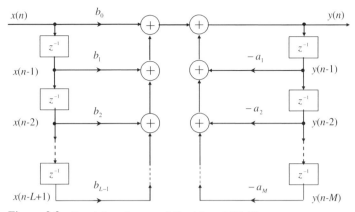

Figure 3.3 Signal flow diagram of direct-form I IIR filter

$$H(z) = \frac{b_0 + b_1 z^{-1} + b_2 z^{-2} + \ldots + b_{L-1} z^{-(L-1)}}{1 + a_1 z^{-1} + a_2 z^{-2} + \ldots + a_M z^{-M}} = \frac{\sum_{i=0}^{L-1} b_i z^{-i}}{\sum_{m=0}^{M} a_m z^{-m}} = \frac{B(z)}{A(z)}, \quad (3.2.13)$$

where $a_0 = 1$. Note that if all of the denominator coefficients a_m, $m = 1, 2, \ldots, M$ are equal to zero, $H(z)$ is identical to $B(z)$, or an FIR filter as defined in Equation 3.2.7. The signal flow diagram given in Figure 3.3 illustrates this I/O equation.

It is important to note that there is a sign change of a_m in the transfer function $H(z)$ given in Equation 3.2.13 and the I/O equation described in Equation 3.2.12, or the signal-flow diagram shown in Figure 3.3. This direct-form I realization of the IIR filter can be treated as two FIR filters. It requires two signal buffers, $\{x(n), x(n-1), x(n-2), \ldots, x(n-L+1)\}$ and $\{y(n), y(n-1), y(n-2), \ldots, y(n-M)\}$. These two buffers can be combined into one by using the direct-form II realization. In addition, the simple direct-form implementation of an IIR filter will not be used in practical applications because of severe sensitivity problems due to coefficient quantization (which are explained in Chapter 6), especially as the order of the filter increases. To reduce this effect, a high-order IIR filter transfer function is factored into second-order sections plus a first-order section if the order of filter is an odd number. These sections are connected in cascade or parallel to form an overall filter. We discuss these issues further in Chapter 4.

EXAMPLE 3.4

Consider the second-order IIR filter expressed as

$$y(n) = b_0 x(n) + b_1 x(n-1) + b_2 x(n-2) - a_1 y(n-1) - a_2 y(n-2). \quad (3.2.14)$$

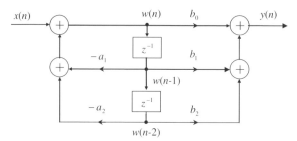

Figure 3.4 Direct-form II realization of second-order IIR filter

The transfer function is expressed as

$$H(z) = \frac{b_0 + b_1 z^{-1} + b_2 z^{-2}}{1 + a_1 z^{-1} + a_2 z^{-2}}. \tag{3.2.15}$$

The direct-form II realization of the second-order IIR filter (also called biquad) is illustrated in Figure 3.4.

EXERCISE 3.4

1. Derive the I/O equation of the second-order IIR filter based on the direct-form II realization shown in Figure 3.4, using the intermediate parameters $w(n)$, $w(n-1)$, and $w(n-2)$.
2. Prove that the direct-form II realization is identical to the direct-form I realization.
3. Identify any difference between direct-form I and direct-form II realizations.

Factoring the numerator and denominator polynomials of $H(z)$ given in Equation 3.2.13, we obtain

$$H(z) = \frac{b_0(z-z_1)\ldots(z-z_i)\ldots(z-z_{L-1})}{(z-p_1)\ldots(z-p_m)\ldots(z-p_M)} = \frac{B(z)}{A(z)}, \tag{3.2.16}$$

where z_i and p_m denote the zero and pole of $H(z)$, respectively. The zeros of the system can be calculated by setting the numerator $B(z) = 0$ in Equation 3.2.13, and the poles can be obtained by setting the denominator $A(z) = 0$. To calculate zeros and poles of a given transfer function $H(z)$, we can use the MATLAB function `roots` on both the numerator and denominator polynomials.

The system will be bounded-input, bounded-output stable if

$$\sum_{n=-\infty}^{\infty} |h(n)| < \infty. \tag{3.2.17}$$

For a causal system, the system is stable if and only if the transfer function has all its poles inside the unit circle. That is,

$$|p_m| < 1, \quad m = 1, 2, \ldots, M. \tag{3.2.18}$$

If any $|p_m| > 1$, the system is unstable. A system is also unstable if the system has multiple-order pole(s) on the unit circle. However, a system is marginally stable if $H(z)$ has a first-order pole on the unit circle.

EXAMPLE 3.5

Consider the second-order IIR filter given as $H(z) = z^{-1}/(1 - 2z^{-1} + z^{-2})$. This system has second-order poles at $z = 1$, and the impulse response of the system is $h(n) = n$. Therefore, this system is unstable because $h(n) \to \infty$ when $n \to \infty$.

Given the first-order IIR filter $H(z) = z/(z + 1)$, this system has a pole on the unit circle, and the impulse response is $h(n) = (-1)^n$, which is oscillated between ±1. The filter with oscillatory bounded impulse response is marginally stable.

EXERCISE 3.5

1. Evaluate the stability of system $H(z) = z/(z - a)$ for different values of coefficient a.
2. Consider the second-order IIR filter defined in Equation 3.2.15. Evaluate the stability of the system in terms of its coefficients a_1 and a_2.

In general, an IIR filter requires fewer coefficients to approximate the desired frequency response than an FIR filter with comparable performance. The primary advantage of IIR filters is that sharper cutoff characteristics are achievable with a relatively low-order filter. This results in savings of processing time and/or hardware complexity. However, IIR filters are more difficult to design and implement on fixed-point processors for practical applications. Stability, finite-precision effects, and nonlinear phase must be considered in IIR filter designs. The issues of filter design, realization of IIR filters in direct-form II, filter design tools, quantization effects, and implementation of IIR filters based on cascade or parallel connection of second-order sections are discussed further in Chapters 4 and 6.

Consider the moving-average filter given in Equation 3.2.10. The transfer function can be rewritten as

$$H(z) = \frac{1}{L} \frac{z^L - 1}{z^{L-1}(z-1)}. \tag{3.2.19}$$

Therefore, there is a pole at $z = 1$ from the solution of $z - 1 = 0$ and $L - 1$ poles at $z = 0$ from the solution of $z^{L-1} = 0$. There are L zeros, $z_l = e^{j\frac{2\pi}{L}l}$, $l = 0, 1, \ldots, L - 1$, that come from solving $z^{L-1} = 0$, which are located on the unit circle and separated by the angle $2\pi/L$. Note that the pole at $z = 1$ is canceled by the zero at $z = 1$. Thus the moving-averaging filter is still an FIR filter even though its transfer function defined in Equation 3.2.10 is in rational form and its I/O equation given in Equation 3.2.11 is in recursive form.

EXAMPLE 3.6

The MATLAB function `zplane(b, a)` computes and plots both zeros and poles on the z-plane with the unit circle as reference for the given numerator vector b and denominator vector a. Each zero is represented with an open circle and each pole with an x on the plot. Multiple zeros and poles are indicated by the multiplicity number shown to the upper right of the zero or pole. The following script (`example3_6.m`) shows the poles and zeros of a moving-average filter with $L = 8$ in Figure 3.5:

```
b = [1/8 0 0 0 0 0 0 0 -1/8];   % numerator vector
a = [1 -1];                      % denominator vector
zplane(b, a)                     % shows poles & zeros
```

As shown in Figure 3.5, the pole at $z = 1$ was canceled by the zero at that location. Therefore, the filter is an FIR filter. Also, it shows that there are seven zeros at angles $\pi l/4$ for $l = 1, 2, \ldots, 7$.

Consider the I/O equation of the moving-average filter given in Equation 3.2.11. The system needs $L + 1$ memory locations for storing $\{x(n), x(n-1), \ldots, x(n-L)\}$. To reduce memory requirements, we assume $x(n - L) \cong y(n - 1)$ because $y(n - 1)$ is the average of $x(n)$ samples; thus Equation 3.2.11 can be simplified to

$$y(n) \cong \left(1 - \frac{1}{L}\right) y(n-1) + \frac{1}{L} x(n) = (1-\alpha) y(n-1) + \alpha x(n), \qquad (3.2.20)$$

where $\alpha = 1/L$. Taking the z-transform of both sides and rearranging terms, the transfer function of this simplified filter can be derived as

$$H(z) = \frac{\alpha}{1 - (1-\alpha) z^{-1}}. \qquad (3.2.21)$$

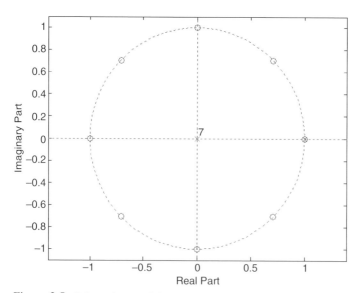

Figure 3.5 Poles and zeros of the moving-average filter, $L = 8$

3.2 The z-Transform

This is a first-order IIR filter with a pole at $z = (1 - \alpha)$ that is equivalent to the moving-average FIR filter with $\alpha = 1/L$. Because $1 - \alpha = (L - 1)/L < 1$, this system is guaranteed to be stable since the pole is always inside the unit circle. As α gets smaller, L becomes larger. When the window is longer, the filter provides better averaging effects. However, a long window is not suitable for time-varying signals.

The MATLAB function y = filter(b, a, x) implements the IIR filtering, where the vectors b and a contain the filter coefficients $\{b_l\}$ and $\{a_m\}$ and the vectors x and y contain the input and output signals. As a special case, the FIR filtering can be implemented as y = filter(b, 1, x).

EXAMPLE 3.7

In Example 2.2, we generated 60 samples of a sine wave with noise as shown in Figure 2.2 and saved the signal in the data file sineNoise.dat. In Example 2.3, we filtered the noisy sine wave with a moving-average filter of lengths $L = 5$, 10, and 20. For $L = 5$, this filter is equivalent to the first-order IIR filter given in Equation 3.2.21 with $\alpha = 0.2$. This IIR filter is implemented with MATLAB code example3.7.m. Both input and output waveforms are displayed in Figure 3.6.

EXERCISE 3.6

i. Compare the first-order IIR filter output shown in Figure 3.6 with the moving-average filter output shown in Figure 2.8 for $L = 5$. Observe and explain the differences.

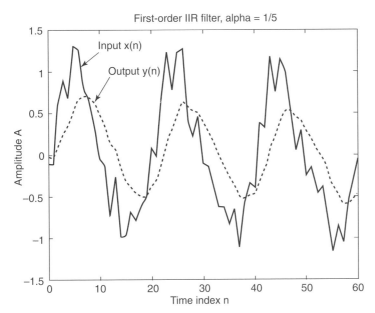

Figure 3.6 Performance of the first-order IIR filter

2. Implement the first-order IIR filter given in Equation 3.2.21 with $\alpha = 0.1$ and 0.05. Compare the results with the moving-average filter outputs shown in Figure 2.8 with $L = 10$ and 20. Pay special attention to the IIR filter with $\alpha = 0.05$. Does the output approach to zero?

3. Evaluate the complexity of moving-average filters for $L = 5, 10, 20$ with their corresponding first-order IIR filters in terms of memory and multiplication requirements.

3.3 FREQUENCY ANALYSIS

Digital signals and systems can be represented and analyzed in the frequency domain. We studied z-domain representations of digital signals and systems in Section 3.2. In this section, we introduce the discrete Fourier transform and the fast Fourier transform and their applications for analyzing digital signals and systems. We revisit some unanswered problems from Chapter 2 on using the moving-average filter for noise reduction.

3.3.1 Frequency Response

The discrete-time Fourier transform (DTFT) of infinite-length digital signal $x(n)$ is defined as

$$X(\omega) = \sum_{n=-\infty}^{\infty} x(n) e^{-j\omega n}. \tag{3.3.1}$$

Compare this equation with the equation for the z-transform defined in Equation 3.2.1; they are equal if the variable z is defined as

$$z = e^{j\omega}. \tag{3.3.2}$$

Thus, evaluating the z-transform on the unit circle, $|z| = 1$, in the complex z-plane shown in Figure 3.1 is equivalent to the frequency-domain representation of the sequence.

Similarly, the transfer function $H(z)$ can be evaluated on the unit circle to yield the frequency-domain representation of the system as

$$H(\omega) = H(z)|_{z=e^{j\omega}} = |H(\omega)| e^{j\phi(\omega)}, \tag{3.3.3}$$

where $H(\omega)$ is called the frequency response of the system, $|H(\omega)|$ is the magnitude (amplitude) response, and $\phi(\omega)$ is the phase response. The value $|H(\omega_0)|$ is called the system gain at a given frequency ω_0.

Consider the transfer function of the moving-average filter given in Equation 3.2.10. From Equation 3.3.3, the frequency response can be expressed as

$$H(\omega) = \frac{1}{L} \left[\frac{1 - e^{-jL\omega}}{1 - e^{-j\omega}} \right].$$

Because $e^{j\omega/2}e^{-j\omega/2} = e^{jL\omega/2}e^{-jL\omega/2} = 1$ and $\sin(\omega) = \frac{1}{2j}(e^{j\omega} - e^{-j\omega})$, this equation becomes

$$H(\omega) = \frac{1}{L}\left[\frac{e^{jL\omega/2}e^{-jL\omega/2} - e^{-jL\omega}}{e^{j\omega/2}e^{-j\omega/2} - e^{-j\omega}}\right] = \frac{1}{L}\left[\frac{e^{jL\omega/2} - e^{-jL\omega/2}}{e^{j\omega/2} - e^{-j\omega/2}}\right]\frac{e^{-jL\omega/2}}{e^{-j\omega/2}}$$

$$= \frac{1}{L}\left[\frac{\sin(L\omega/2)}{\sin(\omega/2)}\right]e^{-j(L-1)\omega/2} \quad (3.3.4)$$

Beause $|e^{-j(L-1)\omega/2}| = 1$, the magnitude response is given by

$$|H(\omega)| = \frac{1}{L}\left|\frac{\sin(L\omega/2)}{\sin(\omega/2)}\right|, \quad (3.3.5)$$

and the phase response is

$$\phi(\omega) = \begin{cases} -\frac{(L-1)\omega}{2}, & |H(\omega)| \geq 0 \\ -\frac{(L-1)\omega}{2} \pm \pi, & |H(\omega)| < 0 \end{cases}. \quad (3.3.6)$$

EXAMPLE 3.8

The Hanning filter defined in Example 2.4 has coefficients {0.1 0.2 0.4 0.2 0.1}. The transfer function of the filter is $H(z) = 0.1 + 0.2z^{-1} + 0.4z^{-2} + 0.2z^{-3} + 0.1z^{-4}$. Therefore, the frequency response is

$$H(\omega) = 0.1 + 0.2e^{-j\omega} + 0.4e^{-2j\omega} + 0.2e^{-3j\omega} + 0.1e^{-4j\omega}$$
$$= e^{-2j\omega}[0.1(e^{2j\omega} + e^{-2j\omega}) + 0.2(e^{j\omega} + e^{-j\omega}) + 0.4]$$
$$= e^{-2j\omega}[0.2\cos(2\omega) + 0.4\cos(\omega) + 0.4].$$

The magnitude response is $|H(\omega)| = 0.4 + 0.2\cos(2\omega) + 0.4\cos(\omega)$, and the phase response is -2ω.

EXERCISE 3.7

1. Find the frequency response of the first-order IIR filter defined in Equation 3.2.21.
2. Find the magnitude response of a moving-average filter with $L = 2$.
3. The bandwidth of a filter represents the value of the frequency at which the squared-magnitude response equals half its maximum value. Find the bandwidth of the filter given in Exercise 2.
4. Find the transfer function and frequency response of a comb filter defined by the I/O equation

$$y(n) = x(n) - x(n - L).$$

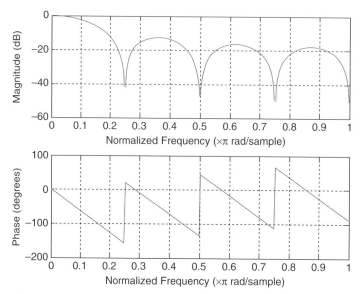

Figure 3.7 Magnitude and phase responses of moving-average filter, $L = 8$

EXAMPLE 3.9

The magnitude and phase responses of the system can be computed and displayed with the MATLAB function `freqz`. The MATLAB script `example3_9.m` plots the responses of the moving-average filter for $L = 8$ and is shown in Figure 3.7. The magnitude response shows that there are dips that occur at radian frequencies $\pi/4$, $\pi/2$, $3\pi/4$, and π (or normalized frequencies 0.25, 0.5, 0.75, and 1), same as the zeros shown in Figure 3.5. Equation 3.3.6 and the phase response of Figure 3.7 imply that $\phi(\omega)$ is a piecewise linear function of ω. The MATLAB script is listed as follows:

```
b=[1/8 0 0 0 0 0 0 0 -1/8];   % numerator vector
a=[1 -1];                      % denominator vector
freqz(b, a)                    % magnitude and phase responses
```

From Equation 3.3.6, the time (group)-delay function is defined as

$$T_d(\omega) = -\frac{d\phi(\omega)}{d\omega} = \frac{L-1}{2}, \quad (3.3.7)$$

which is independent of ω. The systems (or filters) that have constant $T_d(\omega)$ are called linear-phase systems. The constant time-delay function causes all sinusoidal components in the input to be delayed by the same amount, thus avoiding phase distortion. The group delay can be calculated and displayed by using the MATLAB function `grpdelay(b,a)`.

HANDS-ON EXPERIMENT 3.1

The MATLAB *Signal Processing Toolbox* [49] provides a graphical user interface filter visualization tool (FVTool) that allows the user to analyze digital filters. The command fvtool(b, a) launches the FVTool window and computes the magnitude response for the filter defined by numerator and denominator coefficients in vectors b and a, respectively. The command fvtool(b1, a1, b2, a2, ...) performs an analysis of multiple filters.

Similar to example3_9.m, we design three moving-average filters with L = 5, 10, and 20 and launch the FVTool using experiment3_1.m. Using FVTool, we can display the phase response, group delay, impulse response, step response, pole-zero plot, and coefficients of the filters. We can export the displayed response to a file with **Export** on the **File** menu. A window that shows the magnitude responses of three filters is displayed as shown in Figure 3.8. Determine where the nulls of the magnitude response occurred for different L.

As shown in Figure 3.8, the analysis toolbar has the following options:

1. ▨ : Magnitude response of the current filter. See freqz and zerophase for more information.

2. ▨ : Phase response of the current filter. See phasez for more information.

3. ▨ : Superimposes the magnitude response and the phase response of the current filter. See freqz for more information.

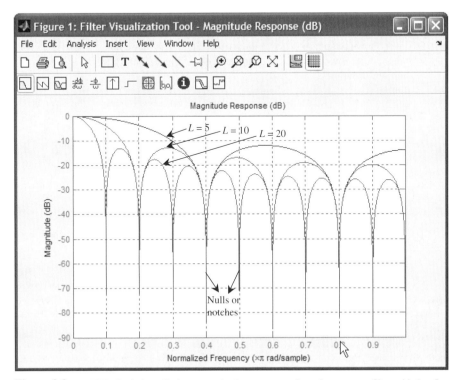

Figure 3.8 A FVTool window displays magnitude responses of moving-average filter with L = 5, 10, and 20

74 Chapter 3 Frequency-Domain Analysis and Processing

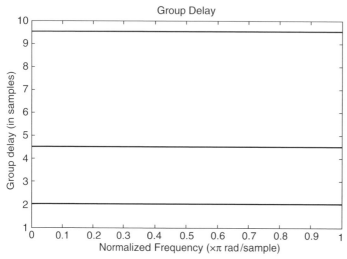

Figure 3.9 Group delays of moving-averaging filters, $L = 5$, 10, and 20 (from bottom to top)

4. $\frac{-d\phi}{d\omega}$: Shows the group delay of the current filter. The group delay is the average delay of the filter as a function of frequency. See `grpdelay` for more information.

5. $\frac{-\phi}{\omega}$: Shows the phase delay of the current filter. The phase delay is the time delay that the filter imposes on each component of the input signal. See `phasedelay` for more information.

6. ⬆ : Impulse response of the current filter. The impulse response is the response of the filter to an impulse input. See `impz` for more information.

7. ⎍ : Step response of the current filter. The step response is the response of the filter to a step input. See `stepz` for more information.

8. ✦ : Pole-zero plot, which shows the pole and zero locations of the current filter on the z-plane. See `zplane` for more information.

9. [b,a] : Filter coefficients of the current filter, which depend on the filter structure (e.g., direct form, etc.) in a text box. For second-order section filters, each section is displayed as a separate filter.

For example, selecting the group-delay response icon (or from the **Analysis** menu) will compute and display group delay. Export the result as shown in Figure 3.9, where constant group delays of 2, 4.5, and 9.5 samples are observed and can be easily obtained from Equation 3.3.7 for $L = 5$, 10, and 20, respectively.

An important characteristic of an LTI system is its steady-state response to a sinusoidal input. Consider a sinusoidal signal defined in Equation 2.2.3 at frequency ω_0 expressed as

$$x(n) = A \sin(\omega_0 n).$$

Applying this sinusoidal signal to the digital filter defined in Equation 3.3.3, the corresponding output in the steady state is

$$y(n) = A|H(\omega_0)|\sin[\omega_0 n + \phi(\omega_0)], \quad (3.3.8)$$

where $|H(\omega_0)|$ is the gain and $\phi(\omega_0)$ is the phase shift of $H(z)$ at frequency ω_0. This equation shows that when sinusoidal input is applied to an LTI system its steady-state output is also a sinusoid having the same frequency ω_0. The amplitude of the input sinusoid A is multiplied by the gain of the system, and the phase is shifted by the phase of the system.

EXAMPLE 3.10

Several questions were asked in Quiz 2.3. These questions may be answered with the knowledge of the magnitude response and the group delay of the system.

1. The outputs of the moving-average filter with different lengths have different phase shifts. This fact can be explained by Equation 3.3.7 and is shown in Figure 3.9. For example, the output sine wave is delayed by 2.5 samples, as shown in Figure 2.7 for $L = 5$.
2. Figure 2.7 uses a 200-Hz sine wave sampled at 4 kHz. From Equation 2.2.4, the digital frequency of the sine wave is $\omega = 0.1\pi$. Figure 3.8 shows that the first notch of a moving-average filter with $L = 20$ is located at $\omega = 0.1\pi$. Therefore, the sine wave will be completely attenuated by the moving-average filter with $L = 20$, as shown in Figure 2.8.
3. Figure 3.8 also shows that the gains $|H(\omega_0)|$ of a filter at $\omega_0 = 0.1\pi$ for filter lengths $L = 5$ and 10 are less than 1 (0 dB). The steady-state sinusoidal response given in Equation 3.3.8 shows that the output will be attenuated, and $L = 10$ will have more attenuation because it has smaller gain.

EXERCISE 3.8

1. Compute the gains $|H(\omega_0)|$ of moving-average filters with lengths $L = 5$ and 10 at frequency $\omega_0 = 0.1\pi$. Compare the results with Figure 3.8.
2. Compute the gains of the first-order IIR filter defined in Equation 3.2.21 for $\alpha = 0.2, 0.1$, and 0.05 at frequency $\omega_0 = 0.1\pi$. Compare the results with the results obtained in Exercise 1.
3. What frequencies of sine waves will be attenuated by the moving-average filter with $L = 5$ if the sampling rate is 8 kHz? Why?
4. A 100-Hz sine wave is corrupted by white noise, and the corrupted sine wave is sampled at 1 kHz. To remove noise, what is the proper length of the moving-average filter? Why?
5. Show the magnitude responses of comb filters defined in Exercise 3.7(4) for $L = 5, 10$, and 20. Compare the results with the corresponding moving-average filters and explain the differences using pole-zero diagrams.

6. Based on the magnitude response of a comb filter, suggest its applications for reducing noise.

3.3.2 Discrete Fourier Transform

The z-transform and DTFT are defined for infinite-length sequences and are functions of continuous frequency variables z and ω, respectively. These transforms are useful for analyzing digital signals and systems in theory. However, they are difficult to implement from the numerical computation standpoint. In this section, we introduce a numerical computable transform, the discrete Fourier transform (DFT). The DFT is a basic operation used in many different DSP applications. It is used to transform a sequence of signal samples from the time domain into the frequency domain, so that spectral information about the signal can be known explicitly.

If the digital signal $x(n)$ is a finite-duration sequence $\{x(0), x(1), \ldots, x(N-1)\}$ of length N, the DTFT given in Equation 3.3.1 can be modified to become the DFT expressed as

$$X(k) = X(\omega_k)|_{\omega_k = 2\pi k/N}$$
$$= \sum_{n=0}^{N-1} x(n) e^{-j\left(\frac{2\pi}{N}\right)kn}, \quad k = 0, 1, \ldots, N-1, \qquad (3.3.9)$$

where k is the frequency index. Usually the signal $x(n)$ is a real-valued sequence, but the DFT coefficients $X(k)$ are complex values. It is important to note that the DFT defined in Equation 3.3.9 assumes that the signal is a periodic signal with period N. The DFT is equivalent to evaluating (or sampling) the DTFT $X(\omega)$ at N equally spaced frequencies $\omega_k = 2\pi k/N$, $k = 0, 1, \ldots, N-1$; thus the DFT is computable with digital computers. The interval between the adjacent frequency samples is called the frequency resolution, expressed as

$$\Delta \omega = \frac{2\pi}{N}. \qquad (3.3.10)$$

EXAMPLE 3.11

Compute the DFT of the impulse response of a moving-average filter with length L. From Equations 2.3.2 and 3.3.9, we have

$$X(k) = \sum_{n=0}^{L-1} x(n) e^{-j\left(\frac{2\pi}{L}\right)kn} = \frac{1}{L}\sum_{n=0}^{L-1} e^{-j\left(\frac{2\pi}{L}\right)kn}, \quad k = 0, 1, \ldots, L-1.$$

For $k = 0$, we obtain $X(0) = \frac{1}{L}\sum_{n=0}^{L-1} e^{-j\left(\frac{2\pi}{L}\right)0n} = \frac{1}{L}\sum_{n=0}^{L-1} 1 = 1$. For $k \neq 0$, we get

$$X(k) = \sum_{n=0}^{L-1} x(n) e^{-j\left(\frac{2\pi}{L}\right)kn} = \frac{1}{L}\sum_{n=0}^{L-1} e^{-j\left(\frac{2\pi}{L}\right)kn}, \quad k = 1, \ldots, L-1.$$

Using the finite geometric series given in Equation 3.2.9, $\sum_{n=0}^{L-1} e^{-j\left(\frac{2\pi}{L}\right)kn} = \frac{1-e^{-j(2\pi/L)kL}}{1-e^{-j(2\pi/L)k}} = 0$.

Therefore, we have $X(k) = \delta(k)$.

EXERCISE 3.9

1. Find the DFT of the unit impulse function defined in Equation 2.3.5. It is interesting to compare the result with the result in Example 3.11 and summarize the observations.
2. If the signal $x(n)$ is a real sequence and N is an even number, show that $X(N/2 + k) = X^*(N/2 - k)$, for $k = 0, 1, \ldots, N/2$, where $X^*(k)$ denotes the complex conjugate of $X(k)$.
3. If the signal $x(n)$ is a real sequence and N is an even number, show that $X(0)$ and $X(N/2)$ are real numbers.

By defining a twiddle factor

$$W_N = e^{-j\left(\frac{2\pi}{N}\right)}, \tag{3.3.11}$$

the DFT given in Equation 3.3.9 can be modified to

$$X(k) = \sum_{n=0}^{N-1} x(n) W_N^{kn}, \quad k = 0, 1, \ldots, N-1. \tag{3.3.12}$$

The inverse DFT can be expressed as

$$x(n) = \frac{1}{N} \sum_{k=0}^{N-1} X(k) e^{j\left(\frac{2\pi}{N}\right)kn}$$

$$= \frac{1}{N} \sum_{k=0}^{N-1} X(k) W_N^{-kn}, \quad n = 0, 1, \ldots, N-1. \tag{3.3.13}$$

Thus the inverse DFT is the same as the DFT except for the sign of the exponent and the scaling factor $1/N$.

The DFT is often used as an analysis tool for determining the spectra of digital signals. The DFT can be broken into magnitude and phase components as follows:

$$X(k) = \text{Re}[X(k)] + j\text{Im}[X(k)] = |X(k)| e^{j\phi(k)}, \tag{3.3.14}$$

where

$$|X(k)| = \sqrt{\{\text{Re}[X(k)]\}^2 + \{\text{Im}[X(k)]\}^2} \tag{3.3.15}$$

is the magnitude spectrum and

$$\phi(k) = \tan^{-1}\left\{\frac{\text{Im}[X(k)]}{\text{Re}[X(k)]}\right\} \tag{3.3.16}$$

is the phase spectrum. It is often preferable to measure the magnitude spectrum in dB scale defined as

$$\text{spectrum in dB} = 20\log_{10}|X(k)|. \tag{3.3.17}$$

If the sequence $x(n)$ is a sampled signal with sampling rate f_s, the frequency index k corresponds to frequency

$$\omega_k = k\left(\frac{2\pi}{N}\right) = k\Delta\omega \quad \text{(radians per sample or radians)} \tag{3.3.18}$$

or

$$f_k = k\left(\frac{f_s}{N}\right) = k\Delta f \quad \text{(Hz)} \tag{3.3.19}$$

for $k = 0, 1, \ldots, N-1$. Thus the frequency components can only be discriminated if they are separated by at least $\Delta f = f_s/N$ Hz. This frequency resolution is a common term used in determining the size of the FFT and sampling frequency required to achieve good frequency analysis. In Section 3.3.3, we use computer simulations to demonstrate the effects of frequency resolution using DFT.

If $x(n)$ is a real-valued sequence and N is an even number, we can show that

$$X(N/2+k) = X*(N/2-k), \quad k = 0, 1, \ldots, N/2, \tag{3.3.20}$$

where $X*(k)$ denotes the complex conjugate of $X(k)$. This complex conjugate property of the DFT demonstrates that only the first $(N/2 + 1)$ DFT coefficients of a real data sequence are independent. From Equation 3.3.20, the even-symmetrical property applies to the magnitude $|X(k)|$ and Re$[X(k)]$, while the odd-symmetrical property occurs in phase $\phi(k)$ and Im$[X(k)]$. Thus it is common to plot the magnitude spectrum only from $k = 0, 1, \ldots, N/2$, because the rest of the spectrum points from $k = N/2 + 1$ to $N - 1$ are symmetrical to the points from $k = N/2 - 1$ to $k = 1$. It is also shown that $X(0)$ and $X(N/2)$ are real valued.

A limitation of DFT is its inability to handle signals extending over all time. It is also unsuitable for analyzing nonstationary signals (such as speech) that have time-varying spectra. For such a signal, it makes more sense to divide the signal into blocks over which it can be assumed to be stationary and estimate the spectrum of each block.

3.3.3 Fast Fourier Transform

The DFT given in Equation 3.3.9 shows that N complex multiplications and additions are needed to produce one output. To compute N outputs, a total of approximately N^2 complex multiplications and additions are required. A 1,024-point DFT requires over a million complex multiplications and additions. The Fast Fourier Transform (FFT) is a family of very efficient algorithms for computing the DFT. The FFT is not a new transform that is different from the DFT; it is simply an efficient algorithm for computing the DFT by taking advantage of the fact that many computations are repeated in the DFT because of the periodic nature of the twiddle factors. The ratio of computing cost in terms of number of multiplications is approximately

$$\frac{\text{FFT}}{\text{DFT}} = \frac{\log_2 N}{2N}, \qquad (3.3.21)$$

which is 10/2,048 when N is equal to 1,024.

The FFT algorithm [16] was first introduced by Cooley and Tukey in 1965. Since then, many variations of FFT algorithms have been developed. Each variant FFT has a different strength and makes different trade-offs between code complexity, memory requirements, and computational speed. The FFT algorithm becomes lengthy when N is not a power of 2. This restriction on N can be overcome by appending zeros at the tail of the sequence to cause N to become a power of 2.

MATLAB provides a function fft for computing the DFT of a vector x. The following command

```
Xk = fft(x);
```

performs N-point DFT, where N is the length of vector x, and the Xk vector contains N samples of $X(k)$, $k = 0, 1, \ldots, N - 1$. If N is a power of 2, an efficient radix-2 FFT algorithm will be used; otherwise, a slower mixed-radix FFT algorithm or the direct DFT will be used. To avoid slow computation, we can use

```
Xk = fft(x, L);
```

where L is a power of 2. If N is less than L, the vector x will be automatically padded with $(L - N)$ zeros at the tail of the sequence to make a new sequence of length L. If N is larger than L, only the first L samples will be used for computing DFT.

EXAMPLE 3.12

Generate a signal that consists of two sinusoids at 400 Hz and 820 Hz with a 4-kHz sampling rate. We can use the MATLAB function fft to compute the DFT coefficients $X(k)$ with DFT length $N = 100$. MATLAB provides the functions abs and angle to calculate the magnitude and phase spectra defined in Equations 3.3.15 and 3.3.16, respectively. The MATLAB script (example3_12.m) plots the magnitude spectra of these two sine waves as shown in Figure 3.10.

QUIZ 3.1

1. What is the frequency resolution for the case given in Example 3.12?
2. Based on Equation 3.3.19, what frequency indexes correspond to sine waves at frequencies 400 Hz and 820 Hz?
3. Because the Fourier transform of a sinusoidal function is a delta function, we expected to see two lines in the plot. In Figure 3.10, we find a spike (not

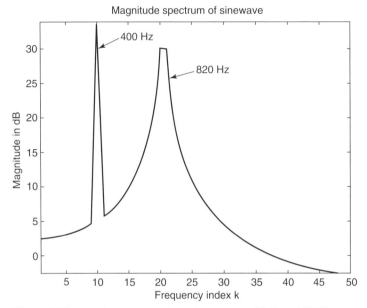

Figure 3.10 Magnitude spectra of two sine waves at 400 Hz and 820 Hz

a line) for the 400-Hz sine wave. Why? (Hint: try to use `stem` instead of the `plot` function provided by MATLAB).

4. Why is the spectrum of the 820-Hz sine wave spread to adjacent frequency bins?

5. Derive a general rule that relates to FFT size N, sampling rate f_s, and sinusoidal frequency f to predict whether the spectrum of a sine wave will have a spike or spread to adjacent bins.

We discuss these problems further in Section 3.3.4 and introduce windowing techniques to solve them.

EXERCISE 3.10

1. Similar to `example3_12.m`, show the magnitude spectrum of two sine waves at frequency 400 Hz and 420 Hz. Are you able to see two separated frequency components? Why not?

2. Modify the program `example3_12.m` by changing $N = 100$ to $N = 200$. Are you able to see two spikes now? Why?

3. Also, use $N = 200$ for Exercise 1. Are you able to see two separated frequency components now? Why?

3.3 Frequency Analysis   81

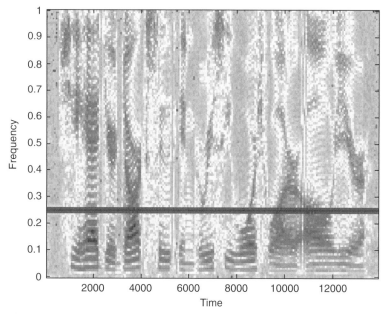

Figure 3.11 Spectrogram of the speech signal, timit1.asc, which is corrupted by a tonal noise at normalized frequency 0.25

EXAMPLE 3.13

In Exercise 2.8, a digitized speech signal (sampling rate 8,000 Hz) stored in file timit1.asc was corrupted by a 1,000-Hz (normalized frequency 0.25) tonal noise and examined with SPTool. The DFT is inadequate for analyzing nonstationary signals such as speech. The short-time Fourier transform (STFT) breaks up the signal sequence into consecutive blocks and performs the FFT of individual blocks over the entire signal. MATLAB provides a function specgram to compute and plot a time-frequency spectrogram of the input signal. For example, the following command

```
specgram(xn_tone,256);
```

will plot the spectrogram of the speech signal timit1.asc with tonal noise, using an FFT of length 256. This plot is shown in Figure 3.11 by MATLAB script example3_13.m. It contains both time and frequency information. Therefore, it can pinpoint the time instance where the signal is active by looking at the color plot. A darker color indicates higher energy.

EXERCISE 3.11

1. Generate a sinusoid embedded in white noise with SNR = 10 dB. Assuming that the variance of white noise is 1, what is the amplitude of sine wave needed for generating the noisy sine wave?

2. Compute and plot the noisy sine wave's magnitude spectrum. Compare the spectrum of the noisy sine wave with the clean sine wave shown in Figure 3.10. The white noise has a flat spectrum over the entire frequency range from 0 to π.
3. Analyze this spectrum using the MATLAB function `specgram`.

HANDS-ON EXPERIMENT 3.2

In this experiment, we use SPTool to perform frequency analysis. As shown in Figure 2.3, the rightmost function, **Spectra**, provides the spectral estimation methods supported by the *Signal Processing Toolbox*. As introduced in Hands-On Experiment 2.1, we can bring signals (with their sampling frequencies) from the MATLAB workspace (or data files) into the SPTool workspace by using **Import** under the **File** menu. As shown in Figure 2.4, the data file xn is imported from the file `sine3sec.mat` with the **Signals** column. The imported data file is named `sig1` in this example. For 3s of data file with 12,000 samples, we have to zoom in on the signal to see it as shown in Figure 2.5.

To use the **Spectrum Viewer** for analyzing the imported signal `sig1`, highlight the signal in the **Signals** column and click on the **Create** button in the **Spectra** column. The new **Spectrum Viewer** window will appear as shown in Figure 3.12. Click on the **Apply** button on the bottom left corner to create the spectrum file `spect1`, and the PSD (power spectrum density) subwindow will be displayed as shown in Figure 3.12. The user can select one of the many spectral estimation methods to implement the spectrum estimation. In addition, the size of FFT, window functions, and overlapping samples can be selected to complete the PSD estimation. We introduce window functions in Section 3.3.4.

Repeat the above process for importing the second data file `sineNoise3sec.mat` and creating the second spectrum file `spect2`. Select both `spect1` and `spect2` in the **Spectra**

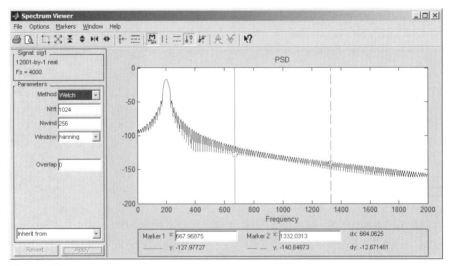

Figure 3.12 Spectrum Viewer window

3.3 Frequency Analysis

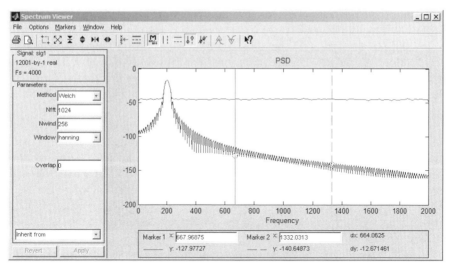

Figure 3.13 Spectra of clean (bottom) and noisy (top) sine waves

column, and click on the **View** button to display the spectra as shown in Figure 3.13. This feature allows us to display several spectra on the same plot, which is very useful for comparing the filter input and output spectra. As shown in Figure 3.13, the spectrum of the pure sine wave has a spike at 200 Hz, but the white noise spectrum is flat in the frequency range from 0 to π.

EXERCISE 3.12

1. Create the spectra of these two import signals using different parameters such as **Method**, **Nfft**, **Nwind**, **Window**, and **Overlap** as shown in the left side of Figure 3.13, and observe the differences.
2. Import the three data files created in Example 2.9 and use the **Spectrum Viewer** to perform frequency analysis.
3. Modify example2_10.m to generate 3 s of square wave, save it using the .mat format, import it to SPTool, and use the **Spectrum Viewer** to perform frequency analysis. How many sinusoidal components are there, and what are their frequencies? Observe the relationship between the frequency of the square wave and the frequencies of those harmonics.

3.3.4 Window Functions

In Example 3.12, Quiz 3.1, and Exercise 3.10, we found some problems with using DFT for spectral analysis. From Equation 3.3.19, the frequency index k

corresponds to frequency kf_s/N(Hz). If the sine wave frequency f can be exactly represented by an integer value of k, we obtain a line spectrum. In Example 3.12, f_s is 4,000 Hz and $N = 100$. The 400-Hz sine wave can be exactly represented by $k = 10$. However, the 820-Hz sine wave is located between $k = 20$ and $k = 21$. Therefore, its spectrum cannot be represented by one spectral bin and thus is spread (or leaks out) to adjacent bins as shown in Figure 3.10. This phenomenon is called spectral leakage.

For the 400-Hz sine wave sampled at 4,000 Hz, one cycle of sine wave consists of 10 samples and $N = 100$ covers exactly 10 cycles. However, for the 820-Hz sine wave, $N = 100$ covers 20.5 cycles. The DFT defined in Equation 3.3.9 assumes that the signal is a periodic signal with period N, which is only valid for the 400-Hz sine wave. To solve this problem, we can change N from 100 to 200. If we only have L samples that are less than N, we can pad $N - L$ zero samples. Unfortunately, in most practical applications, we do not know the sine wave frequency, and thus we are not able to select an appropriate N.

As shown in Figure 3.12, a window is often employed for spectral analysis using DFT. For a long signal sequence, taking only N samples for analysis is equivalent to applying a rectangular window $w(n)$ of length N to the signal. This action can be expressed as

$$x_N(n) = x(n)w(n), \qquad (3.3.22)$$

where

$$w(n) = \begin{cases} 1, & 0 \le n \le N-1 \\ 0, & \text{otherwise} \end{cases}. \qquad (3.3.23)$$

The spectrum of the rectangular function is similar to Equation 3.3.4 without the scaling factor $1/L$. Windowing not only produces leakage effects, it also reduces spectral resolution.

HANDS-ON EXPERIMENT 3.3

The MATLAB *Signal Processing Toolbox* provides two graphical user interface tools, Window Visualization Tool (WVTool) and Window Design and Analysis Tool (WINTool), to design and analyze windows. In the MATLAB command window, wvtool(winname(n)) opens WVTool with the time- and frequency-domain plots of the n-length window specified in winname, which can be any window in the *Signal Processing Toolbox*. For example, the following command will open the WVTool for comparing rectangular and Blackman windows as shown in Figure 3.14:

```
wvtool(rectwin(64), blackman(64))
```

In Figure 3.14, the time-domain values of window coefficients are displayed on the left-hand side and the magnitude spectrum is shown on the right-hand side. As shown in the figure, the Blackman window is tapered to zero at both ends; thus the Blackman window has wider bandwidth but higher attenuation at stopband compared to the rectangular window.

3.3 Frequency Analysis   85

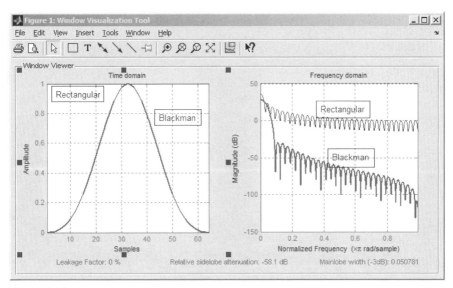

Figure 3.14 WVTool for evaluating rectangular and Blackman windows

A more powerful tool for designing and analyzing window is WINTool, which can be launched as

```
wintool
```

It opens with a default 64-point Hamming window as shown in Figure 3.15.

As shown in the figure, WINTool has three panels: **Window Viewer** displays the time-domain and frequency-domain representations of the selected window(s). Three window measurements are shown below the plots. (1) **Leakage Factor** indicates the ratio of power in the sidelobes to the total window power. (2) **Relative sidelobe attenuation** shows the difference in height from the mainlobe peak to the highest sidelobe peak. (3) **Mainlobe width (-3dB)** is the bandwidth of the mainlobe at 3 dB below the mainlobe peak.

The second panel, called **Window List**, lists the windows available for display in the **Window Viewer**. Highlight one or more windows to display them. The four **Window List** buttons are as follows. (1) **Add a new window** allows the user to add a default Hamming window with length 64 and symmetric sampling. We can change the information for this window by applying changes made in the **Current Window Information** panel. (2) **Copy window** copies the selected window(s). (3) **Save to workspace** saves the selected window(s) as vector(s) to the MATLAB workspace. The name of the window in WINTool is used as the vector name. (4) **Delete** removes the selected window(s) from the window list.

Current Window Information displays information about the currently active window. We can change the current window's characteristics by changing its parameters and clicking **Apply**. The active window name is shown in the **Name** field. We can either select a name from the menu or type the desired name in the edit box. The parameter **Type** presents the algorithm for the window. Select the type from the menu. All windows in the *Signal Processing Toolbox* are available. The parameter **Length** indicates the total number of samples.

For example, click on the **Add a new window** button at the **Window List** panel; select **Kaiser** from the **Type** menu in the **Current Window Information** panel, and click on **Apply**.

86 Chapter 3 Frequency-Domain Analysis and Processing

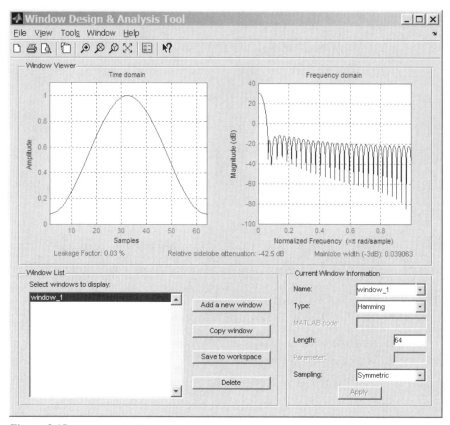

Figure 3.15 A default WINTool window

Both time-domain and frequency-domain representations in the **Window Viewer** panel are updated. Highlight both **window_1** and **window_2** from the **Window List** panel; the time- and frequency-domain representations of both windows are displayed as shown in Figure 3.16. Note that we use window length 100 for both cases and use Beta = 8.96 for the Kaiser window. The Kaiser window has more sidelobe attenuation (−65.7 dB) as compared to the Hamming window (−42.5 dB) that is shown in Figure 3.16. After verifying the designed window, we can use **Export** from the **File** menu to export window coefficients to the MATLAB workspace as a text file or a MAT file.

EXAMPLE 3.14

In Example 3.12, we found undesired spectral leakage effects due to the use of a rectangular window that is not tapered to zeros at both ends as shown in Figure 3.14. In this example, we modify the MATLAB program `example3_12.m` by applying the Kaiser window shown in Figure 3.16 to the signal, using Equation 3.3.22. We then compute the spectra of signals

3.3 Frequency Analysis

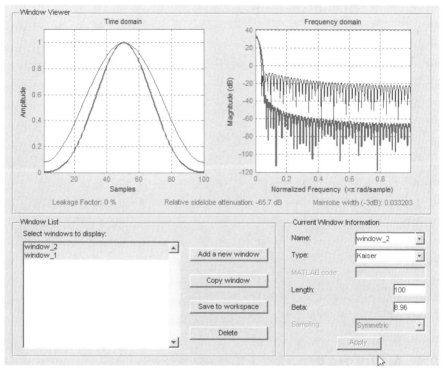

Figure 3.16 Comparison of Hamming window (top) with Kaiser window

with a rectangular window and a Kaiser window, using the MATLAB script example 3_14.m. The results are displayed in Figure 3.17. Compared with Figure 3.10, we observe that the Kaiser window has effectively reduced the spectral leakage of the sine wave at 820 Hz, but at the cost of broadening the peak for 400 Hz because of the wider bandwidth of the Kaiser window compared to the rectangular window shown in Figure 3.14.

EXERCISE 3.13

1. Use WINTool to evaluate and compare different window functions such as barthannwin, bartlett, blackman, blackmanharris, bohmanwin, chebwin, flattopwin, gausswin, hamming, hann, kaiser, nuttallwin, parzenwin, rectwin, triang, and tukeywin.
2. Evaluate the performance of the Kaiser window with different lengths and values of beta.
3. Apply different windows to the signal given in Example 3.14 and observe the differences between the spectral leakage reduction and main lobe broadening.
4. Redo Exercise 3 using the Kaiser window with different values of beta.

88 Chapter 3 Frequency-Domain Analysis and Processing

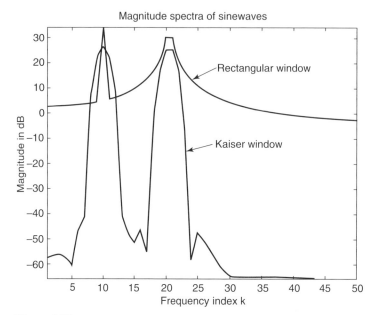

Figure 3.17 Spectral analysis using rectangular and Kaiser windows

3.4 MORE HANDS-ON EXPERIMENTS

In this section, we use the SPTool for filter design and analysis. We apply pole/zero concepts learned in Section 3.2 for designing notch and peak filters and use these filters for reducing noises.

3.4.1 Simple Low-Pass Filters

Assume that the signal $x(n)$ consists of a sine wave corrupted by random noise. Our goal is to develop and implement a digital low-pass filter to reduce noise, thus enhancing the sinusoidal component. For simulation purposes, we use the same data file sineNoise3sec.mat used in Hands-On Experiment 3.2 as $x(n)$, which is a sine wave corrupted by noise. The spectrum of this noisy sine wave is shown as the top line in Figure 3.13. The figure clearly shows that the sinusoidal component (a peak) is located at a low frequency (200 Hz) and can be enhanced by reducing noise components in high-frequency ranges. This objective may be achieved by using a simple low-pass filter.

EXAMPLE 3.15

In Figure 3.8, we show the magnitude responses of a moving-average FIR filters with lengths 5, 10, and 20. The simple IIR low-pass filter implemented in example3_7.m with $\alpha = 0.2$

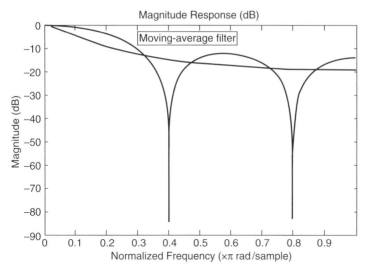

Figure 3.18 Magnitude responses of FIR and IIR filters

is equivalent to the moving-average filter of $L = 5$. We can compute and compare the magnitude responses of these FIR and IIR filters with the MATLAB script `example3_15.m`. The magnitude responses are shown in Figure 3.18. Compared with Figure 3.13, we know that the sinusoidal component (200 Hz or 0.1π rad) will be passed with little attenuation when the moving-average filter is used, but the simple IIR filter has higher attenuation. Both filters will reduce noise at the high-frequency range with, however, only about 10 dB noise reduction. Thus these simple filters will work for the purpose of reducing random noise that corrupts a low-frequency sinusoidal signal with high SNR. The design of higher-performance IIR filters is introduced in Chapter 4.

HANDS-ON EXPERIMENT 3.4

In this experiment, we use SPTool to perform the filtering and analysis. As shown in previous experiments, SPTool provides a rich graphic environment for signal viewing, filter design, and spectral analysis.

We first import the corrupted sine wave `sineNoise3sec.mat` and name it `sig1`. The next task is to design the FIR and IIR low-pass filters as shown in Example 3.15 to filter out the high-frequency noise components. The simplest way is to import the filter from the MATLAB workspace to the SPTool by clicking on **File → Import** and importing the numerator (`b1`) and denominator (`a1`) coefficients as shown in Figure 3.19 (note that after execution of `example3_15.m` both FIR and IIR filter coefficients are available from the workspace). The imported filter is named `FIR1`. Repeat this process to import the first-order IIR filter from workspace and name it `IIR1`. The user can examine the characteristics of the imported filters by highlighting filters `FIR1` and `IIR1`, followed by **View**. The selected display in the **Filter Viewer** is similar to that displayed in Figure 3.18. The **Filter Viewer** also allows the user to view the characteristics of a designed or imported filter, including the magnitude response, phase response, group delay, pole-zero plot, impulse, and step responses of the filter.

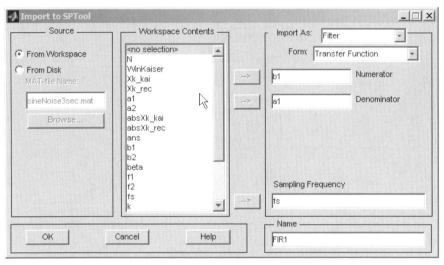

Figure 3.19 Import the moving-average filter's coefficients into SPTool

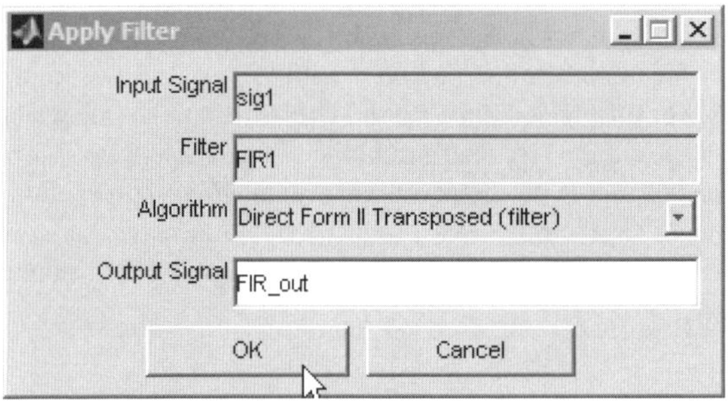

Figure 3.20 **Apply Filter** dialog box for filtering

The imported filter, FIR1 (or IIR1), can be selected and applied to the input signal, sig1, by clicking on the **Apply** button in the SPTool window under **Filters**. A new window as shown in Figure 3.20 is displayed, which allows the user to select the structure of the filter and specify the name of the output signal as FIR_out. Repeat this process for filtering the same sig1 with the IIR1 filter, and name the output signal IIR_out.

The time-domain plots of the input signal and filtered output signals can be viewed by selecting sig1, FIR_out, and IIR_out. This is followed by clicking on the **View** button in the **Signal** column. The third column of the SPTool is the **Spectrum Viewer**, which is used to analyze and view the frequency content of the signals. Following the steps introduced in Hands-On Experiment 3.2, we can create three spectra for the signals. Figure 3.21 shows the magnitude spectra for the input signal and the output signals from both FIR and IIR filters. Finally, the SPTool session can be saved in the file experiment3_4.spt for later reference.

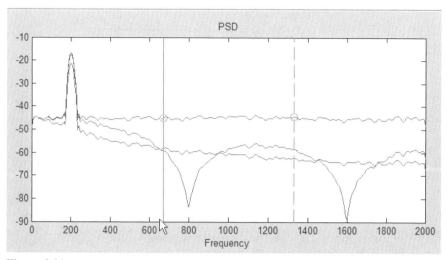

Figure 3.21 Spectra of input signal (top) and filtered signals (the one with notches is the output of the FIR moving-average filter)

3.4.2 Design and Applications of Notch Filters

A notch filter contains one or more deep notches in its magnitude response. To create a notch at frequency ω_0, we simply introduce a pair of complex-conjugate zeros on the unit circle at angle ω_0 as

$$z = e^{\pm j\omega_0}. \tag{3.4.1}$$

The transfer function of this FIR notch filter is

$$\begin{aligned}H(z) &= \left(1 - e^{j\omega_0} z^{-1}\right)\left(1 - e^{-j\omega_0} z^{-1}\right) \\ &= 1 - 2\cos(\omega_0) z^{-1} + z^{-2}.\end{aligned} \tag{3.4.2}$$

This is the FIR filter of order 2 because there are two zeros in the system. Note that a pair of complex-conjugate zeros guarantees that the filter will have real-valued coefficients as shown in Equation 3.4.2.

The magnitude response of the second-order FIR notch filter described in Equation 3.4.2 has a relatively wide bandwidth, which means that other frequency components around the null are severely attenuated. To reduce the bandwidth of the null, we may introduce poles into the system. Suppose that we place a pair of complex-conjugate poles at

$$z_p = re^{\pm j\theta_0}, \tag{3.4.3}$$

where r and θ_0 are the radius and angle of poles, respectively. The transfer function for the resulting filter is

$$H(z) = \frac{\left(1 - e^{j\omega_0} z^{-1}\right)\left(1 - e^{-j\omega_0} z^{-1}\right)}{\left(1 - re^{j\theta_0} z^{-1}\right)\left(1 - re^{-j\theta_0} z^{-1}\right)} = \frac{1 - 2\cos(\omega_0) z^{-1} + z^{-2}}{1 - 2r\cos(\theta_0) z^{-1} + r^2 z^{-2}}. \tag{3.4.4}$$

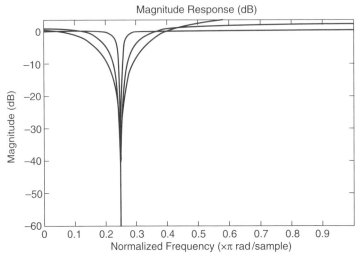

Figure 3.22 Magnitude responses of notch filter for different values of r (from top to bottom, $r = 0.95, 0.75,$ and 0.5).

The notch filter expressed in Equation 3.4.4 is the second-order IIR filter because there are two poles in the system.

EXAMPLE 3.16

The magnitude response of the filter defined by Equation 3.4.4 is plotted in Figure 3.22. We use the fixed frequency ($\theta_0 = \omega_0 = 0.25\pi$), but change the value of r. The magnitude responses of the filter are shown for $r = 0.5, 0.75,$ and 0.95, using the MATLAB script `example3_16.m`. Compared with the magnitude response of the FIR filter given in Equation 3.4.2, we note that the effect of the pole is to reduce the bandwidth of the notch. Obviously, the closer the r value to 1 (poles are closer to the unit circle), the narrower the bandwidth.

EXERCISE 3.14

In MATLAB program `exercise2_8.m`, a speech file, `timit1.asc`, is corrupted by 1-kHz tone, and the corrupted signal with a sampling rate of 8 kHz is saved in the file `speech_tone.mat`.

1. Implement the notch filter with $r = 0.95$ designed in Example 3.16 for attenuating the tonal noise. Play both input and output signals with the MATLAB function `soundsc` and display the spectrogram of output speech (see Fig. 3.23) with the function `specgram`. Compared with Figure 3.11, we found that the tonal noise at 1 kHz (0.25π rad) was removed and the sound quality of the output speech is high. See `exercise3_14.m` for reference.

2. Redo Exercise 1, using different values of r, and evaluate the results by playing the outputs and viewing the spectrograms.

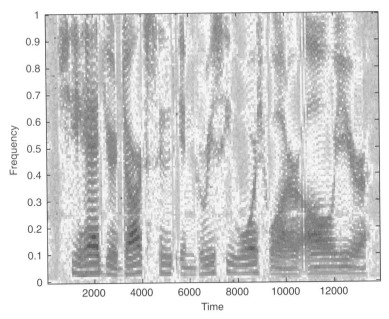

Figure 3.23 Output spectrogram with the tone removed by the notch filter

3. Compute the magnitude response of a notch filter with only two complex-conjugate zeros on the unit circle as expressed in Equation 3.4.2. Compare it with the notch filter with poles as shown in Figure 3.22.

4. Modify the MATLAB script `exercise2_8.m` that adds a 60-Hz hum (tone) into the speech file `timit1.asc` with different SNRs. Play the corrupted speech with `soundsc`. Note that many real-world speeches are corrupted by 60-Hz tone and its harmonics.

5. Design a notch filter to attenuate this undesired 60-Hz hum. Evaluate the filter's performance by examining the spectrogram and playing the output speech.

HANDS-ON EXPERIMENT 3.5

In Example 2.9, we used the MATLAB code `mulSineGen.m` for generating signal consisting of a desired sine wave that is corrupted by another sine wave at different frequency and saved it in a data file. We are not able to use simple low-pass filters shown in Figure 3.18 to remove those interferences because both desired and undesired sine waves are at the low-frequency range. In this experiment, we design a notch filter to attenuate the undesired tone. We use the first data file, `sine200plus400at4k.mat`, which is a 200-Hz sine wave corrupted by a 400-Hz sine wave with a sampling rate of 4 kHz, as an example. Note that the digital frequencies of 200 Hz and 400 Hz are equivalent to 0.1π and 0.2π, respectively.

94 Chapter 3 Frequency-Domain Analysis and Processing

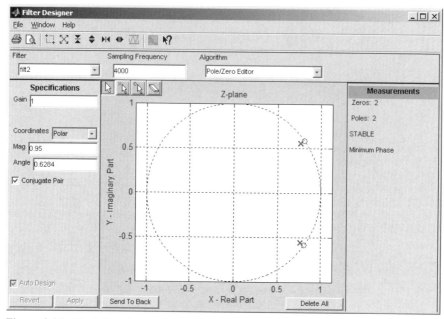

Figure 3.24 Notch filter design using **Pole/Zero Editor**

In this experiment, we use the **Pole/Zero Editor** in SPTool to specify the poles and zeros of the IIR filter. In the SPTool window, click on the **New** button under **Filters**; a **Filter Designer** dialog box will be displayed. Select the **Pole/Zero Editor** in the **Algorithm** menu; the default filter with a pole/zero diagram is displayed. Clear this filter by clicking on the **Delete All** button. The window that appears in Figure 3.24 allows the user to drag- and-drop poles and zeros in the z-plane with the icons ![] and ![], respectively. To reduce 400-Hz tone, the complex-conjugate poles and zeros are positioned at frequency 0.2Π (or 0.6284) rad. We use these two icons to drop poles and zeros into the desired position and use the **Specifications** panel at the left-hand side to fine-tune the position as shown in Figure 3.24. Note that we also change the default **Sampling Frequency** to 4000 Hz. Once the pole and zero positions are specified, we name the filter filt2.

We can evaluate the filter characteristics by highlighting the filter filt2 and clicking on the **View** button under **Filters** in the SPTool window. A Filter Visualization Tool (FVTool) window is displayed, which was introduced in Hands-On Experiment 3.1. For example, we select the **Magnitude and Phase Responses** option from the **Analysis** menu, and Figure 3.25 is displayed. This figure clearly shows that the 400-Hz tone is attenuated by the designed notch filter and the desired 200-Hz tone passes without attenuation.

Import the file sine200plus400at4k.mat into the SPTool and name it sig1. Perform IIR filtering to produce the output sig2. We can view both input (mixing of two sine waves) and output (only 400-Hz tone) by the **Signal Browser**. We also create spectra of both input and output signals and display both spectra in Figure 3.26. From the figure, we find that the 400-Hz interference was attenuated by about 60 dB. We also can play input and output signals separately and evaluate their differences.

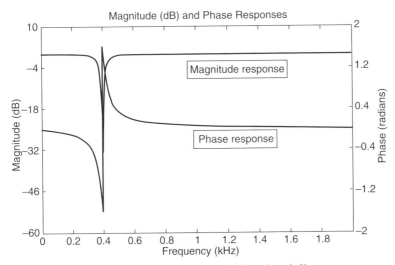

Figure 3.25 Magnitude and phase responses of designed notch filter

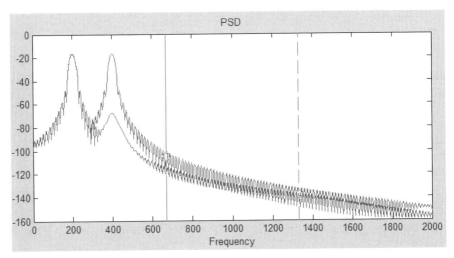

Figure 3.26 Input and output spectra

EXERCISE 3.15

1. Redesign the notch filter shown in Figure 3.25, using $r = 0.99$. Apply this filter to the data file `sine200plus400at4k.mat` and show both input and output spectra.

2. Design a notch filter to remove the 400-Hz sine wave generated in Example 2.9(2) and show the results.

3. Design a notch filter to remove the 60-Hz hum generated in Example 2.9(3) and show the results.

3.4.3 Design and Applications of Peak Filters

In Hands-On Experiment 3.4, we used the moving-average FIR filter and the first-order IIR filter for enhancing a sine wave that was corrupted by white noise. As shown in Figure 3.18, these filters have undesired attenuation of signals because the gain is less than 1. They also only provide about 10–15 dB of noise reduction, as shown in Figure 3.21. In Section 3.4.2, we introduced notch filters that have very narrow bandwidth for attenuation of sinusoidal interference. In this section, we extend a similar technique to enhance a sinusoidal signal that is corrupted by broad-band noises.

To create a peak (or narrow passband) at frequency ω_0, we may think we can simply follow the example of designing a notch filter by introducing a pair of complex-conjugate poles on the unit circle at angle ω_0. However, as discussed in Section 3.2.3, the resulting second-order IIR filter will be unstable. To solve this problem, we have to move the poles slightly inside the unit circle (i.e., use $r_p < 1$) as

$$z_p = r_p e^{\pm j\omega_0}. \tag{3.4.5}$$

The transfer function of this second-order IIR peak filter is

$$H(z) = \frac{1}{\left(1 - r_p e^{j\omega_0} z^{-1}\right)\left(1 - r_p e^{-j\omega_0} z^{-1}\right)} = \frac{1}{1 - 2r_p \cos(\omega_0) z^{-1} + r_p^2 z^{-2}}. \tag{3.4.6}$$

Similar to the second-order IIR notch filter introduced in Section 3.4.2, the magnitude response of the second-order IIR peak filter described in Equation 3.4.6 has a relatively wide bandwidth, which means that other frequency components around the peak will also be amplified. To reduce the bandwidth of the peak, we may introduce zeros into the system. Suppose that we place a pair of complex-conjugate zeros at radius r_z with the same angle as the poles; the transfer function for the resulting second-order IIR filter is

$$H(z) = \frac{\left(1 - r_z e^{j\omega_0} z^{-1}\right)\left(1 - r_z e^{-j\omega_0} z^{-1}\right)}{\left(1 - r_p e^{j\omega_0} z^{-1}\right)\left(1 - r_p e^{-j\omega_0} z^{-1}\right)} = \frac{1 - 2r_z \cos(\omega_0) z^{-1} + r_z^2 z^{-2}}{1 - 2r_p \cos(\omega_0) z^{-1} + r_p^2 z^{-2}}. \tag{3.4.7}$$

It is important to note that for designing a peak filter, the poles must be closer to the unit circle as

$$r_p > r_z. \tag{3.4.8}$$

The IIR filter defined in Equation 3.4.7 can be applied as a simple parametric equalizer for boosting ($r_p > r_z$) or cutting ($r_p < r_z$) an audio signal. The amount of boost or cut is determined by the difference between r_p and r_z. The bandwidth of peaks or notches is determined by the value of r_z.

EXAMPLE 3.17

The peak filter with a pair of complex-conjugate poles only as defined in Equation 3.4.6 is implemented in MATLAB code example3_17.m. The magnitude responses for radius 0.99,

0.95, and 0.9 are shown in Figure 3.27. This figure shows that the second-order IIR filter with poles only has wide bandwidth. Also, when the poles are closer to the unit circle, the filter has higher gain.

The peak filter with both complex-conjugate poles and zeros as defined in Equation 3.4.7 is also implemented in the same MATLAB code with the radius of poles fixed at 0.99, but the radius of zero is varied at 0.5, 0.75, and 0.95 in order to satisfy Equation 3.4.8. The magnitude responses for these three different zero positions are shown in Figure 3.28.

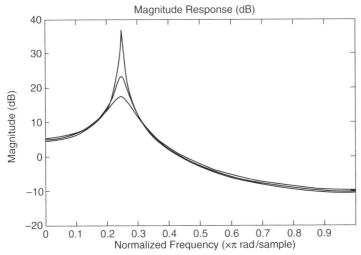

Figure 3.27 Magnitude responses of peak filter with poles only; radius $r = 0.99$ (top), 0.95 (middle), and 0.9 (bottom)

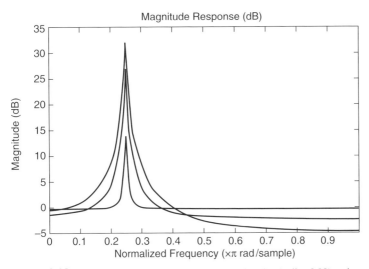

Figure 3.28 Magnitude responses of peak filter with poles (radius 0.99) and zeros; radius of zeros $r = 0.5$ (top), 0.75 (middle), and 0.95 (bottom)

EXERCISE 3.16

1. In Hands-On Experiment 3.4, we used an FIR moving-average filter with $L = 5$ and its equivalent first-order IIR filter for enhancing the 200-Hz sine wave (sampled at 4 kHz) in data file `sineNoise3sec.mat`. Design a peak filter using MATLAB with poles radius 0.99 and zeros radius 0.95 to enhance the sine wave, and compare the results with Figure 3.21.

2. Redo Exercise 1 by using **Pole/Zero Editor** to design different peak filters, analyze their characteristics, filter the noisy sine wave, and compare input and output spectra.

3. In Exercise 2.8, we added tonal and white noises into the speech signal. In Exercise 3.14, we used a notch filter to remove tonal noise. Try all the filters we have learned so far to reduce random noise. You may find that this is a very difficult task. Why?

EXERCISE 3.17

1. If the impulse response of an IIR filter is $h(n) = \sin(\omega_0 n)$, show that the transfer function of the filter is expressed as

$$H(z) = \frac{\sin(\omega_0) z^{-1}}{1 - 2\cos(\omega_0) z^{-1} + z^{-2}}. \quad (3.4.9)$$

2. Find the poles of this filter by comparing the transfer function with Equations 3.4.5 and 3.4.6. Is this a stable IIR filter?

3. If the sampling rate is 8 kHz, compute the values of IIR filter coefficients for generating sinewaves at frequencies of 697, 770, 852, 941, 1,209, 1,336, 1,477, and 1,633 Hz.

4. Write a MATLAB program to implement Equation 3.4.9. Apply an impulse function to the filter to generate sine waves at frequency defined in Exercise 3.

3.5 FREQUENCY ANALYSIS WITH BLACKFIN SIMULATOR

This section introduces additional features of the VisualDSP++ simulator for the Blackfin (BF533 and BF537) processors. We use optimized functions provided in the C run time library and DSP run time library. These library functions greatly ease the development of efficient C code for embedded applications.

HANDS-ON EXPERIMENT 3.6

This experiment uses a simple program, main.c, to compute the magnitude spectrum of the input signal. The project file, exp3_6.dpj, is provided in directory c:\adsp\chap3\exp3_6. Double-click on main.c to display it, and examine the program carefully. This C main program includes the following header files:

```
#include <stdio.h>      // standard input and output
#include <fract.h>      // support fractional values
#include <complex.h>    // basic complex arithmetic
#include <filter.h>     // filters and transformations
#include <math.h>       // math functions
```

The main program reads in data from the input file in1.dat, which contains 512 samples of sine wave. The program then calls the library functions to compute FFT, square root, and magnitude spectrum. The result is stored in the output file out1.dat. The critical portion of program is listed as follows:

```
twidfft_fr16(w, VEC_SIZE); // initialize twiddle factors rfft_
fr16(in, t, out, w, 1, VEC_SIZE, 0, 0); // FFT
mag[0] = sqrt(out[0].re*out[0].re+out[0].im*out[0].im);
fprintf(fid, "%d,\n", (short)mag[0]);
for (i=1; i<VEC_SIZE/2; i++) {
  mag[i] = 2*sqrt(out[i].re*out[i].re+out[i].im*out[i].im);
  fprintf(fid, "%d,\n", (short)mag[i]);
}
mag[VEC_SIZE/2]=sqrt(out[VEC_SIZE/2].re*out[VEC_SIZE/2].re
          +out[VEC_SIZE/2].im*out[VEC_SIZE/2].im);
fprintf(fid, "%d,\n", (short)mag[VEC_SIZE/2]);
fclose(fid);
```

Build the project and load the program into the VisualDSP++ simulator. Before running the program, open two previously created debug windows, input.vps and output.vps, to display the input signal waveform and the output spectrum, respectively. Click on **View** → **Debug Windows** → **Plot** → **Restore**, and search for input.vps in directory c:\adsp\chap3\exp3_6. A graphic window will appear in the right column, as shown in Figure 3.29. Right-click on the displayed window, click on **Configure...** → **Settings...**, and select the **2D-Axis** tab. Note that the increment value of the x-axis should be equal to the sampling period (1/48,000 in this case).

The main program, main.c, uses the FFT routine provided in the library to compute the magnitude spectrum of the input signal. Open the output window by clicking on **View** → **Debug Windows** → **Plot** → **Restore** and search for output.vps in directory c:\adsp\chap3\exp3_6. Right-click on the output window and select **Auto Refresh**. Run the program, and the output will be updated in the window. Again, right-click and activate the **Data Cursor** to measure the frequency of the peak magnitude. Instead of measuring the frequency in hertz, the magnitude spectrum plot can also be measured in term of the frequency index, k. The relationship between the frequency in hertz and the frequency index is defined in Equation 3.3.19. Because we are using 512-point FFT and the sampling frequency f_s is 48 kHz, the frequency resolution Δf is 93.75 Hz. The peak frequency is measured at 3,000 Hz, which corresponds to the frequency index $k = 32$. Note that the index k starts from 0 to 511, and only 257 points are displayed from the frequency index $k = 0$ (corresponds to DC or 0 Hz) to $k = 257$ (corresponds to the Nyquist frequency, which is 24 kHz in this

100 Chapter 3 Frequency-Domain Analysis and Processing

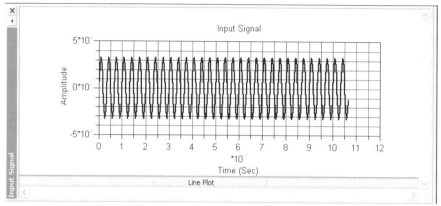

Figure 3.29 Display of input signal

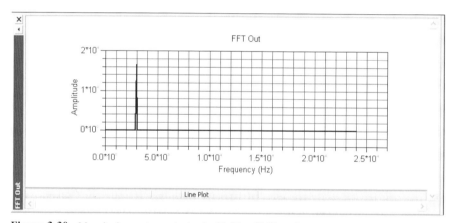

Figure 3.30 Magnitude spectrum obtained with VisualDSP++ data process

example). Users can also export outl.dat to the MATLAB workspace for observing the magnitude spectrum. It is important to note that the MATLAB index starts from 1, so that $k = 1$ corresponds to 0 Hz, and so on.

Instead of writing an FFT routine to compute the magnitude spectrum, users can also display the magnitude spectrum of the signal by right-clicking on the input window and selecting **Modify Settings.** Click on the **Data Processing** tab, and selecting **FFT Magnitude** under **Data Process**. Click on **OK**, and the magnitude spectrum is displayed as shown in Figure 3.30. Examine the position of the peak and its frequency index and magnitude. The peak should occur at 3,000 Hz.

We can also display the spectrogram by using the VisualDSP++ graphical feature. Right-click on the input signal window and select **Configure....** Select **Spectrogram Plot** under **Plot Type**. Because we have 512 samples of input signal, we can divide the time axis (y-axis) into 5 parts and the frequency axis (x-axis) into 102 parts. Figure 3.31(a) shows the

3.5 Frequency Analysis with Blackfin Simulator   **101**

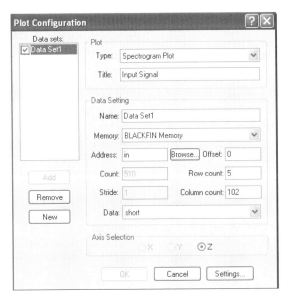

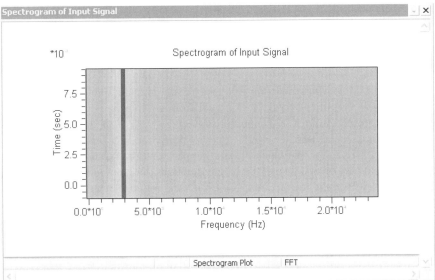

Figure 3.31 (a) Configuration for plotting spectrogram (b) Spectrogram plot of the input signal, in1.dat

settings used for the spectrogram plot. The new plot is shown in Figure 3.31(b). It shows a vertical red line at the 3,000-Hz mark, which indicates that the 3,000-Hz sine wave is present in the signal at all times. The spectrogram plot provides both frequency and time information for the signal being analyzed. This plot is particularly useful for analyzing nonstationary signals like speech.

EXERCISE 3.18

Use the display tool given in Hands-On Experiment 3.6 to perform the following tasks:

1. The digitized speech plus tonal signal (speech_sine.dat) given in Example 3.13 can be imported into the VisualDSP++ simulator for analysis. However, the data file is too long to fit into the internal memory of the processor. This problem can be overcome by placing the data in the external memory. To demonstrate this method, a project file, exercise3_18.dpj, is given in directory c:\adsp\chap3\exercise3_18. Build the project and perform signal analysis with VisualDSP++.

2. Display the time-domain waveform of the corrupted speech signal and listen to this signal with the computer sound card. Set the sampling rate to 8,000 Hz. Export data to the sound card can be enabled by right-clicking on the display window and selecting the **Export...** option.

3. Note the length of the speech signal and plot its spectrogram with VisualDSP++. What is the frequency of the sine wave?

3.6 FREQUENCY ANALYSIS WITH BLACKFIN BF533/BF537 EZ-KIT

This section implements real-time frequency analysis of the input signal with the Blackfin BF533 or BF537 EZ-KIT. A signal from a CD player or computer sound card is connected to the ADC1 input pair (see Fig. 2.23) of the BF533 EZ-KIT or the stereo input of the BF537 EZ-KIT. The Blackfin processor on the EZ-KIT computes the FFT based on the block of data samples, and the result is displayed with the graphical plot of the VisualDSP++ environment. To continuously analyze and display the spectrum of the incoming signal, a background telemetry channel (BTC) is required to facilitate data exchange between the VisualDSP++ hosted on the computer and the Blackfin processor without interrupting the processor. This feature allows the frequency spectrum to be updated as soon as the block of signal is being calculated. In the following exercises, we use BTC to change the parameters of the program (such as the window type, window overlapping, and number of data samples per block) on the fly.

HANDS-ON EXPERIMENT 3.7

This experiment uses the frequency analyzer programs provided in the project exp3_7_533.dpj for the BF533 EZ-KIT (or exp3_7_537.dpj for the BF537), which can be found in directory c:\adsp\chap3\exp3_7_533 (or c:\adsp\chap3\exp3_7_537). This project includes four source files summarized as follows:

1. init.c initializes the EZ-KIT for analog-to-digital conversion. This program also sets up the direct memory access (DMA) controller and interrupts. A more detailed explanation of this program is postponed to Chapter 7.
2. isr.c is the interrupt service routine (ISR) that performs the real-time frequency analysis after the CODEC has acquired a block of data samples.
3. main.c is the main program that declares all the variables and their memory allocation. It also initializes the EZ-KIT and forces the processor into an infinite loop for frequency analysis.
4. process.c performs windowing and window overlapping and computes FFT and magnitude spectrum. This code is called isr.c.

Build the project. Open the graphical window by clicking on **View** → **Debug Windows** → **Plot** → **Restore**. Select FFT In.vps and FFT Out.vps in directory c:\adsp\chap3\exp3_7. Right-click on the plots and make sure that **Auto Refresh** is enabled. Select the wave file chirp.wav from directory c:\adsp\audio_files and play this wave file continuously with an audio player on the computer. Run the project and observe the two graphical windows as shown in Figure 3.32. Note that the **FFT Out** spectrum will be swept from left to right and back from right to left along the frequency axis. This plot shows that the chirp waveform is a linear swept-frequency signal.

The updates of the signal waveform and the magnitude spectrum (shown in Fig. 3.32) are performed by the BTC via the USB interface. Users can adjust the parameters of the program with the BTC. Five steps needed to include the BTC in the source code are listed as follows:

1. Add the btc.h header file to the source code.
2. Define the channels in the source code (main.c) as follows:

   ```
   BTC_MAP_BEGIN
   // channel name, starting address, length
   BTC_MAP_ENTRY("FFT_INPUT",(long)&BTC_CHAN0,sizeof(BTC_CHAN0))
   BTC_MAP_ENTRY("FFT_OUTPUT",(long)&BTC_CHAN1,sizeof(BTC_CHAN1))
   BTC_MAP_ENTRY("INPUT SIZE",(long)&new_size,sizeof(new_size))
   BTC_MAP_ENTRY("WINDOW TYPE",(long)&new_win_select,sizeof(new_
     win_select))
   BTC_MAP_ENTRY("OVERLAP",(long)&overlap, sizeof(overlap))
   BTC_MAP_END
   ```

   In this case, five channels are defined to capture the user's data entries when the program is running.
3. Define the BTC polling loop with the btc_poll( ) command in the interrupt service routine. The polling loop checks for incoming commands from the host computer.
4. Initialize the BTC with the btc_init( ) command in main.c.
5. Add the BTC library libbtc532.dlb into the project.

To activate the BTC, click on **View** → **Debug Windows** → **BTC Memory**. The **BTC Memory** window is displayed as shown in Figure 3.33, which indicates five channels that are defined in the source code. The first two channels (FFT_INPUT and FFT_OUTPUT) are used to display the real-time data, while the remaining three channels (INPUT_SIZE, WINDOW_TYPE, and OVERLAP) are used for user-defined parameters. Refer to Table 3.1 for the options provided for these three channels. Change the parameter values during the program execution and observe the changes to the plots.

104 Chapter 3 Frequency-Domain Analysis and Processing

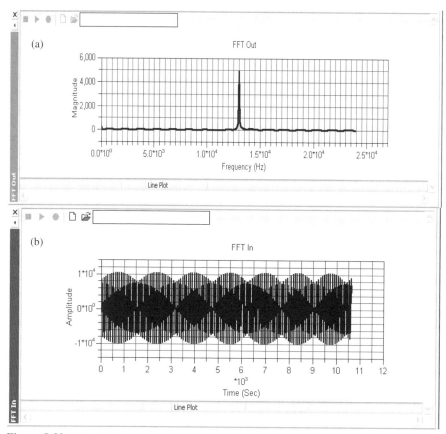

Figure 3.32 Magnitude spectrum and waveform of input signal, spectrum (a) and time-domain signal (b)

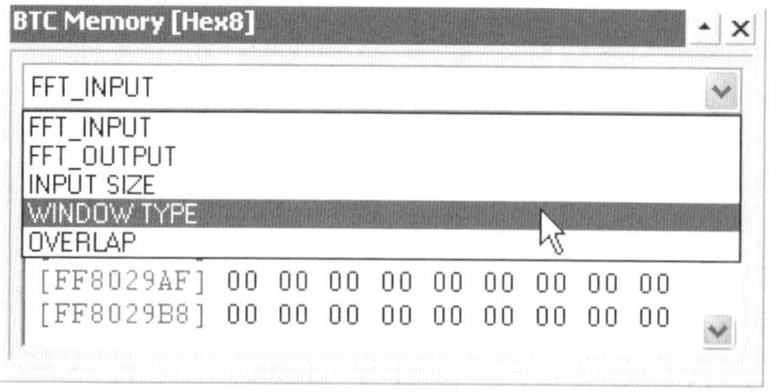

Figure 3.33 BTC memory window

3.7 Frequency Analysis with LabVIEW Embedded Module for Blackfin Processors

Table 3.1 Parameters Used to Adjust the Frequency Analysis

INPUT_SIZE (Hex16)	
0×0200	512-point FFT
0×0100	256-point FFT
0×0080	128-point FFT
0×0040	64-point FFT
WINDOW_TYPE (Hex8)	
0×00	Rectangular
0×01	Bartlett
0×02	Hanning
0×03	Hamming
0×04	Blackman
OVERLAP (Hex8)	
0×00	No overlap between frames
0×01	50% overlap between frames

EXERCISE 3.19

Replace `chirp.wav` with the wave file, `liutm_48k_mono.wav`, and perform the following tasks:

1. Change `INPUT_SIZE` of the FFT and observe the resulting magnitude spectrum. Users have to adjust the resolution settings of the **FFT Out** plot by changing the **Increment value** of the x-axis and the **Count** value of the display.
2. Change `WINDOW_TYPE` for windowing the input signal and note the spectrum differences by using different windows.
3. Change `OVERLAP` of the input blocks and note any change in the spectrum.
4. Modify the source code to process stereo input. Display the magnitude spectra in real time for both the left and right channels.
5. Instead of using the given code to perform FFT, users can use the graphical plot feature to display the magnitude spectrum. Display the spectrum of the input signal and compare it with that obtained in Hands-On Experiment 3.7.

3.7 FREQUENCY ANALYSIS WITH LABVIEW EMBEDDED MODULE FOR BLACKFIN PROCESSORS

Frequency analysis allows a designer to extract useful signal information in the presence of noise or to detect a physical phenomenon like failing bearings in a motor. Accurate time and frequency information is often needed before the embedded design process begins, allowing the designer to choose sampling rates and filters to accurately design and implement the system.

The following examples explore frequency-domain properties in LabVIEW, enabling the designer to simulate, prototype, and deploy an application that detects frequency information from time-domain signals. First, we explore frequency-domain concepts through an interactive simulation. Then we implement similar functionality with the FFT through the LabVIEW Embedded Module for Blackfin Processors, using the LabVIEW graphical interface to view and interact with signals streaming into the Blackfin EZ-KIT.

HANDS-ON EXPERIMENT 3.8

This hands-on experiment introduces the effects of windowing on frequency analysis. Understanding the frequency content of signals within a system is key to development of effective filters and other system components. We accomplish this by using the FFT algorithm, which is based on having a time-domain signal that continues on to infinity. In real-world applications, finite-length signals will cause leakage in the frequency domain that distorts the result of the FFT. Therefore, windowing is used to minimize the leakage introduced at the beginning and end of a time-domain finite-length signal.

Open the program `Window_FFT_Sim.exe`, located in directory `c:\adsp\chap3\exp3_8`. This LabVIEW application studies different window topologies applied to an input signal and their effects on the performance of the FFT. Figure 3.34 shows the user interface for `Window_FFT_Sim.exe`. Note that the **Time & Frequency** tab shows the **Time Domain** signal and the **Frequency Response**, whereas the **Window** tab shows the time representation of both the window and the input signal after the window has been applied.

The default input signal is a 1-kHz sine wave sampled at 48 kHz with a buffer length of 512 samples. When no window is applied, as shown in Figure 3.34, leakage is expected in the frequency domain because the **Time Domain** plot shows that the sine wave stops short of completing a full period. The effects of leakage can be seen in the **Frequency Response** plot because the power at high frequencies rests above −60 dB. The **Window** pull-down menu can be used to change the type of windowing to be implemented. Applying a Hanning window provides improved performance, as seen in Figure 3.35, where the power at high frequencies drops below −120 dB.

Now select the `Bartlett` window. How do the results change? Experiment with the other window types and pay attention to the corresponding frequency response of the sinusoidal input. What trade-offs are apparent in both the time and frequency domains when using the different windows?

Now load and experiment with the audio file `speech_tone_48k.wav`, which was used in Hands-On Experiment 2.5. Can you see the tonal noise in the frequency-response graph? Experiment with different windows and parameters to see the FFT results. How does windowing affect the non-periodic speech file differently from the periodic sinusoid? How would you expect the results to vary if you were to break up the speech tone into 512-sample buffers? Explain.

3.7 Frequency Analysis with LabVIEW Embedded Module for Blackfin Processors 107

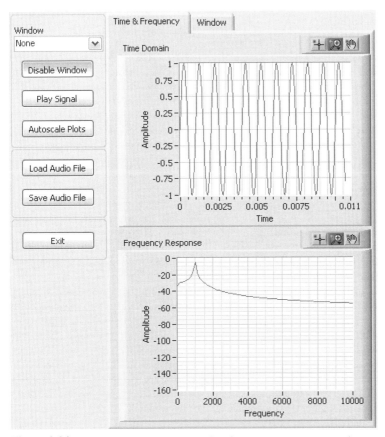

Figure 3.34 Frequency analysis (user interface for `Window_FFT_Sim.exe`)

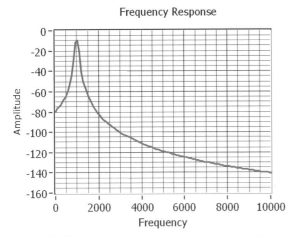

Figure 3.35 Frequency analysis of the input signal with a Hanning window

HANDS-ON EXPERIMENT 3.9

This experiment implements the FFT algorithm in the LabVIEW Embedded Module for Blackfin Processors for execution on the Blackfin EZ-KIT to explore the advantages of various windowing implementations. Executing the project in debug mode on the Blackfin processor allows the user to interact with the LabVIEW front panel and to see the results of the FFT.

Open the `Audio FFT - BF5xx.lep` project appropriate for your Blackfin processor in directory `c:\adsp\chap3\exp3_9`. Open the block diagram to see how the FFT is implemented. The block diagram shown in Figure 3.36 is intuitive, allowing the programmer to easily see how to integrate the analog input and processing algorithm. The **Window Type** control is wired to a **Case Structure** allowing the user to specify the window to be applied to the acquired input signal. The windowed time-domain result is plotted to the front panel indicator and passed to the **FFT subVI**. The FFT result is then converted from complex representation to magnitude response, which is then plotted to a front panel indicator as well. The data type is converted before plotting to limit the amount of data transferred between the code running on the Blackfin target and the front panel interface running on the computer.

Customize the debugging mode and parameters within the LabVIEW Embedded Module for Blackfin Processors to enhance the experience and value of this exercise. First, modify the target debugging parameters to increase the amount of data to be downloaded to graph indicators. Then choose the debugging method. Finally, execute the application with debugging support. These steps are described in more detail below.

Modify the target debugging parameters by navigating to the LabVIEW **Embedded Project Manager** window and selecting **Target → Configure Target → Debug Options**, change **Max array elements** to 256, and click on **OK**. This is necessary to view all of the data on the graph because the buffer length on the block diagram is specified as 512 samples, which is divided into 256 for each channel. Therefore, the time-domain signal and FFT result will have 256 samples.

These three debugging methods offer different advantages depending on the application and resources. By default, this project is configured to the standard JTAG debugging

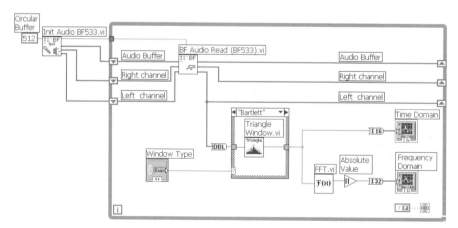

Figure 3.36 Frequency analysis block diagram

3.7 Frequency Analysis with LabVIEW Embedded Module for Blackfin Processors

method using JTAG over USB. JTAG is the standard communication method between the Blackfin processor and VisualDSP++. In this debug mode, data transfers interrupt the processor when transmitting data to and from the computer, which disrupts real-time processing. JTAG debugging is adequate for this and many other applications as long as real-time processing is not necessary. Another option, unique to the LabVIEW Embedded Module for Blackfin Processors, is called instrumented debugging, which requires the USB cable and an additional cable connection between the Blackfin and the computer. The additional cable can be either serial or TCP/IP and allows for quicker update rates, providing a faster, more interactive experience. Instrumented debugging adds extra code to the project to initiate debug data transfers while the LabVIEW code is running on the processor, adding as much as 40% to the size of the embedded executable code. If you are interested in using serial or TCP/IP debug, attach the appropriate cables and change the debug mode for the project by navigating to **Target → Build Options** and changing **Debug Mode** to **Instrumented (via serial port)** or **Instrumented (via TCP port)**.

Run the project on the Blackfin processor. Close the block diagram window and navigate back to the project window. Run the VI using debug mode to explore the various types of windowing. Different windows can be applied during run time by changing the **Window Type** control on the front panel as shown in Figure 3.37. Change the window type to see how it affects the time-domain signal and the performance of the FFT. How does windowing improve the resulting frequency-domain calculation?

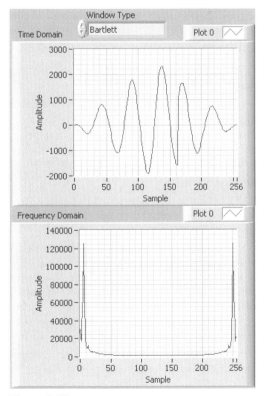

Figure 3.37 Frequency analysis front panel

Probe the values passed through wires on the block diagram using probes. Note that the resulting vector from the FFT is complex and the **Absolute Value** function converts the complex values to the magnitude response. Modify the experiment to show the real part of the frequency response in one graph and the imaginary part in the other.

3.8 MORE EXERCISE PROBLEMS

1. The digital signal $x(n)$ consists of two real-valued sinusoidal components expressed as
   $$x(n) = A_1\sin(\omega_1 n) + A_2\sin(\omega_2 n) + v(n),$$
   where $v(n)$ is zero-mean and unit-variance white noise, $\omega_1 = 0.1\pi$, and $\omega_2 = 0.2\pi$.
   (a) Compute A_1 and A_2 such that $A_1 = 2A_2$ and the signal-to-noise ratio is 10 dB. Generate 512 samples $x(n)$ with MATLAB.
   (b) Design a moving-average filter of $L = 10$. Filter the signal $x(n)$ and plot both $x(n)$ and $y(n)$ with MATLAB. Discuss the results and explain why.
   (c) Use MATLAB functions `fft`, `abs`, and `log10` to compute and plot the magnitude spectra of $x(n)$ and $y(n)$ with a dB scale. Plot the spectra from frequencies 0 to π.
   (d) Explain why the sinusoidal component at frequency $\omega_2 = 0.2\pi$ has been attenuated.
   (e) Design a peak filter with two peaks to enhance these two sinusoids.
2. Redo Problem 1 with SPTool.
3. Given a speech file `timit1.asc`, perform the following task:
   (a) Use different frequency analysis techniques to obtain its spectrum. Other than `specgram`, do other techniques really work?
   (b) Add white noise with a mean of 0 and a variance of 20,000 into the original speech. Replay the original and corrupted speech with the MATLAB function `soundsc`, and show its frequency contents with the MATLAB function `specgram`.
   (c) Can you design a digital filter to reduce the noise?
4. The convolution function `conv` in MATLAB can be used to perform FIR filtering. Redo Example 3.7 with the `conv` function. Is the output derived from `conv` identical to that from `filter`? If not, how can the two outputs be the same?
5. Generate a 3-s sine wave of 2,000 Hz in MATLAB, using a sampling frequency of 16,000 Hz. Use MATLAB function `soundsc` to play the generated sine wave. Save the sine wave as a wave file with extension .wav by using the `wavwrite` function. The wave file can be saved as 16 bits per sample. Record the file size of the saved wave file. Examine the size of the file saved if 8 bits per sample is used. Finally, the wave file can be read back into the MATLAB workspace by using the `wavread` function. Plot the wave file and compare it with the original data file generated by MATLAB.
6. Using the property of z-transform of an N-periodic sequence given in Exercise 3.2(3), find the z-transform of square wave $x(n) = \{1\ 1\ 1\ -1\ -1\ -1\ 1\ 1\ 1\ \ldots\}$.

7. A digital system is defined by the I/O equation $y(n) = x(n) + x(n-1)$.
   (a) Sketch the signal flow diagram of the system.
   (b) Find the magnitude and phase responses of the system.

8. Find the transfer function of the systems described by the following difference equations:
   (a) $2y(n) + y(n-1) + 0.75y(n-2) = x(n-1) + 2x(n-2)$
   (b) $y(n) - 0.5y(n-1) + 2y(n-2) = x(n) - x(n-1) + 0.75x(n-2)$
   (c) Derive Equation 3.2.21 from Equation 3.2.20.

9. A digital system is defined by the I/O equation $y(n) = x(n) + ay(n-1)$.
   (a) Sketch the signal-flow diagram of the system.
   (b) Find the magnitude response of the system and sketch it for $a = 0.9$ and $a = 0.5$.

10. A three-point moving-average filter is defined by the I/O equation
    $$y(n) = \tfrac{1}{3}[x(n) + x(n-1) + x(n-2)].$$
    (a) Find the frequency response of the system.
    (b) Compute the magnitude and phase responses of the system.

11. An analog sine wave with frequency $f_0 = 100\,\text{Hz}$ is sampled at a sampling rate of 1,000 Hz. What is the normalized frequency F_0 in cycles per sample and the digital frequency ω_0 in radians per sample?

12. Similar to Equation 3.2.20, a simple power estimator can be expressed as
    $$P(n) \equiv \left(1 - \frac{1}{L}\right) P(n-1) + \frac{1}{L} x^2(n) = (1-\alpha)P(n-1) + \alpha x^2(n).$$
    Use this equation to estimate the power of speech signal `timit1.asc`, using different values of L. For example, try $L = 32$, 128, and 1,024. Plot these power estimation results, compare with the speech waveform, and summarize the results.

13. The N-point radix-2 inverse FFT function `ifft` is used to transform the frequency-domain complex sequence into the time-domain signal. Using the VisualDSP++ simulator, extend the source code in `exp3_6.dpj` to include the IFFT routine after FFT to recover the original time-domain signal.

14. Extend Hands-On Experiment 3.6 to plot the phase spectrum of the input signal.

15. Play the wave file `speech_tone_48k.wav` with the computer and use the frequency analyzer program (given in Hands-On Experiment 3.7) running on the EZ-KIT to detect the occurrence of the sinusoidal tone. Design a simple moving-average filter or bandstop filter to remove the tone. Observe the magnitude spectrum after the filter is applied to the input signal. Did you still see the tone?

16. Modify Hands-On Experiment 3.9 to include the moving-average filter in Problem 15. Show the magnitude spectrum of both the filtered and the unfiltered signal.

Chapter 4

Digital Filtering

As discussed in Chapters 2 and 3, a filter can be designed to alter the spectral content of input signals in a specified manner to achieve the desired objectives, and digital filtering is widely used for embedded systems. This chapter introduces the design, analysis, application, and implementation of time-invariant FIR and IIR filters and time-varying adaptive filters.

4.1 INTRODUCTION

As discussed in Section 3.2.3, digital filters can be divided into two categories: FIR filters and IIR filters. These filters can be represented by difference (or I/O) equations, system transfer functions, and signal flow diagrams. The I/O equations of FIR and IIR filters are defined in Equations 2.3.4 and 3.2.12, the transfer functions are defined in Equations 3.2.7 and 3.2.13, and the signal flow diagrams are illustrated in Figure 2.16 and Figure 3.3, respectively.

The process of deriving the digital filter transfer function $H(z)$ that satisfies a given specification is called digital filter design. Although some applications require only simple filters such as the moving-average, notch, and peaking filters introduced in Chapters 2 and 3, the design of more sophisticated filters requires the use of more advanced techniques. A number of computer-aided design tools (such as MATLAB) are available for designing digital filters. In this chapter, we focus on using the Filter Design and Analysis Tool (FDATool) for designing FIR and IIR filters in Sections 4.2 and 4.3, respectively.

Linear, time-invariant filters are characterized by magnitude response, phase response, stability, rise time, settling time, and overshoot. As introduced in Section 3.3, magnitude response specifies the gains of the filter at different frequencies, whereas phase response indicates the amount of phase changed by the filter. Magnitude and phase responses determine the steady-state response of the filter. Rise

---

Embedded Signal Processing with the Micro Signal Architecture. By Woon-Seng Gan and Sen M. Kuo
Copyright © 2007 John Wiley & Sons, Inc.

time, settling time, and overshoot specify the transient response of the filter in the time domain.

4.1.1 Ideal Filters

As discussed in Section 3.3.1, a filter passes certain frequency components in a signal through the system and attenuates others based on the magnitude response of the filter. The range of frequencies that is allowed to pass through the filter is called the passband, and the range of frequencies that is attenuated by the filter is called the stopband. If a filter is defined in terms of its magnitude response, there are four different types of filters: low-pass, high-pass, bandpass, and bandstop filters. The magnitude response of an ideal filter is given by $|H(\omega)| = 1$ in the passband and $|H(\omega)| = 0$ in the stopband. This two-level shape of the magnitude response helps in analyzing and visualizing actual filters used in DSP systems. However, achieving an ideal characteristic is not feasible, and ideal filters are only useful for conceptualizing the impact of filters on signal frequency components.

The frequency response $H(\omega)$ of a digital filter is a periodic function of frequency ω, and the magnitude response $|H(\omega)|$ of a digital filter with real coefficients is an even function of ω. Therefore, digital filter specifications are given only in the range of $0 \leq \omega \leq \pi$, or $0 \leq F \leq 1$, where F is the normalized frequency defined in Equation 2.2.6. The magnitude response of an ideal low-pass filter is illustrated in Figure 4.1. The regions $0 \leq \omega \leq \omega_c$ and $\omega_c \leq \omega \leq \pi$ are referred to as the passband and the stopband, respectively. The frequency that separates the passband and the stopband is called the cutoff frequency ω_c. An ideal low-pass filter has magnitude response $|H(\omega)| = 1$ in the frequency range $0 \leq \omega \leq \omega_c$ and has $|H(\omega)| = 0$ for $\omega_c \leq \omega \leq \pi$. Thus a low-pass filter passes low-frequency components below the cutoff frequency and attenuates high-frequency components above ω_c. For an ideal high-pass filter, the regions $\omega_c \leq \omega \leq \pi$ and $0 \leq \omega \leq \omega_c$ are referred to as the passband and the stopband, respectively. A high-pass filter passes frequency components above the cutoff frequency ω_c and attenuates frequency components below ω_c.

The magnitude response of an ideal bandpass filter is illustrated in Figure 4.2. The frequencies ω_l and ω_u are called the lower and upper cutoff frequencies, respectively. The region $\omega_l \leq \omega \leq \omega_u$ is called the passband, and the regions $0 \leq \omega \leq \omega_l$

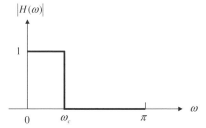

Figure 4.1 Magnitude response of an ideal low-pass filter

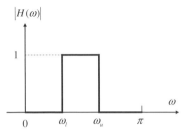

Figure 4.2 Magnitude response of an ideal bandpass filter

and $\omega_u \leq \omega \leq \pi$ are referred to as the stopband. A bandpass filter passes frequency components between the two cutoff frequencies ω_l and ω_u and attenuates frequency components below the frequency ω_l and above the frequency ω_u. If the passband is narrow, the center frequency and bandwidth are commonly used to specify a bandpass filter. A narrow bandpass filter also can be called a resonator (or peaking filter) as introduced in Section 3.4.3 and shown in Figures 3.27 and 3.28.

For an ideal bandstop (or band reject) filter, the region $\omega_l < \omega < \omega_u$ is called the stopband, and the regions $0 \leq \omega \leq \omega_l$ and $\omega_u \leq \omega \leq \pi$ are referred to as the passband. A bandstop filter attenuates frequency components between the cutoff frequencies ω_l and ω_u and passes frequency components below the frequency ω_l and above the frequency ω_u. A narrow bandstop filter designed to attenuate a single frequency component is called a notch filter, as introduced in Section 3.4.2 and illustrated in Figure 3.22.

EXAMPLE 4.1

Similar to Example 3.12, we use a MATLAB script (example4_1.m) to generate a signal $x(n)$ that consists of four sinusoids at $f_1 = 400\,\text{Hz}$, $f_2 = 800\,\text{Hz}$, $f_3 = 1{,}200\,\text{Hz}$, and $f_4 = 1{,}600\,\text{Hz}$. The magnitude spectrum is shown in Figure 4.3. If the sampling frequency is 4 kHz, Equation 2.2.4 shows that the digital frequencies of these four sine waves are $\omega_1 = 0.2\pi$, $\omega_2 = 0.4\pi$, $\omega_3 = 0.6\pi$, and $\omega_4 = 0.8\pi$. In addition, Equation 2.2.6 indicates that the normalized digital frequencies of these four sine waves are 0.2, 0.4, 0.6, and 0.8.

As shown in Figure 4.1, we can pass only a 400-Hz sine wave and attenuate others by using a low-pass filter with cutoff frequency $\omega_c = 0.3\pi$. Figure 4.2 shows that we can pass only a 1,200-Hz sine wave by using a bandpass filter with $\omega_l = 0.5\pi$ and $\omega_u = 0.7\pi$.

EXERCISE 4.1

Determine the filter types and cutoff frequencies for the following tasks:

1. Pass a 1,600-Hz sine wave only and attenuate others.
2. Attenuate a 800-Hz sine wave only.

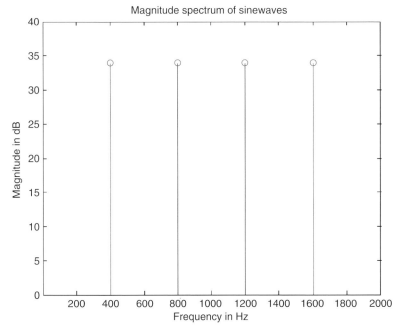

Figure 4.3 Magnitude spectrum of four sine waves

3. Pass 800-Hz and 1,200-Hz sine waves and attenuate others.
4. Pass 400-Hz and 1,600-Hz sine waves and attenuate others.

4.1.2 Practical Filter Specifications

In practice, it is very difficult to achieve the sharp cutoff implied by the ideal filters shown in Figures 4.1 and 4.2. We must accept a more gradual roll-off between the passband and the stopband. A transition band is introduced to permit the smooth magnitude drop-off between the passband and the stopband. In addition, the deviation from $|H(\omega)| = 1$ (0 dB) in the passband is called magnitude distortion. For frequency-selective filters, the magnitude specifications of a digital filter are often given in the form of tolerance (or ripple) schemes. A typical magnitude response of a low-pass filter is shown in Figure 4.4. The dashed horizontal lines in the figure indicate the tolerance limits. In the passband the magnitude response has a peak deviation δ_p, and in the stopband it has a maximum deviation δ_s. The frequencies ω_p and ω_s are the passband edge frequency and the stopband edge frequency, respectively.

As shown in Figure 4.4, the magnitude of the passband in the range $0 \leq \omega \leq \omega_p$ approximates unity with an error of $\pm\delta_p$. That is,

$$1 - \delta_p \leq |H(\omega)| \leq 1 + \delta_p, \quad 0 \leq \omega \leq \omega_p. \tag{4.1.1}$$

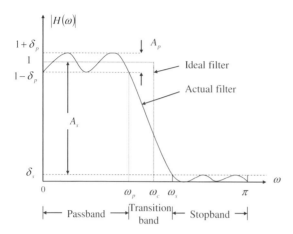

Figure 4.4 Practical low-pass filter specifications

The passband ripple, δ_p, is a measure of the allowed variation in magnitude response in the passband of the filter. Note that the gain of the magnitude response is normalized to 1 (0 dB). In practical applications, it is easy to scale the filter output by multiplying the output with a constant, which is equivalent to multiplying the whole magnitude response by the same constant gain.

In the stopband, the magnitude approximates zero with an error δ_s. That is,

$$|H(\omega)| \leq \delta_s, \quad \omega_s \leq \omega \leq \pi. \qquad (4.1.2)$$

The stopband ripple describes the minimum attenuation for signal components in the stopband.

Passband and stopband deviations may be expressed in decibels. The peak passband ripple, δ_p, and the minimum stopband attenuation, δ_s, in decibels are given as

$$A_p = 20 \log_{10}\left(\frac{1+\delta_p}{1-\delta_p}\right) \text{ dB} \qquad (4.1.3)$$

and

$$A_s = -20 \log_{10}\delta_s \text{ dB}. \qquad (4.1.4)$$

EXAMPLE 4.2

Consider a filter specified as having a magnitude response in the passband within ±0.001. That is, $\delta_p = 0.001$. From Equation 4.1.3, we have

$$A_p = 20\log_{10}\left(\frac{1.001}{0.999}\right) = 0.0174 \text{ dB}.$$

The minimum stopband attenuation is given as $\delta_s = 0.001$. From Equation 4.1.4, we have

$$A_s = -20\log_{10}(0.001) = 60\,\text{dB}.$$

The transition band is the frequency range between the passband edge frequency ω_p and the stopband edge frequency ω_s. The magnitude response decreases monotonically from the passband to the stopband in this region. The width of the transition band determines how sharp the filter is. Generally, the smaller δ_p and δ_s, and the narrower the transition band, the more complicated (higher order) is the required filter.

HANDS-ON EXPERIMENT 4.1

This experiment uses FDATool to introduce filter types and their design specifications. FDATool is a graphical user interface (GUI) to design (or import) and analyze digital filters. For general information about using FDATool, refer to the *Signal Processing Toolbox User's Guide* [49].

FDATool can be activated by typing

```
fdatool
```

in the MATLAB command window. The FDATool window is shown in Figure 4.5, which has tools similar to the **Filter Designer** window in the SPTool, as shown in Figure 2.11. The design steps and features to view the filter characteristics are also similar. However, FDATool is a more advanced filter design tool. FDATool can be used to (1) design filters, (2) quantize

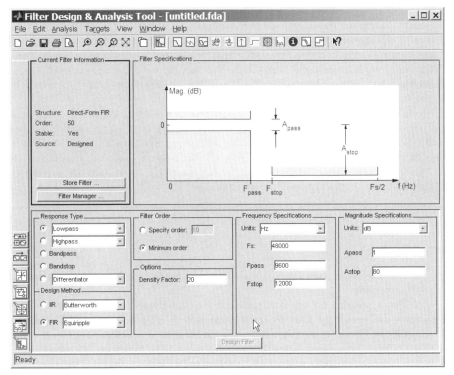

Figure 4.5 FDATool window

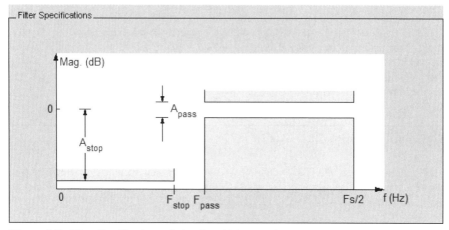

Figure 4.6 **Filter Specifications** window for a high-pass window

filters, (3) analyze filters, (4) modify existing filter designs, (5) create multirate filters, (6) realize Simulink models of quantized, direct-form, FIR filters, (7) import filters into FDATool, and (8) perform digital frequency transformations of filters. Note that capabilities (2), (5), (6), and (8) are only available when the *Filter Design Toolbox* [50] is installed, which integrates advanced filter design methods and supports quantized filters.

As shown in Figure 4.5, we can choose from several response (filter) types: **Lowpass**, **Highpass**, **Bandpass**, **Bandstop**, and **Differentiator**. The filter design specifications vary according to response types and design methods. As shown in Figure 4.5, we can enter (1) **Filter Order**, (2) **Options**, (3) **Frequency Specifications**, and (4) **Magnitude Specifications**. For example, to design a high-pass filter, select the radio button next to **Highpass** in the **Response Type** region on the GUI. The **Filter Specifications** window is shown in Figure 4.6.

It is important to compare the **Filter Specifications** window in Figure 4.6 with Figure 4.4. The parameters F_{stop}, F_{pass}, A_{stop}, and A_{pass} correspond to ω_s, ω_p, A_s, and A_p, respectively. These parameters can be entered in the **Frequency Specifications** and **Magnitude Specifications** regions at the bottom of Figure 4.5. The frequency units are **Normalized (0 to 1)**, **Hz** (default), **kHz**, **MHz**, or **GHz**, and the magnitude options are **dB** (default) or **Linear**. Note that Fs/2 corresponds to $\omega = \pi$ or $F = 1$.

EXAMPLE 4.3

For the four sine waves shown in Figure 4.3, design a bandstop FIR filter to attenuate the 1,200-Hz sine wave.

As shown in Figure 4.5, select the radio button next to **Bandstop** in the **Response Type** region. We can design the filter by entering parameters in **Frequency Specifications** as shown in Figure 4.7 and using default settings in the **Magnitude Specifications** region. Press the **Design Filter** button to compute the filter coefficients. The **Filter Specifications** region will change to **Magnitude Response (dB)** as shown in Figure 4.8. Comparing the magnitude

4.1 Introduction 119

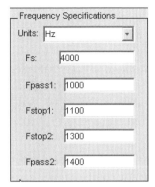

Figure 4.7 **Frequency Specifications** for a bandstop filter

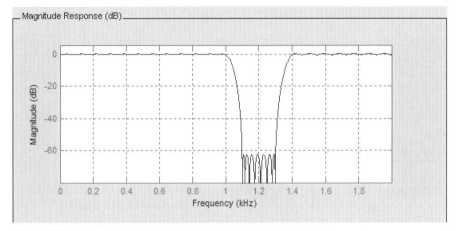

Figure 4.8 Magnitude response of the designed bandstop filter

response of the designed filter with the signal spectrum shown in Figure 4.3; we know this designed filter can attenuate the 1,200-Hz sine wave by 60 dB.

EXERCISE 4.2

Given the signal defined in Example 4.1, use FDATool to design FIR filters for the following tasks:

1. A low-pass filter that is passing the 400-Hz sine wave only.
2. A bandpass filter that is passing the 1,200-Hz sine wave only.
3. A high-pass filter that is passing the 1,600-Hz sine wave only.

4. A bandstop filter that is attenuating the 800-Hz sine wave only.
5. Redo 1–4 by assuming that the sampling rate is 8 kHz, and use normalized frequency for frequency specifications.
6. Compare the required filter order for the filter designed in 1–4. Why do they need different orders?
7. Redo 1–4 with IIR filters by selecting the radio button next to **IIR** in the **Design Method** region on the GUI. Pay attention to the required IIR filter orders as compared with the FIR filters.

When a signal passes through a filter, its amplitude and phase are modified. The phase response is an important filter characteristic because it affects time delays of different frequency components passing through the filter. If we consider a signal that consists of several frequency components, the phase delay of the filter is the time delay the composite signal suffers at each frequency. A filter is said to have a linear phase if its phase response satisfies

$$\phi(\omega) = -\alpha\omega, \quad -\pi \leq \omega \leq \pi. \tag{4.1.5}$$

This equation shows that for a filter with a linear phase, the group delay $T_d(\omega)$ given in Equation 3.3.7 is a constant α for all frequencies. As shown in Equation 3.3.8, this filter avoids phase distortion because all sinusoidal components in the input are delayed by the same amount. A filter with a nonlinear phase will cause a phase distortion in the signal that passes through it.

4.2 FINITE IMPULSE RESPONSE FILTERS

As discussed in Section 3.2.3, an FIR filter of length L can be represented by its impulse response $h(n)$, which has at most L nonzero samples. That is, $h(n) = 0$ for all $n \geq L$. An FIR filter is also called a transversal filter. Some characteristics of FIR filters are summarized as follows:

1. Because there is no feedback of past output samples as defined in Equation 2.3.4, the FIR filters are always stable. This inherent stability is also manifested in the absence of poles in the transfer function as defined in Equation 3.2.7, except possibly at the origin.
2. The filter has finite memory from $x(n)$ to $x(n - L + 1)$, as shown in Figure 2.16.
3. The design of linear-phase filters can be guaranteed. In many real-world applications such as audio signal processing, linear-phase filters are preferred because they avoid phase distortion.
4. The finite-precision errors (discussed in Chapter 6) are less severe in FIR filters than in IIR filters.
5. FIR filters can be easily implemented on most DSP processors such as the Blackfin processor.

6. A relatively higher-order FIR filter is required to obtain the same characteristics as compared with an IIR filter, and this may result in higher computational cost.

4.2.1 Characteristics and Implementation of FIR Filters

The signal flow diagram of the FIR filter is shown in Figure 2.16. The general I/O equation of FIR filter is defined in Equation 2.3.4. This equation describes the output of the FIR filter as a convolution sum of the input with the impulse response of the system. An example of linear convolution of two sequences is given in Example 2.5. Note that the convolution of the length M input with the length L impulse response results in a length $L + M - 1$ output.

As shown in Example 2.5, the input sequence is flipped around (folding) and then shifted to the right to overlap with the filter coefficients. At each time instant, the output value is the sum of products of overlapped coefficients with the corresponding input data aligned below it. At time $n = 0$, the only nonzero product comes from b_0 and $x(0)$, which are time aligned. It takes the filter L iterations to completely overlap with the input sequence. Therefore, the first $L - 1$ outputs correspond to the transient behavior of the FIR filter. For $n \geq L - 1$, the filter aligns over the nonzero portion of the input sequence. That is, when the signal buffer of FIR filter is full, the filter is operated in steady state.

An FIR filter has linear phase if its coefficients satisfy the following symmetric condition

$$b_l = b_{L-1-l}, \quad l = 0, 1, \ldots, L-1, \tag{4.2.1}$$

or the antisymmetric (negative symmetry) condition

$$b_l = -b_{L-1-l}, \quad l = 0, 1, \ldots, L-1. \tag{4.2.2}$$

There are four types of linear-phase FIR filters, depending on whether L is an even or an odd number and whether coefficients have positive or negative symmetry. These four linear-phase FIR filters are summarized as follows:

Type I—Positive symmetry and L is even.

Type II—Positive symmetry and L is odd.

Type III—Negative symmetry and L is even.

Type IV—Negative symmetry and L is odd.

The frequency response of the Type I filter is always zero at the Nyquist frequency (or $f_s/2$). This type of filter is unsuitable for a high-pass filter. Type III and IV filters introduce a 90° phase shift; thus they are often used for designing Hilbert transformers. The frequency response is always zero at DC frequency, making them unsuitable for low-pass filters. In addition, Type III response is always zero at the Nyquist frequency, making it also unsuitable for a high-pass filter.

The symmetry (or antisymmetry) property of a linear-phase FIR filter can be exploited to reduce the total number of multiplications almost in half. Consider the realization of an FIR filter with an even number L and positive symmetric impulse response as given in Equation 4.2.1; Equation 3.2.7 can be combined as

$$H(z) = b_0(1 + z^{-L+1}) + b_1(z^{-1} + z^{-L+2}) + \ldots + b_{L/2-1}(z^{-L/2+1} + z^{-L/2}). \quad (4.2.3)$$

The I/O equation given in Equation 2.3.10 can be generalized as

$$y(n) = \sum_{l=0}^{L/2-1} b_l[x(n-l) + x(n-L+1+l)]. \quad (4.2.4)$$

For an antisymmetric FIR filter, the addition of two signals is replaced by subtraction. That is,

$$y(n) = \sum_{l=0}^{L/2-1} b_l[x(n-l) - x(n-L+1+l)]. \quad (4.2.5)$$

EXERCISE 4.3

1. Realize $H(z)$ defined in Equation 4.2.3 as a signal-flow diagram of a symmetric FIR filter where L is even.
2. Redo Exercise 1 with L an odd number.
3. Redo Exercise 1 with an antisymmetric FIR filter.

As shown in Equation 4.2.4, the number of required multiplications is reduced in half by first adding the pair of samples and then multiplying the sum by the corresponding coefficient. The trade-off is that instead of addressing data linearly in the signal buffer with a single circular pointer, we need two address pointers.

An FIR filter can be realized either with a block-by-block- or a sample-by-sample-based operation. In block processing, the input sequence is segmented into multiple blocks. Filtering is performed one block at a time, and the resulting output blocks are recombined to form the overall output. The filtering of each block can be implemented with the linear convolution technique or with fast convolution using FFT. In the sample-by-sample operation, the input samples are processed at every sampling period after the current input $x(n)$ is available.

As shown in Equation 2.3.4, the filter output $y(n)$ is a linear combination of L input samples $\{x(n), x(n-1), \ldots, x(n-L+1)\}$ with the corresponding L coefficients $\{b_l, l = 0, 1, \ldots, L-1\}$, which can be represented as two separate tables (arrays or vectors) in memory. To compute the output at any given time, we simply multiply the corresponding values in each table and sum the results. The coefficient values are constant, but the data in the signal buffer change every sampling period T. For example, the sample $x(n)$ at time n becomes $x(n-1)$ in the next sampling period, then becomes $x(n-2)$, etc., until it simply drops off the end of the delay chain. The signal buffer is refreshed in every sampling period, where the oldest sample $x(n-L+1)$ is discarded and other signals are shifted one location in the

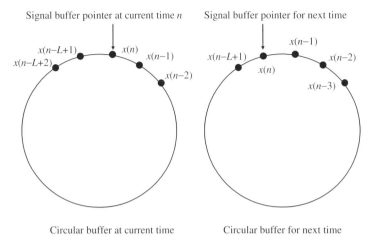

Figure 4.9 Circular buffer for storing signal samples

buffer. A new sample is inserted to the memory location labeled as $x(n)$. The FIR filtering operation that computes $y(n)$ with Equation 2.3.4 is then repeated. The process of refreshing the signal buffer requires intensive processing time if the operation is not implemented by the hardware.

The most efficient method for refreshing a signal buffer is to load the signal samples into a circular buffer, as illustrated in Figure 4.9. Instead of shifting the data samples while holding the buffer address fixed, the data are kept fixed and the address is shifted backwards (counterclockwise) in the circular buffer. The current signal buffer is arranged as the left-hand circle in Figure 4.9. After calculating the output $y(n)$, the pointer is moved counterclockwise one position pointing at $x(n - L + 1)$, which is no longer needed, and we wait for the next input signal. As shown in the right-hand circle, the new input is written to the $x(n - L + 1)$ position pointed by the circular pointer, and this new sample is referred to as $x(n)$ for the next iteration. The circular buffer implementation of a signal buffer (or a tapped-delay line) is very efficient. The update is carried out by adjusting the address pointer without physically shifting any data in memory.

A circular buffer also can be used for addressing FIR filter coefficients. The circular buffer allows the coefficient pointer to wrap around when it reaches the end of the coefficient buffer. That is, the pointer moves from b_{L-1} back to the adjacent b_0 such that the filtering will always start from the first coefficient. Details of how these pointers can be programmed in the Blackfin processor for efficient FIR filtering are discussed in Chapter 8.

4.2.2 Design of FIR Filters

The objective of FIR filter design is to determine a set of filter coefficients $\{b_l, l = 0, 1, \ldots, L - 1\}$ such that the filter performance is close to the given specifications.

A variety of techniques have been proposed for the design of FIR filters. The Fourier series method offers a very simple and flexible way of computing FIR filter coefficients, but it does not allow the designer to control the filter parameters. With the availability of an efficient and easy-to-use filter design software package such as MATLAB, the Park–McClellan algorithm is now widely used in industry for most practical applications.

MATLAB filter design functions operate with the normalized frequencies defined in Equation 2.2.6, so they do not require the sampling rate as an extra input argument. The *Signal Processing Toolbox* uses the convention that unit frequency is the Nyquist frequency, defined as half the sampling frequency. The normalized frequency, therefore, is always in the interval $0 \leq F \leq 1$. For a system with a 1,000-Hz sampling frequency, 300 Hz is equivalent to a normalized frequency $F = 300/500 = 0.6$. To convert normalized frequency to angular frequency ω around the unit circle, multiply it by π. To convert normalized frequency back to hertz, multiply it by half the sampling frequency.

The FIR filter design functions provided by the *Signal Processing Toolbox* are listed in Table 4.1. For example, fir1 function implements the classic method of windowed linear-phase FIR digital filter design. It designs filters in standard low-pass, high-pass, bandpass, and bandstop types. By default, the magnitude response of the filter at the center frequency of the passband is normalized to 0 dB. An example syntax of using this function is

    b = fir1(n,Wn);

This returns row vector b containing n + 1 coefficients of low-pass FIR filter with order n. This is a Hamming window-based linear-phase filter with normalized cutoff frequency Wn between 0 and 1, where 1 corresponds to the Nyquist frequency. If Wn is a two-element vector such as Wn = [w1 w2], the function fir1 returns

Table 4.1 FIR Filter Design Functions Provided by *Signal Processing Toolbox*

Methods	Functions	Description
Windowing	fir1, fir2, kaiserord	Apply window to truncated inverse Fourier transform of desired filter
Multiband with transition bands	firls, firpm, firpmord	Equiripple or least-squares approach over subbands of the frequency range
Constrained least squares	fircls, fircls1	Minimize squared integral error over entire frequency range subject to maximum error constraints
Arbitrary response	cfirpm	Arbitrary responses, including non-linear-phase and complex filters
Raised cosine	firrcos	Low-pass response with smooth, sinusoidal transition

Adapted from **Help** menu in MATLAB.

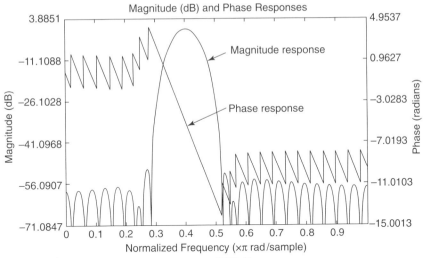

Figure 4.10 Magnitude and phase responses of FIR filter

a bandpass filter with passband w1 < F < w2. The MATLAB *Filter Design Toolbox* [50] provides an additional 10 FIR filter design functions.

EXAMPLE 4.4

Given a signal consisting of four sinusoidal components as shown in Figure 4.3, design an FIR filter to pass only the 800-Hz sine wave and attenuate others.

Because the sampling rate is 4 kHz, the normalized frequency of 800 Hz is 0.4. Therefore, we choose w1 = 0.35 and w2 = 0.45. The MATLAB program (example4_4.m) is listed as follows:

```
w1 = 0.35; w2 = 0.45;   % define edge frequencies
b = fir1(48,[w1 w2]);   % design FIR filter
fvtool(b,1)             % activate FVTool
```

The magnitude and phase responses of the designed filter are shown in Figure 4.10.

EXERCISE 4.4

1. Modify example4_4.m to design a bandpass filter with different edge frequencies and filter lengths.
2. Design FIR filters with different design functions listed in Table 4.1.
3. Design an FIR filter to attenuate the 800-Hz sine wave and pass other components.
4. Design an FIR filter to pass the 1,600-Hz sine wave only, using a high-pass filter.

4.2.3 Hands-On Experiments

In this section, we expand on Hands-On Experiment 4.1 for designing FIR filters. We use FDATool to design FIR filters for enhancing a 1,000-Hz tonal signal that is corrupted by broadband speech (`timit1.asc`) as shown in Figure 3.11.

HANDS-ON EXPERIMENT 4.2

In the FDATool window shown in Figure 4.5, select the radio button next to **Bandpass** in the **Response Type** region. The **Filter Specifications** window is shown as Figure 4.11. In the **Frequency Specification** region, enter 8000 in F_s, 900 in F_{stop1}, 950 in F_{pass1}, 1050 in F_{pass2}, and 1100 in F_{stop2}. In the **Magnitude Specifications** region, enter 60 in A_{stop1}, 1 in A_{pass}, and 60 in A_{stop2}. The magnitude response of the designed bandpass filter is shown in Figure 4.12.

In addition to the magnitude response shown in Figure 4.12, we can analyze the phase response, group delay response, phase delay, impulse response, step response, pole/zero plot, filter coefficients, etc. from the toolbar icons or from the **Analysis** pull-down menu. We can

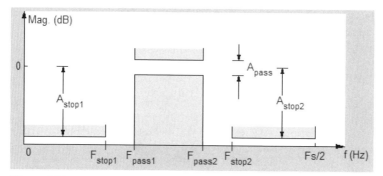

Figure 4.11 Filter Specifications window for a bandpass filter

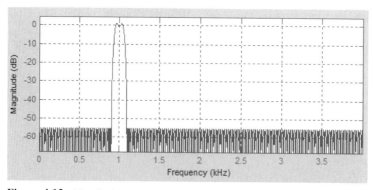

Figure 4.12 Magnitude response of the designed bandpass filter

4.2 Finite Impulse Response Filters

Figure 4.13 An **Export** window

export the coefficients of the filter with the **Export** option from the **File** menu. The **Export** window is displayed as shown in Figure 4.13. From the **Export To** pull-down menu, we have options of **Workspace**, **Coefficient File** (ASCII), **MAT-File**, and **SPTool**. From the **Export As** menu, we can choose **Coefficients** or **Objects**. We also can enter **Variable Names**. Using the default settings, and clicking on **Export** tab, the designed FIR filter coefficients are saved in the default vector Num under the current workspace.

EXERCISE 4.5

1. Design different bandpass filters with different frequency and magnitude specifications and compare the required filter order, which is displayed in the **Current Filter Information** region. Also, analyze the designed filter performance, using features provided in the **Analysis** menu.
2. Save the design filter coefficients to an ASCII file and in the format that can be imported by SPTool.
3. The default FIR filter design method is Equiripple, as shown in the **Design Method** region. From the pull-down menu, there are 11 options. Use different methods such as Window for designing FIR filters and evaluate the performance.

After the filter design process has generated the filter coefficient vectors, two functions are available in the *Signal Processing Toolbox* for implementing the filter: dfilt and filter. The dfilt function is a discrete-time filter object that allows users to specify its structure and values. The fitler function filters a data sequence with a digital filter that works for both real and complex inputs. The syntax

```
y = filter(b, a, x);
```

128 Chapter 4 Digital Filtering

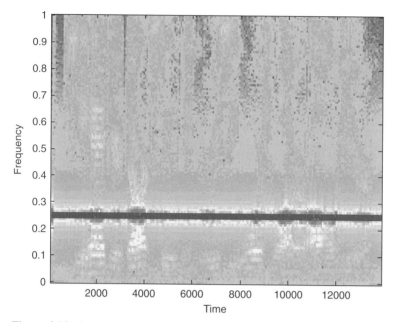

Figure 4.14 Spectrogram of tone enhanced with the designed bandpass filter

filters the data in vector x with the filter described by numerator coefficient vector b and denominator coefficient vector a. For an FIR filter, a = 1.

EXAMPLE 4.5

We use the FIR filter designed in Hands-On Experiment 4.2 (save as vector Num in Workspace) to enhance a tonal signal corrupted by speech in file timitl.asc. The MATLAB code is given in example4_5.m. The program plays the original speech, the mixed tone with speech, and the enhanced tone after the bandpass filtering. It also displays the spectrogram as shown in Figure 4.14. Comparing this spectrum with Figure 3.11, we verify that the bandpass filter shown in Figure 4.12 enhances the tonal signal.

EXERCISE 4.6

1. Use FDATool to design an FIR filter that can enhance the speech signal (timitl.asc) by attenuating the tonal noise at 1,000 Hz. Using different design methods and frequency and magnitude specifications, compare the filter order and performance.
2. Export the designed filter coefficients for SPTool. Similar to hands-on experiments given in Chapter 2, import the signal and filter into SPTool, perform filtering, and examine waveforms and spectra for both input and output signals.

3. Instead of designing a notch filter to attenuate the tonal noise at 1,000 Hz, we can simply subtract the output of the bandpass filter in Hands-On Experiment 4.2 from the original noisy speech signal. Compare the performance of this method with that from Exercise 1. Explain why it works.

4.3 INFINITE IMPULSE RESPONSE FILTERS

This section introduces the design, realization, and implementation of digital IIR filters. We discuss the basic characteristics and structures of digital IIR filters and focus on the techniques used for the design and implementation of these filters.

4.3.1 Design of IIR Filters

Digital IIR filters can be designed by beginning with the design of an analog filter in the s-domain and using mapping technique to transform it into the z-domain. The one-sided Laplace transform of analog signal $x(t)$ can be expressed as

$$X(s) = \int_0^\infty x(t)e^{-st}dt, \tag{4.3.1}$$

where

$$s = \sigma + j\Omega \tag{4.3.2}$$

is a complex variable. The z-transform can be viewed as the Laplace transform of the sampled function $x(nT)$ with the change of variable

$$z = e^{sT}. \tag{4.3.3}$$

This relationship represents the mapping of a region in the s-plane to the z-plane because both s and z are complex variables. The portion of the $j\Omega$-axis between $\Omega = -\pi/T$ and $\Omega = \pi/T$ in the s-plane is mapped onto the unit circle in the z-plane from $-\pi$ to π. As Ω varies from 0 to ∞, there are an infinite number of encirclements of the unit circle in the counterclockwise direction. Similarly, there are an infinite number of encirclements of the unit circle in the clockwise direction as Ω varies from 0 to $-\infty$. Each strip of width $2\pi/T$ in the left half of the s-plane is mapped onto the unit circle. This mapping occurs in the form of concentric circles in the z-plane as σ varies from 0 to $-\infty$. Also, each strip of width $2\pi/T$ in the right half of the s-plane is mapped outside of the unit circle. This mapping also occurs in concentric circles in the z-plane as σ varies from 0 to ∞. In conclusion, the mapping from the s-plane to the z-plane is not one-to-one because many points in the s-plane are mapped to a single point in the z-plane.

Because analog filter design is a mature and well-developed field, we begin the design of digital IIR filters in the analog domain and then convert the designed analog filter transfer function $H(s)$ into the digital domain. The problem is to determine a digital filter $H(z)$ that will approximate the performance of the desired analog filter $H(s)$. There are two commonly used methods, the impulse-invariant and the bilinear transform, for designing digital IIR filters based on existing analog IIR filters.

The impulse-invariant method preserves the impulse response of the original analog filter by digitizing the impulse response of the analog filter but not its frequency (magnitude) response. Because of inherent aliasing, this method is inappropriate for high-pass or bandstop filters. The bilinear-transform method yields very efficient filters and is well suited for the design of frequency-selective filters. The bilinear transform is defined as

$$s = \frac{2}{T}\left(\frac{z-1}{z+1}\right) = \frac{2}{T}\left(\frac{1-z^{-1}}{1+z^{-1}}\right), \quad (4.3.4)$$

or, equivalently,

$$z = \frac{1+(T/2)s}{1-(T/2)s}. \quad (4.3.5)$$

As discussed above, the $j\Omega$-axis of the s-plane ($\sigma = 0$) maps onto the unit circle in the z-plane. The left ($\sigma < 0$) and right ($\sigma > 0$) halves of the s-plane map onto the inside and outside of the unit circle, respectively. There is a direct relationship between the s-plane frequency Ω and the z-plane frequency ω. It can be easily shown that the corresponding mapping of frequencies is obtained as

$$\Omega = \frac{2}{T}\tan\left(\frac{\omega}{2}\right), \quad (4.3.6)$$

or, equivalently,

$$\omega = 2\tan^{-1}\left(\frac{\Omega T}{2}\right). \quad (4.3.7)$$

Thus the entire $j\Omega$-axis is compressed into the interval $-\pi/T \leq \omega \leq \pi/T$ in a one-to-one manner. Each point in the s-plane is uniquely mapped onto the z-plane.

The relationship in Equation 4.3.7 between the frequency variables Ω and ω is illustrated in Figure 4.15. The bilinear transform provides a one-to-one mapping of the points along the $j\Omega$-axis onto the unit circle, that is, the entire $j\Omega$-axis is mapped uniquely onto the unit circle or onto the Nyquist band $|\omega| \leq \pi$. However, the mapping is highly nonlinear. The entire band $\Omega T \geq 1$ is compressed onto $\pi/2 \leq \omega \leq \pi$. This frequency compression effect associated with the bilinear transform is known as frequency warping because of the nonlinearity of the arctangent function given in Equation 4.3.7. This nonlinear frequency warping phenomenon must be taken into consideration when designing digital filters with the bilinear transform. This can be done by prewarping the critical frequencies and using frequency scaling.

The classic IIR filters, Butterworth, Chebyshev Types I and II, elliptic, and Bessel, all approximate the ideal filter in different ways. The MATLAB *Signal Processing Toolbox* provides functions to create these types of IIR filters in both the analog and digital domains (except Bessel), and in low-pass, high-pass, bandpass, and bandstop types. Table 4.2 summarizes the various filter design methods

4.3 Infinite Impulse Response Filters

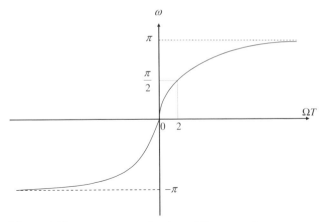

Figure 4.15 Frequency mapping due to bilinear transform

Table 4.2 IIR Filter Design Methods and Functions

Methods	Functions	Description

Analog prototyping	Filter design functions:
	  besself, butter,
	  cheby1, cheby2,
	  ellip
	Order estimation functions:
	  buttord, cheb1ord,
	  cheb2ord, ellipord
		Using lowpass prototype filter in the *s*-domain, obtain a digital filter through frequency transformation.

Direct design	yulewalk	Design digital filter directly by approximating a magnitude response.
Generalized Butterworth design	maxflat	Design low-pass Butterworth filters with more zeros than poles.

Parametric modeling	Time-domain modeling functions:
	  lpc, prony, stmcb
	Frequency-domain modeling functions:
	  invfreqs, invfreqz
		Find a digital filter that approximates a prescribed time- or frequency-domain response.

Adapted from **Help** menu in MATLAB.

provided in the toolbox and lists the functions available to implement these methods. The direct filter design function yulewalk finds a filter with magnitude response approximating a desired function. This is one way to create a multiband bandpass filter.

EXAMPLE 4.6

We design simple Butterworth and Chebyshev I low-pass IIR filters, using `example4_6.m`. The following function

    [b, a] = butter(N,Wn);

designs an Nth-order low-pass Butterworth filter with the cutoff frequency Wn and returns the filter coefficients in vectors b (numerator) and a (denominator). The function

    [b, a] = cheby1(N,R,Wn);

designs a Chebyshev I filter with R decibels of peak-to-peak ripple in the passband. The magnitude responses of these two filters are shown in Figure 4.16.

As shown in Figure 4.16, the magnitude response of the Butterworth filter is monotonically decreasing in both the passband and the stopband. The Butterworth filter has a flat magnitude response over the passband and the stopband. This flat passband is achieved at the expense of the transition region, which has a slower roll-off. For a given transition band, the order of the Butterworth filter required is often higher than that of other types. Chebyshev filters permit a certain amount of ripples in the passband or the stopband but have a much steeper roll-off near the cutoff frequency. Type I Chebyshev filters are all-pole filters that exhibit equiripple behavior in the passband and a monotonic characteristic in the stopband. The family of type II Chebyshev filters contains both poles and zeros and exhibits a monotonic behavior in the passband and an equiripple behavior in the stopband. In general, the Chebyshev filter meets the specifications with a lower number of poles than the corresponding Butterworth filter; however, it has a poorer phase response. The sharpest transition from passband to stopband can be achieved with the elliptic design. Elliptic filters exhibit equiripple behavior in both the passband and the stopband. In addition, the phase response of an elliptic filter is extremely nonlinear in the passband, especially near the cutoff frequency. However, for a given transition band, the order of the elliptic filter required is the lowest among the classic IIR filters.

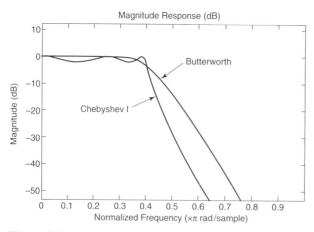

Figure 4.16 Magnitude responses of Butterworth and Chebyshev I filters

EXERCISE 4.7

1. Examine frequency responses of Chebyshev II and elliptic filters by modifying `example4_6.m` and compare the required filter orders for the same specifications.
2. Evaluate filter performance for different orders and different R values.

4.3.2 Structures and Characteristics of IIR Filters

Given an IIR filter described by Equation 3.2.13, the direct-form I realization is defined by the I/O equation (Eq. 3.2.12) and is illustrated as the signal-flow diagram shown in Figure 3.3. In DSP implementation, we have to consider the required operations, memory storage, and finite-wordlength effects. A given transfer function $H(z)$ can be realized in several forms or configurations. As discussed in Section 3.2.3, a high-order IIR filter is factored into second-order sections and connected in cascade or parallel to form an overall filter. In this section, we discuss direct-form I, direct-form II, cascade, and parallel realizations.

The transfer function of the second-order IIR filter is defined in Equation 3.2.15, and the I/O equation is given in Equation 3.2.14. The direct-form II realization is illustrated in Figure 3.4, where two signal buffers are combined into one. This realization requires three memory locations for the second-order IIR filter, as opposed to six memory locations required for the direct-form I realization given in Figure 3.3. Therefore, the direct-form II realization is called the canonical form because it realizes the given transfer function with the smallest possible numbers of delays, adders, and multipliers.

The cascade realization of an IIR filter assumes that the transfer function is the product of first-order and/or second-order IIR sections. By factoring the numerator and the denominator polynomials of the transfer function $H(z)$ as a product of lower-order polynomials, an IIR filter can be realized as a cascade of low-order filter sections. Consider the transfer function $H(z)$ given in Equation 3.2.16, it can be expressed as

$$H(z) = GH_1(z)H_2(z)\ldots H_k(z) = G\prod_{k=1}^{K} H_k(z), \qquad (4.3.8)$$

where each $H_k(z)$ is a first- or second-order IIR filter and K is the total number of sections. In this form, any complex-conjugate roots must be grouped into the same section to guarantee that all the coefficients of $H_k(z)$ are real numbers. The realization of Equation 4.3.8 in cascade form is illustrated in Figure 4.17.

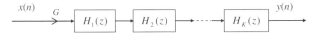

Figure 4.17 Cascade realization of IIR filter

The transfer function for each section of filter can be expressed as

$$H_k(z) = \frac{1 + b_{1k} z^{-1}}{1 + a_{1k} z^{-1}}, \qquad (4.3.9)$$

for the first-order filter or

$$H_k(z) = \frac{b_{0k} + b_{1k} z^{-1} + b_{2k} z^{-2}}{1 + a_{1k} z^{-1} + a_{2k} z^{-2}} \qquad (4.3.10)$$

for the second-order section. Assuming that every $H_k(z)$ is the second-order IIR filter, the I/O equations describing the time-domain operations of the cascade realization are expressed as

$$w_k(n) = x_k(n) - a_{1k} w_k(n-1) - a_{2k} w_k(n-2), \qquad (4.3.11)$$

$$y_k(n) = b_{0k} w_k(n) + b_{1k} w_k(n-1) + b_{2k} w_k(n-2), \qquad (4.3.12)$$

$$x_{k+1}(n) = y_k(n), \qquad (4.3.13)$$

for $k = 1, 2, \ldots, K$ sections and

$$x_1(n) = G x(n), \qquad (4.3.14)$$

$$y(n) = y_K(n). \qquad (4.3.15)$$

The signal flow diagram of the second-order IIR filter is illustrated in Figure 4.18.

By different ordering and pairing, it is possible to obtain many different cascade realizations for the same transfer function $H(z)$. Ordering means the order of connecting $H_k(z)$, and pairing means the grouping of poles and zeros of $H(z)$ to form a section. Each cascade realization behaves differently from others because of the finite-wordlength effects. The best ordering is the one that generates the minimum overall roundoff noise.

In the direct-form realization shown in Figure 3.3, the variation of one parameter will affect the locations of all the poles of $H(z)$. In the cascade realization illustrated in Figure 4.17, the variation of one parameter will affect only pole(s) in that section. Therefore, the cascade realization is less sensitive to parameter variation (due to coefficient quantization, etc.) than the direct-form structure. In practical implementations of digital IIR filters, the cascade form is often preferred.

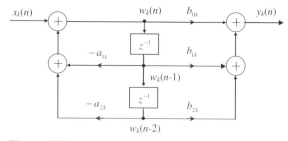

Figure 4.18 Block diagram of second-order IIR filter section

EXAMPLE 4.7

Given the second-order IIR filter

$$H(z) = \frac{3 + 1.5z^{-1} + 0.75z^{-2}}{1 + 0.86z^{-1} + 0.78z^{-2} + 0.3z^{-3}},$$

realize it using the cascade form in terms of first-order and second-order sections.

By factoring the numerator and denominator polynomials of $H(z)$, we obtain

$$H(z) = \frac{3(1 + 0.5z^{-1} + 0.25z^{-2})}{(1 + 0.5z^{-1})(1 + 0.36z^{-1} + 0.6z^{-2})}.$$

By different pairings of poles and zeros, there are different realizations of $H(z)$. For example, we choose $G = 3$,

$$H_1(z) = \frac{1}{1 + 0.5z^{-1}} \quad \text{and} \quad H_2(z) = \frac{1 + 0.5z^{-1} + 0.25z^{-2}}{1 + 0.36z^{-1} + 0.6z^{-2}}.$$

The expression of $H(z)$ in a partial-fraction expansion leads to another canonical structure called the parallel form. It is expressed as

$$H(z) = c + H_1(z) + H_2(z) + \ldots + H_k(z), \tag{4.3.16}$$

where c is a constant. The variation of parameters in a parallel form affects only the poles of the $H_k(z)$ associated with the parameters. Therefore, the pole sensitivity of a parallel realization is less than that of the direct form.

The cascade realization of an IIR transfer function $H(z)$ involves its factorization in the form of Equation 3.2.16. This can be done in MATLAB by using the function roots. From the computed roots, the coefficients of each section can be determined by pole-zero pairings. A much simpler approach is to use the function tf2zp in the *Signal Processing Toolbox*, which finds the zeros, poles, and gains of systems in transfer functions of single-input or multiple-output form. For example, the statement

```
[z, p, c] = tf2zp(b, a);
```

returns the zero locations in p, the pole locations in p, and the gains for each numerator transfer function in c. Vector a specifies the coefficients of the denominator, and b indicates the numerator coefficients. Some related linear system transformation functions are listed in Table 4.3.

As listed in Table 4.3, MATLAB provides a useful function, tf2sos (or zp2sos), to convert a given system transfer function to an equivalent representation of second-order sections. The function

```
[sos, G] = tf2sos(b, a);
```

finds a matrix sos in second-order section form and a gain G that represents the same system $H(z)$ as the one with numerator b and denominator a. The poles and zeros of $H(z)$ must be in complex conjugate pairs. The matrix sos is a $K \times 6$ matrix as follows:

$$sos = \begin{bmatrix} b_{01} & b_{11} & b_{21} & 1 & a_{11} & a_{21} \\ b_{02} & b_{12} & b_{22} & 1 & a_{12} & a_{22} \\ \vdots & \vdots & \vdots & \vdots & \vdots \\ b_{0K} & b_{1K} & b_{2K} & 1 & a_{1K} & a_{2K} \end{bmatrix}, \tag{4.3.17}$$

Table 4.3 List of Linear System Transformation Functions

Functions	Description
residuez	z-transform partial-fraction expansion.
sos2tf	Convert digital filter second-order section data to transfer function form.
sos2zp	Convert digital filter second-order section parameters to zero-pole-gain form.
ss2sos	Convert digital filter state-space parameters to second-order section form.
tf2sos	Convert digital filter transfer function data to second-order section form.
tf2zp	Convert continuous-time transfer function filter parameters to zero-pole-gain form.
tf2zpk	Convert discrete-time transfer function filter parameters to zero-pole-gain form.
zp2sos	Convert digital filter zero-pole-gain parameters to second-order section form.
zp2ss	Convert zero-pole-gain filter parameters to state-space form.
zp2tf	Convert zero-pole-gain filter parameters to transfer function form.

Adapted from **Help** menu in MATLAB.

whose rows contain the numerator and denominator coefficients, b_{ik} and a_{ik}, $i = 0, 1, 2$ of the kth second-order section $H_k(z)$. The overall transfer function is expressed as

$$H(z) = G\prod_{k=1}^{K} H_k(z) = G\prod_{k=1}^{K} \frac{b_{0k} + b_{1k}z^{-1} + b_{2k}z^{-2}}{1 + a_{1k}z^{-1} + a_{2k}z^{-2}}. \quad (4.3.18)$$

The parallel realizations can be realized with the function residuez listed in Table 4.3.

As discussed above, MATLAB function filter supports the implementation of direct-form IIR filters. To implement the cascade of second-order IIR sections, we can use

    y = sosfilt(sos,x);

which applies the second-order section filter sos to the vector x. The output y has the same length as x.

The transfer function of IIR filter $H(z)$ can be factored into the pole-zero form defined in Equation 3.2.16. The system is stable if and only if all its poles are inside the unit circle. For the cascade structure given in Equation 4.3.8, the stability can be guaranteed if every filter section $H_k(z)$ defined in Equation 4.3.8 or Equation 4.3.10 is stable. Consider the second-order IIR filter defined by Equation 4.3.10, the poles will lie inside the unit circle if the coefficients satisfy the following conditions:

$$|a_{2k}| < 1 \quad (4.3.19)$$

and

$$|a_{1k}| < 1 + a_{2k} \quad (4.3.20)$$

for all k. Therefore, the stability of the IIR filter realized as a cascade of first- or second-order sections is easy to examine.

4.3.3 Hands-On Experiments

This section uses FDATool and SPTool for designing IIR filters to enhance the speech signal (timit1.asc) corrupted by a 1,000-Hz tone. As shown in Figure 3.11, a bandstop filter centered at 1,000 Hz (with sampling frequency 8 kHz) is needed to attenuate the narrowband noise.

HANDS-ON EXPERIMENT 4.3

In the FDATool window shown in Figure 4.5, select the radio button next to **Bandstop** in the **Response Type** region and **IIR** in the **Design Method** region, and choose Elliptic from the pull-down menu. In the **Frequency Specification** region, enter 8000 in **Fs**, 900 in **Fpass1**, 950 in **Fstop1**, 1050 in **Fstop2**, and 1100 in **Fpass2**. In the **Magnitude Specifications** region, enter 1 in **Apass1**, 60 in **Astop**, and 1 in **Apass2**. Click on the **Design Filter** button; the magnitude response of the designed bandstop filter is shown in Figure 4.19.

Note that the **Current Filter Information** region shows that it required 10th-order IIR filter, realized as five direct-form II second-order sections. It is interesting to design an FIR filter with the same specifications, and we need an FIR filter with order 342. We can use the **Analysis** menu for comparing the IIR filter with the FIR filter. In the **Edit** menu, there are options of **Convert Structure**, **Reorder and Scale Second-Order Sections**, and **Convert to Single Section**. We are able to realize the designed IIR filter with different structures for implementation.

After becoming satisfied with the designed filter, we select **Export** from the **File** menu. Select **SPTool** from the **Export To** menu; the SPTool startup window as shown in Figure 2.3 is displayed. The designed filter is displayed in the **Filters** column as Hd[imported].

Following the procedure given in Hands-On Experiment 2.1, we can import speech_tone.mat generated by example3_13.m. This file is named sig1 in the **Signals** region. View and play the imported speech that is corrupted by a 1,000-Hz tone. Following the steps given in Hands-On Experiment 2.2, select the input signal (sig1) and the imported filter (Hd) and click on the **Apply** button to perform IIR filtering. The **Apply Filter** window is displayed, which allows us to specify the file name of the output signal (default sig2). Click on **OK** to perform IIR filtering, which will generate the output file in the **Signals** region. We can evaluate the filter's performance by viewing and playing the input and output waveforms. We should verify that the undesired 1,000-Hz tonal noise was attenuated by the designed bandstop IIR filter.

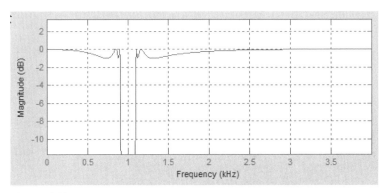

Figure 4.19 Magnitude response of the designed bandstop IIR filter

EXERCISE 4.8

1. Design IIR filters with other design methods such as Butterworth, Chebyshev I, Chebyshev II, etc. supported by FDATool. Compare the required filter orders and filter performance.
2. Design IIR filters with different frequency and magnitude specifications. Compare the required filter orders and filter performance. Find the relationship between the width of transition band and the required filter order.
3. Following the steps given in Hands-On Experiment 2.2, design IIR filters with SPTool, perform IIR filtering, and compare the performance of the designed filters.

EXAMPLE 4.8

Figure 2.2 shows that the narrowband signal is corrupted by a broadband random noise. It can be enhanced by using a simple moving-average filter. We also can design bandpass FIR and IIR filters for reducing random noise. However, if the desired narrowband signal has changing frequency as the chirp signal shown in Figure 4.20 (see `example4_8.m`), can we use bandpass filters to enhance a time-varying signal with changing frequency?

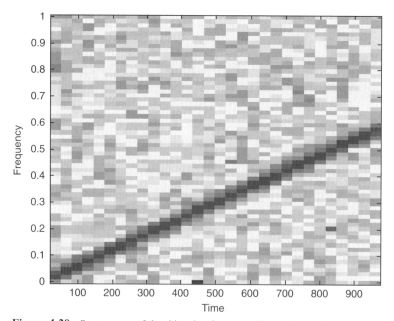

Figure 4.20 Spectrogram of the chirp signal corrupted by random noise

4.4 ADAPTIVE FILTERS

We have introduced techniques for designing and implementing FIR and IIR filters. The characteristics of these filters are time invariant because they have fixed coefficients. As shown in Example 4.8, these filters cannot be applied for time-varying signals and noises. In contrast, adaptive filters' coefficients are updated automatically by adaptive algorithms [6, 22]. The characteristics of adaptive filters are time varying and can adapt to an unknown and/or changing environment. Therefore, coefficients of adaptive filters cannot be determined by filter design software such as FDATool. This section introduces structures, algorithms, design, and applications of adaptive filters.

4.4.1 Structures and Algorithms of Adaptive Filters

In this book, we use many practical applications that involve the reduction of noises. The signal in some physical systems is time varying like the chirp signal given in Example 4.8, unknown, or possibly both. Adaptive filters provide a useful approach for these applications. For example, a modem needs a channel equalizer for transmitting and receiving data over telecommunication channels. Because the dial-up channel has different characteristics on each connection and is time varying, the channel equalizers must be adaptive.

As illustrated in Figure 4.21, an adaptive filter consists of two functional blocks—a digital filter to perform the desired filtering and an adaptive algorithm to automatically adjust the coefficients (or weights) of that filter. In the figure, $d(n)$ is a desired signal, $y(n)$ is the output of a digital filter driven by an input signal $x(n)$, and error signal $e(n)$ is the difference between $d(n)$ and $y(n)$. The adaptive algorithm adjusts the filter coefficients to minimize a predetermined cost function that is related to $e(n)$.

The FIR and IIR filters presented in Sections 4.2 and 4.3 can be used for adaptive filtering. The FIR filter is always stable and can provide a linear-phase response. The poles of the IIR filter may move outside the unit circle during adaptation of

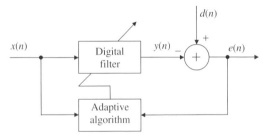

Figure 4.21 Block diagram of adaptive filter

filter coefficients, thus resulting in an unstable filter. Because the filter is adaptive, the stability problem is very difficult to handle. Therefore, adaptive FIR filters are widely used for real-world applications.

Assuming the adaptive FIR filter used in Figure 4.21 with L coefficients, $w_l(n)$, $l = 0, 1, \ldots, L - 1$, the filter output signal is computed as

$$y(n) = \sum_{l=0}^{L-1} w_l(n)x(n-l), \qquad (4.4.1)$$

where the filter coefficients $w_l(n)$ are time varying and updated by an adaptive algorithm. It is important to note that the length of filter is L and the order of the filter is $L - 1$. We define the input vector at time n as

$$\mathbf{x}(n) \equiv [x(n)x(n-1)\ldots x(n-L+1)]^T, \qquad (4.4.2)$$

and the weight vector at time n as

$$\mathbf{w}(n) \equiv [w_0(n)w_1(n)\ldots w_{L-1}(n)]^T. \qquad (4.4.3)$$

The output signal $y(n)$ given in Equation 4.4.1 can be expressed with the vector form as follows:

$$y(n) = \mathbf{w}^T(n)\mathbf{x}(n), \qquad (4.4.4)$$

where T denotes the transpose operation of the vector. The filter output $y(n)$ is compared with the desired response $d(n)$, which results in the error signal

$$e(n) = d(n) - y(n). \qquad (4.4.5)$$

The objective of the adaptive algorithm is to update the filter coefficients to minimize some predetermined performance criterion (or cost function). The most commonly used cost function is based on the mean square error (MSE) defined as

$$\xi(n) \equiv E[e^2(n)], \qquad (4.4.6)$$

where E denotes the expectation operation.

The MSE defined in Equation 4.4.6 is a function of filter coefficient vector $\mathbf{w}(n)$. For each coefficient vector $\mathbf{w}(n)$, there is a corresponding (scalar) value of MSE. Therefore, the MSE values associated with $\mathbf{w}(n)$ form an $(L+1)$-dimensional space, which is called the MSE surface or the performance surface. The steepest-descent method is an iterative (recursive) technique that starts from an initial (arbitrary) weight vector. The weight vector is updated at each iteration in the direction of the negative gradient of the error surface.

The concept of the steepest-descent algorithm can be described as

$$\mathbf{w}(n+1) = \mathbf{w}(n) - \frac{\mu}{2}\nabla\xi(n), \qquad (4.4.7)$$

where μ is a convergence factor (or step size) that controls stability and the rate of descent. The larger the value of μ, the faster the speed of convergence. The vector

$\nabla \xi(n)$ denotes the gradient of the error function with respect to $\mathbf{w}(n)$, and the negative sign increments the weight vector in the negative gradient direction. The successive corrections to the weight vector in the direction of the steepest descent of the performance surface should eventually lead to the minimum MSE, at which point the weight vector reaches its optimum value.

The method of steepest descent cannot be used directly because it requires the exact gradient vector. Many adaptive filter structures and adaptation algorithms have been developed for different applications. This section presents the most widely used adaptive FIR filter with the least mean square (LMS) algorithm, or stochastic gradient algorithm, which uses the instantaneous squared error, $e^2(n)$, to estimate the MSE. The LMS algorithm can be expressed as

$$\mathbf{w}(n+1) = \mathbf{w}(n) + \mu \mathbf{x}(n)e(n). \tag{4.4.8}$$

This equation can be expressed in scalar form as

$$w_l(n+1) = w_l(n) + \mu x(n-l)e(n), \quad l = 0, 1, \ldots, L-1. \tag{4.4.9}$$

Adaptive FIR filters using the LMS algorithm are relatively simple to design and implement. They are well understood with regard to stability, convergence speed, steady-state performance, and finite-precision effects. To effectively use the LMS algorithm, we must determine parameters L, μ, and $\mathbf{w}(0)$, where $\mathbf{w}(0)$ is the initial weight vector at time $n = 0$.

The convergence of the LMS algorithm from an initial condition $\mathbf{w}(0)$ to the optimum filter must satisfy

$$0 < \mu < \frac{2}{LP_x}, \tag{4.4.10}$$

where P_x denotes the power of $x(n)$. Because the upper bound on μ is inversely proportional to L, a smaller μ is used for large-order filters. In addition, μ is inversely proportional to the input signal power. One effective approach is to normalize μ with respect to the input signal power P_x. The resulting algorithm is called the normalized LMS algorithm, which is expressed as

$$\mathbf{w}(n+1) = \mathbf{w}(n) + \frac{\beta}{L\hat{P}_x(n)}\mathbf{x}(n)e(n), \tag{4.4.11}$$

where $\hat{P}_x(n)$ is an estimate of the power of $x(n)$ at time n and β is a normalized step size that satisfies the criterion $0 < \beta < 2$.

The commonly used method to estimate $\hat{P}_x(n)$ sample by sample is similar to the first-order IIR filter described in Equation 3.2.20. This effective technique can be expressed as

$$\hat{P}_x(n) = (1-\alpha)\hat{P}_x(n-1) + \alpha x^2(n). \tag{4.4.11}$$

Because it is not desirable that the power estimate $\hat{P}_x(n)$ be zero or very small, a software constraint is required to ensure that the normalized step size is bounded even if $\hat{P}_x(n)$ is very small.

Convergence of the MSE toward its minimum value is a commonly used performance measurement in adaptive systems because of its simplicity. A plot of the MSE versus time n is called the learning curve. Because the MSE is the performance criterion of LMS algorithms, the learning curve describes its transient behavior. The MSE time constant can be approximated as

$$\tau_{MSE} \propto \frac{1}{\mu} \text{ and } \tau_{MSE} \leq \frac{\max|X(\omega)|^2}{\min|X(\omega)|^2}, \quad (4.4.12)$$

where $|X(\omega)|^2$ is the magnitude-square spectrum of $x(n)$ and the maximum (max) and minimum (min) are calculated over the frequency range $0 \leq \omega \leq \pi$. Therefore, input signals with a flat (white) spectrum have the fastest convergence speed. In addition, τ_{MSE} is inversely proportional to μ. If we use a large value of μ, the time constant is small, which implies faster convergence.

The steepest-descent algorithm defined in Equation 4.4.7 requires the true gradient $\nabla\xi(n)$. The use of estimated gradient $\hat{\nabla}\xi(n)$ produces the gradient estimation noise. After the algorithm converges, perturbing the gradient will cause the weight vector $\mathbf{w}(n+1)$ to move away from the optimum solution. When $\mathbf{w}(n)$ moves away from the optimum weight vector, it causes $\xi(n)$ to be larger than its minimum value, thus producing excess noise called the excess MSE at the filter output. For the LMS algorithm, the excess MSE is directly proportional to μ. The larger the value of μ, the worse the steady-state performance after convergence. However, Equation 4.4.12 shows that a larger μ results in faster convergence. Therefore, there is a design trade-off between the excess MSE and the speed of convergence.

Insufficient spectral excitation of the LMS algorithm may result in divergence of the adaptive algorithm. Divergence can be avoided by means of a "leaking" mechanism used during the weight update. This is called the leaky LMS algorithm, expressed as

$$\mathbf{w}(n+1) = \nu\mathbf{w}(n) + \mu\mathbf{x}(n)e(n), \quad (4.4.13)$$

where ν is the leakage factor with $0 < \nu \leq 1$. The leakage factor introduces a bias on the long-term coefficient estimation. The error due to the leakage is proportional to $[(1-\nu)/\mu]^2$. Therefore, $(1-\nu)$ should be kept smaller than μ in order to maintain an acceptable level of performance. The leaky LMS algorithm not only prevents unconstrained weight overflow, but also limits the output power in order to avoid nonlinear distortion.

4.4.2 Design and Applications of Adaptive Filters

The MATLAB *Filter Design Toolbox* provides the function `adaptfilt` for implementing adaptive filters. The function

```
h = adaptfilt.algorithm(...);
```

returns an adaptive filter h of type `algorithm`. MATLAB supports many adaptive algorithms. For example, the LMS-type adaptive algorithms based on FIR filters are

4.4 Adaptive Filters

summarized in Table 4.4. The *Filter Design Toolbox* also provides many advanced algorithms such as recursive least squares, affine projection, and frequency-domain algorithms.

For example, we can construct an adaptive FIR filter with the LMS algorithm object as follows:

```
h = adaptfilt.lms(l,step,leakage,coeffs,states);
```

Table 4.5 describes the input arguments for `adaptfilt.lms`.

The adaptive filter is able to operate in an unknown environment and to track time variations of the input signals. The essential difference between various applications of adaptive filtering is where the signals $x(n)$, $d(n)$, $y(n)$, and $e(n)$ shown in Figure 4.21 are connected.

Adaptive system identification is illustrated in Figure 4.22, where $P(z)$ is an unknown system to be identified and $W(z)$ is an adaptive filter used to model $P(z)$. The adaptive filter adjusts itself to cause its output to match that of the unknown system. If the input signal $x(n)$ provides sufficient spectral excitation, the adaptive filter output $y(n)$ will approximate $d(n)$ in an optimum sense after convergence. When the difference between the physical system response $d(n)$ and the adaptive model response $y(n)$ has been minimized, the adaptive model $W(z)$ approximates $P(z)$ from the input/output viewpoint. When the plant is time varying, the adaptive

Table 4.4 Summary of LMS-Type Adaptive Algorithms for FIR Filters

Algorithms	Description
lms	Direct-form LMS
nlms	Direct-form normalized LMS
dlms	Direct-form delayed LMS
blms	Block LMS
blmsfft	FFT-based block LMS
ss	Direct-form sign-sign LMS
se	Direct-form sign-error LMS
sd	Direct-form sign-data LMS
filtxlms	Filtered-X LMS
adjlms	Adjoint LMS

Table 4.5 Input Arguments for `adaptfilt.lms`

Input Arguments	Description
l	Filter length L (defaults to 10)
step	Step size μ (defaults to 0.1)
leakage	Leakage factor ν ($0 < \nu \leq 1$, defaults to 1)
coeffs	Initial filter coefficients (defaults to 0)
states	Initial filter states (defaults to 0)

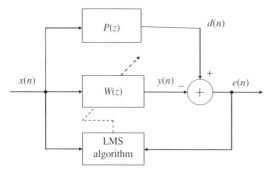

Figure 4.22 Block diagram of adaptive system identification

algorithm will keep the modeling error small by continually tracking time variations of the plant dynamics.

EXAMPLE 4.9

In Figure 4.22, an unknown system $P(z)$ is an FIR filter designed by the following function:

```
p = fir1(15,0.5);
```

The excitation signal $x(n)$ used for system identification is generated as follows:

```
x = randn(1,400);
```

Use an adapting filter $W(z)$ with the LMS algorithm to identify $P(z)$. The length of the adaptive filter is 16, and the step size is 0.01. The adaptive filtering is conducted as follows:

```
ha = adaptfilt.lms(16,mu);
[y,e] = filter(ha,x,d);
```

The MATLAB script for implementing Figure 4.22 is given in example4_9.m (adapted from **Help** menu). Signals $d(n)$, $y(n)$, and $e(n)$ are shown in the top plot of Figure 4.23, which shows that $y(n)$ gradually approximates $d(n)$ and the error signal $e(n)$ is minimized. The bottom plot shows that the coefficients of adaptive filter $W(z)$ after convergence are identical to the corresponding coefficients of unknown system $P(z)$.

EXERCISE 4.9

Modify example4_9.m for the following simulations:

1. Design $P(z)$ with different orders. Set $W(z)$ with the same, higher, and lower orders and compare the identification results. Note that the step size value should be changed according to the filter length.
2. Using a filter length of 16, change the step size value from 0.01 to 0.05, 0.1, 0.005, and 0.001. Evaluate the performance of the adaptive filter.

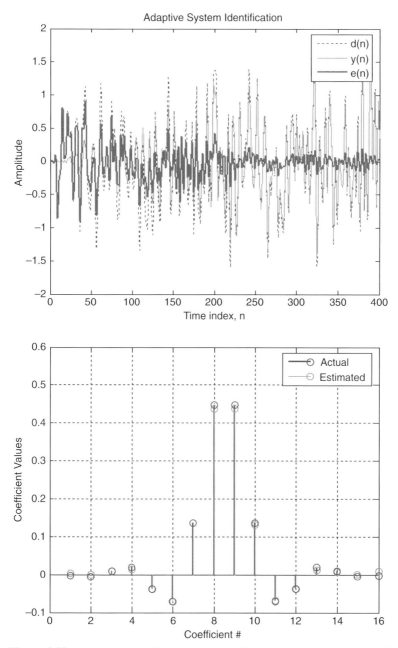

Figure 4.23 Adaptive system identification results. Top: signals $d(n)$, $y(n)$, and $e(n)$. Bottom: coefficients of $P(z)$ and $W(z)$

146 Chapter 4 Digital Filtering

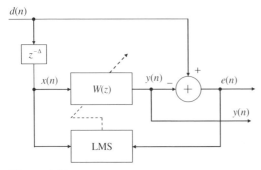

Figure 4.24 Block diagram of adaptive prediction

3. Change the adaptive algorithm from lms to other functions defined in Table 4.4, and evaluate the performance of the system identification results.
4. Table 4.5 shows that leakage is the leakage factor. It must be a scalar between 0 and 1. If it is less than 1, the leaky LMS algorithm is implemented. It defaults to 1 (no leakage). Try different leaky factors and compare the performance with Figure 4.23.
5. Plot $e^2(n)$ instead of $e(n)$. In addition, use the following equation to smooth the curve:

$$\hat{P}_e(n) = (1 - \alpha)\hat{P}_e(n-1) + \alpha e^2(n). \tag{4.4.14}$$

Try different values of α and initial values of $\hat{P}_e(0)$.

Linear prediction has been successfully applied to a wide range of applications such as speech coding and separating signals from noise. As illustrated in Figure 4.24, the adaptive predictor consists of a digital filter in which the coefficients $w_l(n)$ are updated by the LMS algorithm. For example, consider the adaptive predictor for enhancing multiple sinusoids embedded in white noise. In this application, the structure shown in Figure 4.24 is called the adaptive line enhancer (ALE), which provides an efficient means for the adaptive tracking of the sinusoidal components of a received signal $d(n)$ and separates these narrowband signals from broadband noise. In this figure, delay $\Delta = 1$ is adequate for decorrelating the white noise component between $d(n)$ and $x(n)$, and longer delay may be required for other broadband noises. This technique has been shown effective in practical applications when there is insufficient a priori knowledge of the signal and noise parameters.

EXAMPLE 4.10

As shown in Figure 4.24, assume that signal $d(n)$ consists of a desired sine wave that is corrupted by white noise. An adaptive filter will form a bandpass filter to pass the sine wave in $y(n)$. As shown in Figure 4.25, adaptive filter output $y(n)$ gradually approaches a clean sine wave, while the error signal $e(n)$ is reduced to the noise level. The MATLAB script for simulating adaptive line enhancement is given in example4_10.m.

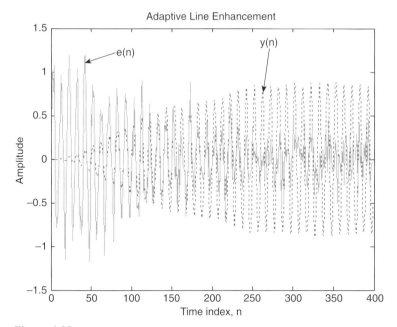

Figure 4.25 Adaptive line enhancement of narrowband signal

EXERCISE 4.10

Modify `example4_10.m` for the following simulations:

1. Add more sine waves into $d(n)$ and redo the simulation with different values of μ and L. We may need a higher order of filter when the sinusoidal frequencies are close. Accordingly, it requires a smaller value of the step size and results in slow convergence.
2. Plot the magnitude response of converged filter $W(z)$ and confirm that it will approximate an bandpass filter. What is the center frequency of the passband?
3. Redo Exercise 1 with different delay values Δ.
4. Use the adaptive line enhancer to enhance the chirp signal that is corrupted by random noise as shown in Figure 4.20.

Adaptive noise cancellation employs an adaptive filter to cancel the noise components in the primary signal picked up by the primary sensor. As illustrated in Figure 4.26, the primary sensor is placed close to the signal source to pick up the desired signal. The reference sensor is placed close to the noise source to sense the noise only. The primary input $d(n)$ consists of signal plus noise, which is highly correlated with $x(n)$ because they are derived from the same noise source. The reference input consists of noise $x(n)$ alone. The adaptive filter uses the reference input $x(n)$ to estimate the noise picked up by the primary sensor. The filter output $y(n)$ is

148 Chapter 4 Digital Filtering

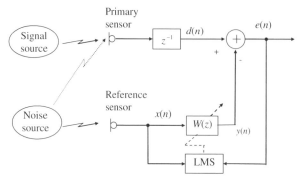

Figure 4.26 Block diagram of adaptive noise cancellation

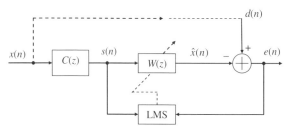

Figure 4.27 Block diagram of adaptive channel equalizer

then subtracted from the primary signal $d(n)$, producing $e(n)$ as the desired signal plus reduced noise. To apply the adaptive noise cancellation effectively, it is critical to avoid the signal components from the signal source being picked by the reference sensor. This "cross talk" effect will degrade the performance because the presence of the signal components in the reference signal will cause the adaptive filter to cancel the desired signal.

The transmission of high-speed data through a channel is limited by intersymbol interference caused by distortion in the transmission channel. This problem can be solved by using an adaptive equalizer in the receiver that counteracts the channel distortion. As illustrated in Figure 4.27, the received signal $s(n)$ is different from the original signal $x(n)$ because it was distorted by the overall channel transfer function $C(z)$, which includes the transmit filter, the transmission medium, and the receive filter. To recover the original signal $x(n)$, we need to process $s(n)$ with the equalizer $W(z)$, which is the inverse of the channel's transfer function $C(z)$, to compensate for the channel distortion. That is, $C(z)W(z) = 1$ such that $\hat{x}(n) = x(n)$. Note that $d(n)$ may not be available during data transmission.

4.4.3 More Hands-On Experiments

As discussed in Section 4.4.1, the optimum step size μ is difficult to determine. Improper selection of μ might make the convergence speed unnecessarily slow or

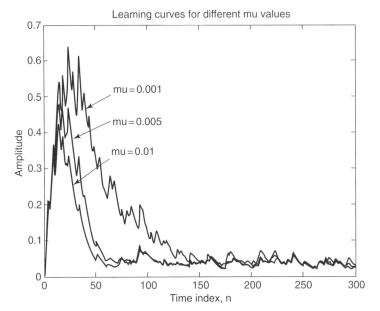

Figure 4.28 Learning curves for different step size values

introduce excess MSE. If the signal is changing and real-time tracking capability is crucial, we can use a larger μ. If the signal is stationary and convergence speed is not important, we can use a smaller μ to achieve better performance in a steady state. In some practical applications, we can use a larger μ at the beginning for faster convergence and use a smaller μ after convergence to achieve better steady-state performance.

EXAMPLE 4.11

Here we examine how step size affects the performance of the algorithm. Similar to Example 4.10, we fix the filter length to $L = 64$; however, we set μ values to 0.001, 0.005, and 0.01. In addition, we use Example 4.4.14 to estimate learning curves, which are shown in Figure 4.28. The MATLAB script is given in example4_11.m. The simulation results confirm that faster convergence can be achieved by using a larger step size.

EXERCISE 4.11

1. Based on example4_11.m, use different values of μ that are larger than 0.01. What values will make the algorithm diverge? Note that the learning curve shown in Figure 4.28 starts from 0. Why? How can we obtain a learning curve that starts from the correct value and converge to the minimum value?

150 Chapter 4 Digital Filtering

2. Based on `example4_11.m`, generate three sine waves that are corrupted by random noise. Find the relationship between the required filter length and the distance between adjacent sine waves.

3. The excess MSE is also proportional to the filter length L, which means that a larger L results in larger algorithm noise. From Equation 4.4.10, a larger L implies a smaller μ, resulting in slower convergence. Design simulations to verify these facts.

As illustrated in Figure 4.26, we can use $P(z)$ to represent the transfer function between the noise source and the primary sensor. The noise canceller has two inputs: the primary signal input $d(n)$ and the reference noise input $x(n)$. The primary input $d(n)$ consists of signal $s(n)$ plus noise $x'(n)$, which is $x(n)$ filtered by $P(z)$. To minimize the residual error $e(n)$, the adaptive filter $W(z)$ will generate an output $y(n)$ that is an approximation of $x'(n)$. Therefore, the adaptive filter $W(z)$ will converge to the unknown plant $P(z)$.

EXAMPLE 4.12

In Figure 4.26, assume that the signal $s(n)$ is speech from file `timit1.asc`. The noise $x(n)$ is a tone at frequency 1,000 Hz, which is also picked up by the primary sensor as $x'(n)$. Assuming that $P(z)$ is the FIR filter defined in Example 4.9, $x'(n)$ can be obtained by filtering $x(n)$ with $P(z)$. An adaptive filter will form a bandpass filter to pass $x(n)$, but change its amplitude and phase to match with $x'(n)$ so it can be canceled. Therefore, the error signal $e(n)$ will consist of speech $s(n)$ as shown in Figure 4.29. The MATLAB script for this

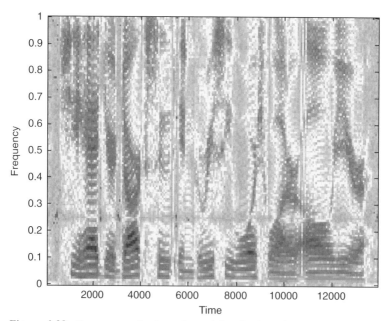

Figure 4.29 Spectrogram of enhanced speech by adaptive noise cancellation

experiment is given in example4_12.m. In the code, the step size used is 0.0001/32,767. Why it is so small? Hint: For DSP simulations, signals are normalized between ±1, which is discussed in Chapter 6. However, many real-world signals (such as timit1.asc) sampled by 16-bit A/D converters usually have amplitude between ±32,767.

EXERCISE 4.12

1. It is difficult to determine an optimum step size for a given application. A trial-and-error method is commonly used. Is it possible to use the normalized LMS algorithm to simplify the process of determining step size?
2. In Example 4.12, if the noise $x(n)$ is a random noise, is the adaptive noise cancellation able to reduce this random noise?
3. What is the difficulty of applying the adaptive noise cancellation technique to reduce noise pickup by the primary microphone of hands-free conversation inside a noisy automobile compartment?

Having analyzed the properties of adaptive filtering in MATLAB, we now examine the implementation of an adaptive filter using the Blackfin simulator and perform real-time experiments using the BF533 (or BF537) EZ-KIT.

4.5 ADAPTIVE LINE ENHANCER USING BLACKFIN SIMULATOR

This section verifies the code ported from MATLAB to C for the adaptive line enhancer shown in Figure 4.24 using the Blackfin simulator. We will use some new functions provided in the C run time library to generate random noise and sine wave.

HANDS-ON EXPERIMENT 4.4

This experiment uses a project file, exp4_4.dpj located in directory c:\adsp\chap4\exp4_4 to simulate the adaptive line enhancer (ALE) given in Example 4.10. The C main program ale_demo.c generates a sine wave that is corrupted by a random noise and calls the adaptive FIR filter with the LMS algorithm to remove the undesired noise. The process.c performs the ALE based on the LMS algorithm. The sin function in the C run time library is used to generate a sine wave of 256 samples as follows:

```
*out = (fract16) (0.5*sin(2*PI*f*step)*32768.0);
```

The function rand is called to generate pseudo-random numbers. Because the rand function generates numbers in the range of [0, $2^{30} - 1$], we subtract the generated numbers by 0x20000000 and shift right by 15 bits to obtain 16-bit zero-mean random integers as follows:

```
(fract16)((rand() - 0x20000000)>>15)
```

152 Chapter 4 Digital Filtering

The generated sine wave samples and the random numbers are added to form a noisy sine wave, which will be enhanced by the ALE. This experiment uses a 32-tap adaptive FIR filter and a delay $\Delta = 1$ for ALE. Because no adaptive filtering function is available in the Blackfin DSP run time library, the process.c file consists of a C routine (LMS_filter) to perform the adaptive filtering. The input arguments for the LMS algorithm are declared as follows:

```
LMS_filter(fract16 in[],fract16 out1[], fract16 out2[], fract16
          d[], fract16 c[], int taps, int size, int delaysize,
          fir_state_fr16* s)
```

Note that the FIR filter function, fir_fr16, is embedded inside the adaptive filtering routine. What is the step size used in the program?

Build and load the project into the Blackfin compiled simulator. The reason for using the compiled simulator is to allow quick evaluation of the ALE's performance. Three signals can be viewed by clicking on **View → Debug Windows → Plot → Restore...** and selecting inp.vps (input signal), out1.vps (output signal), and out2.vps (error signal) as shown in Figure 4.30 (a), (b), and (c), respectively.

Note that the signal out1 is an enhanced version of the input noisy signal and the error signal out2 contains the random noise. Thus the ALE has enhanced the sine wave successfully. Users can plot the magnitude spectra of these signals by using the frequency analysis features provided in the VisualDSP++.

EXERCISE 4.13

1. Use different step sizes for the LMS algorithm and examine the output signal out1. Observe the effects of out1 when increasing and decreasing the value of the step size for adaptive filtering.
2. Increase the length of the adaptive filter and adjust the step size to obtain a good result.
3. Modify the program to implement the normalized LMS algorithm by using the Calc_mu function in process.c.
4. Set up a new window to show the variation of the normalized step size using the normalized LMS algorithm.

4.6 ADAPTIVE LINE ENHANCER USING BLACKFIN BF533/BF537 EZ-KIT

This section tests the real-time performance of the ALE using the BF533 (or BF537) EZ-KIT. Several noisy signals can be played with the sound card or CD player, which is connected to the analog input of the EZ-KIT. The processed output signal from the EZ-KIT is connected to a pair of headphones or external loudspeakers. Users can experiment with different values of step size, delay, and adaptive filter length to determine the optimum set of parameters for the ALE.

4.6 Adaptive Line Enhancer Using Blackfin BF533/BF537 EZ-KIT   153

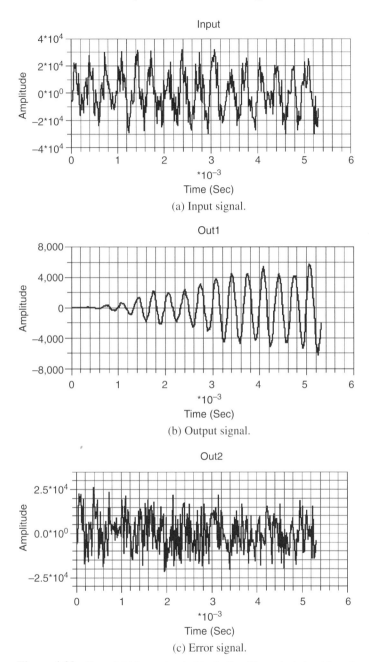

Figure 4.30 Plots of (a) input signal, (b) adaptive filter output signal (out1), and (c) error signal (out2)

154 Chapter 4 Digital Filtering

Table 4.6 Switch Settings and LED Indicators for Listening to Different Modes of Adaptive Filtering

Modes	BF533 EZ-KIT	BF537 EZ-KIT
Pass-through	SW4 (LED4)	SW10 (LED1)
Output from adaptive filter	SW5 (LED5)	SW11 (LED2)
Error of adaptive filter (default mode after loading)	SW6 (LED6)	SW12 (LED3)

HANDS-ON EXPERIMENT 4.5

Activate the VisualDSP++ for the BF533 (or BF537) EZ-KIT and open exp4_5_533.dpj in directory c:\adsp\chap4\exp4_5_533 for the BF533 (or exp4_5_537.dpj in c:\adsp\chap4\exp4_5_537 for the BF537) EZ-KIT. Build the project, and the executable file exp4_5_533.dxe or exp4_5_537.dxe will be automatically loaded into the EZ-KIT. Select speech_tone_48k.wav in directory c:\adsp\audio_files and play this wave file continuously with the computer sound card. Run the program and listen to the original wave file from the headphone (or loudspeakers). The switch settings and LED modes for the BF533 and BF537 EZ-KITs are listed in Table 4.6.

Experiment with different settings (step size, delay, and filter length) of the adaptive filter. Change the adaptive algorithm from the LMS to the normalized LMS and test the performance of the ALE, using the noisy musical signal sineandmusic.wav. Change the delay (DELAY_SIZE) to 1 for decorrelating the noise. Do you notice any distortion to the error signal? Why?

EXERCISE 4.14

1. Generate multiple sine waves at 1, 2, and 3 kHz, and mix them with a musical signal sampled at 48 kHz. Use either the LMS or normalized LMS algorithm to remove the sine waves. Note that we may change the filter length and the corresponding step size for this application.

2. Generate a chirp signal and mix it with the musical signal sampled at 48 kHz. Compare the performance of the LMS and normalized LMS algorithms for removing the chirp signal. Increase the length of the adaptive filter and compare its performance with the filter of shorter length.

3. In the preceding experiments, two adaptive filters are required for the ALE to operate separately on the left and right channels. Devise a way to reduce computation load by using only one filter. Compare the performance of ALEs using one and two filters on removing sine wave in stereo channels.

4.7 ADAPTIVE LINE ENHANCER USING LABVIEW EMBEDDED MODULE FOR BLACKFIN PROCESSORS

Is this section, we design and implement an ALE with the graphical system design approach. First, we simulate the ALE in LabVIEW to gain an intuitive understanding of its behaviors, and then we design and implement the ALE application with the LabVIEW Embedded Module for Blackfin Processors for execution on the Blackfin EZ-KIT.

HANDS-ON EXPERIMENT 4.6

This experiment focuses on the simulation and implementation of a simple adaptive FIR filter with the LMS algorithm. We use the simulation created in LabVIEW to explore the different input parameters that affect the behaviors of the LMS algorithm. Simulation is a crucial part of system design because it allows the designer to test and refine parameters before moving to the real embedded target. This can effectively reduce errors, faulty assumptions, and overall development time.

Open the program LMS_Filter_Sim.exe in directory c:\adsp\chap4\exp4_6. The user interface for this application is shown in Figure 4.31. Some input parameters can be altered to study the effects on the LMS algorithm and its outputs. There are two tabs: **LMS Filter Results (Time Domain)** and **LMS Filter Results (Frequency Domain)**. The

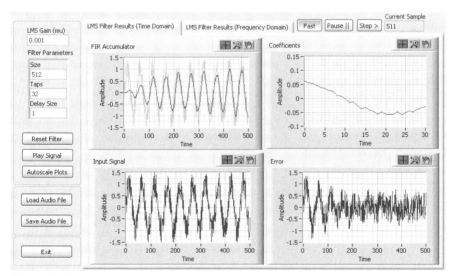

Figure 4.31 Time-domain plots of the signals and coefficients of the filter using the LMS algorithm (LMS_Filter_Sim.exe)

156 Chapter 4 Digital Filtering

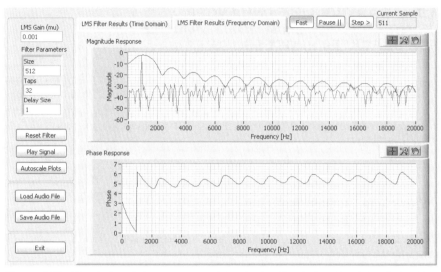

Figure 4.32 Magnitude and phase responses of the adaptive filter designed by the LMS algorithm (LMS_Filter_Sim.exe)

time-domain plots show the noisy sine wave, the filtered output signal, the filter coefficients, and the error signal. Click on the **LMS Filter Results (Frequency Domain)** tab to evaluate the magnitude and phase responses for the FIR filter coefficients as they adapt to reduce the noise.

The most interesting part of the simulation is to see how the LMS algorithm learns and adapts its coefficients. Click on the **Reset Filter** button to reset the filter coefficients to zero and restart this learning process. Additional features have been added to the top right corner for pausing and single-stepping through the algorithm. We can single-step through the calculation while viewing either the time-domain or the frequency-domain plots. These graphical features greatly aid the visualization and understanding of adaptive filtering, which is an iterative algorithm. Select the **LMS Filter Results (Frequency Domain)** tab to see how the FIR filter adapts to suppress the noise while preserving the desired 1-kHz input signal as shown in Figure 4.32.

The input parameters for the LMS algorithm that can be changed include the **LMS Gain** (step size μ), the number of FIR filter taps (length L), and the delay size Δ. As discussed above, the step size determines the amount of correction to be applied to the filter coefficients as the filter adapts. The **Taps** parameter determines the length of the FIR filter. **Delay Size** indicates the interval in samples between the current acquired sample and the newest sample used by the filter.

As the filter designer, optimize the LMS algorithm by fine-tuning the step size for fast learning, while minimizing coefficient fluctuations once the filter converges to a steady state. How do the input parameters affect the resulting filter? Can the LMS algorithm lose stability? Explain.

HANDS-ON EXPERIMENT 4.7

This experiment implements the LMS algorithm with the LabVIEW Embedded Module for Blackfin Processors using C programming. Algorithms written in C can be quickly prototyped in the LabVIEW Embedded Module for Blackfin Processors, allowing them to be implemented as a subcomponent of the graphical system. Graphical system implementation allows other parts of the system to be abstracted, thus allowing the programmer to focus on the algorithm.

Open the `LMS Adaptive Filter - BF5xx.lep` project in directory `c:\adsp\chap2\exp4_7`. When viewing the block diagram of the `LMS Adaptive Filter.vi`, note that it is similar to the audio talk-through example discussed in Chapter 2, but has been modified to include buffer initializations, specification of taps and delay size, and an **Inline C Node** as seen in Figure 4.33. The parameters `Taps` and `Delay Size` can be customized to change the behavior of the LMS algorithm. The **Inline C Node** contains nearly the same LMS algorithm found in the C implementation in earlier experiments with an alternate FIR filter implementation. In this experiment, the FIR filter is implemented with a multiply-accumulate operation within a **For Loop** to reduce the amount of C programming needed. The original implementation using the `fir_fr16` library function required additional programming for initialization, which linked global buffers, coefficients, and state variables to the filter. Push-button input was also added, allowing the user to choose which component of the algorithm to output.

The LMS algorithm is computationally intensive, making it necessary to configure Blackfin optimizations to run the LMS algorithm as fast as possible. These settings optimize the C code generated from LabVIEW and are found in the **Build Options** menu by selecting **Tools → Build Options** from the **Embedded Project Manager** window. First, the **Build Configuration** is changed from **Debug** to **Release**. This option removes all extra debugging code from the generated C. **Stack Variables** and **Disable Parallel Execution** are both enabled to opti-

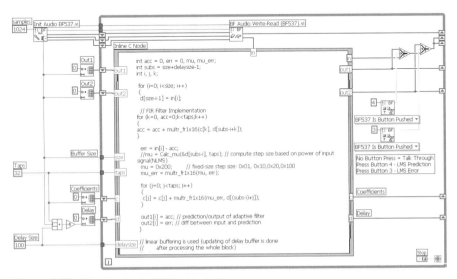

Figure 4.33 The adaptive LMS filter block diagram

mize the way in which memory is managed and parallel programming structures are handled. These optimizations are ideal for implementing fast, serialized algorithms that perform many memory operations, such as the LMS algorithm that performs FIR filtering and coefficient updating for each of the 512 samples in the acquired audio buffer.

Run the project on the Blackfin processor. Play the `speech_tone_48k.wav` that was used in Hands-On Experiment 2.4 and feed the wave output to the audio input of the BF533/BF537 EZ-KIT. When the **SW4** of BF533 (or **SW10/PB4** of BF537) is pressed, the LMS algorithm adapts to suppress the speech. When the **SW5** of BF533 (or **SW11/PB3** of BF537) is pressed, the error signal is heard. What do you notice about the error signal? When can the error signal be used as the desired filter output?

4.8 MORE EXERCISE PROBLEMS

1. In FDATool, the default magnitude specifications in decibels (dB) are $A_p = 1$ and $A_s = 80$. Convert these units to linear scale.

2. Redo Exercise 4.2 with SPTool. In addition to designing those filters, import the signal given in Example 4.1, perform FIR filtering using SPTool, and analyze input and output spectra.

3. Use FIR filter design functions listed in Table 4.1 to design bandstop filters for attenuating tonal noise at 1,000 Hz embedded in speech signal `timit1.asc` as shown in Figure 3.11. Also, perform FIR filtering with MATLAB function `filter`.

4. Use the IIR filter design functions listed in Table 4.2 to design filters specified in Exercise 4.2 and compare the required orders of IIR filters with the corresponding FIR filters.

5. Use the transformation functions listed in Table 4.3 to convert the filters designed in Problem 4 to different realizations.

6. Use the IIR filter design functions listed in Table 4.2 to design IIR bandstop filters specified in Exercise 4.6(1). Realize the designed filters in cascade second-order structure with the functions given in Table 4.3 and perform IIR filtering with MATLAB function `sosfilt`.

7. Examine the stability of the IIR filter given in Example 4.7.

8. Implement the adaptive channel equalization technique illustrated in Figure 4.27, using MATLAB or C program. The input signal $x(n)$ is a zero-mean, unit-variance white noise. The unknown channel $C(z)$ is simulated by the FIR filter used in Example 4.9. Note that a delay of $L/2$ is used to produce $d(n)$ from $x(n)$.

9. Redo Examples 4.9 and 4.10 with different step sizes. Instead of plotting $e(n)$, use Equation 4.4.14 to compute learning curves.

10. Modify the adaptive filtering program in Hands-On Experiment 4.4 to implement the adaptive noise cancellation, described in Section 4.4.2.

11. Repeat Problem 10 by implementing the adaptive channel equalization.

12. The ALE can enhance the detection of DTMF tones in phone dialing. Mix the speech signal `timit.wav` with the dial tone `dtmf_tone.wav`. Play the combined wave file

4.8 More Exercise Problems

and connect it to the input channel of the BF533 (or BF537) EZ-KIT. Choose suitable parameters (algorithm, filter length, step size, and delay) for the ALE and run the program, using the EZ-KIT. Is the ALE program working?

13. A stereo signal is corrupted with a sine wave (speech_tone_48k_stereo.wav). Implement the ALE algorithm in the Blackfin processor to remove the sine wave. To reduce the computational cost of performing ALE to the left and right input channels separately, a single ALE algorithm is performed on the common input of $(x_L + x_R)/2$, where x_L and x_R are the left and right input channels. The sine wave derived from the ALE can be subtracted from the original left and right channels to form a clean stereo signal.

14. In the ALE experiment given in Hands-On Experiment 4.5, what is the smallest step size that can be used for the 16-bit implementation? What is the largest step size for a 32-tap adaptive filter?

Part B

Embedded Signal Processing Systems and Concepts

Chapter 5

Introduction to the Blackfin Processor

This chapter examines the architecture of the Blackfin processor, which is based on the MSA jointly developed by Analog Devices and Intel. We use assembly programs to introduce the processing units, registers, and memory and its addressing modes. At the end of the chapter, we design, simulate, and implement an eight-band graphic equalizer and use this application to explain some of the practical implementation issues. An in-depth discussion of the real-time processing concepts, number representations, peripheral programming, code optimization, and system design is given in Chapters 6, 7, and 8.

5.1 THE BLACKFIN PROCESSOR: AN ARCHITECTURE FOR EMBEDDED MEDIA PROCESSING

This section introduces the architecture of the Blackfin processor and its internal hardware units, memory, and peripherals using assembly instructions. In particular, we use the BF533 processor [23] for explaining the Blackfin processor's architecture. The BF537 processor [24] has core and system architectures identical to those of the BF533, but slightly different on-chip peripherals.

5.1.1 Introduction to Micro Signal Architecture

As introduced in Chapter 1, the MSA core was designed to achieve high-speed DSP performance and best power efficiency. This core combines the best capabilities of microcontroller and DSP processor into a single programming model. This is

---

Embedded Signal Processing with the Micro Signal Architecture. By Woon-Seng Gan and Sen M. Kuo
Copyright © 2007 John Wiley & Sons, Inc.

different from other cores that require separate DSP processor and microcontroller. The main advantage of the MSA core is the integrated feature that combines multimedia processing, communication, and user interface on a single, easy-to-program platform. This highly versatile MSA core performs DSP tasks as well as executing user commands and control tasks. The programming environment has many features that are familiar to both microcontroller and DSP programmers, thus greatly speeding up the development of embedded systems.

The MSA architecture is also designed to operate over a wide range of clock speeds and operating voltages and includes circuitry to ensure stable transitions between operating states. A dynamic power management circuit continuously monitors the software running on the processor and dynamically adjusts both the voltage delivered to the core and the frequency at which the core runs. This results in optimized power consumption and performance for real-time applications.

5.1.2 Overview of the Blackfin Processor

The ADSP-BF5xx Blackfin processor is a family of 16-bit fixed-point processors that are based on the MSA core. This processor targets power-sensitive applications such as portable audio players, cell phones, and digital cameras. Low cost and high performance factors also make Blackfin suitable for computationally intensive applications including video equipment and third-generation cell phones.

The first generation of the BF5xx family is the BF535, which achieves a clock speed up to 350 MHz at 1.6 V. Analog Devices introduced three processor families (BF532, BF533, and BF561) in 2003. These processors can operate up to 750 MHz at 1.45 V. The clock speed and operating voltages can be switched dynamically for given tasks via software for saving power. The BF561 processor incorporates two MSA cores to improve performance using parallel processing. A recent release of the BF5xx family consists of BF534, BF536, and BF537. These processors add embedded Ethernet and controller area network connectivity to the Blackfin processor.

The Blackfin core combines dual multiply-accumulate (MAC) engines, an orthogonal reduce-instruction-set computer (RISC)-like instruction set, single instruction, multiple data (SIMD) programming capabilities, and multimedia processing features into a unified architecture. As shown in Figure 5.1, the Blackfin BF533 processor [23] includes system peripherals such as parallel peripheral interface (PPI), serial peripheral interface (SPI), serial ports (SPORTs), general-purpose timers, universal asynchronous receiver transmitter (UART), real-time clock (RTC), watchdog timer, and general-purpose input/output (I/O) ports. In addition to these system peripherals, the Blackfin processor also has a direct memory access (DMA) controller that effectively transfers data between external devices/memories and the internal memories without processor intervention. Blackfin processors provide L1 cache memory for quick accessing of both data and instructions.

In summary, Blackfin processors have rich peripheral supports, memory management unit (mmu), and RISC-like instructions, which are typically found in

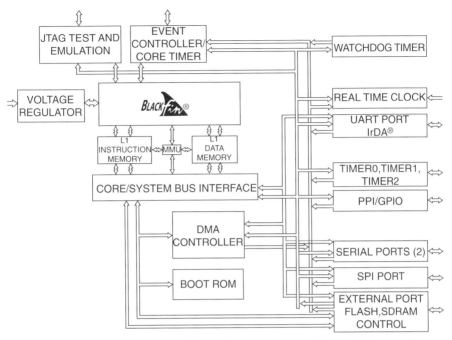

Figure 5.1 Block diagram of the Blackfin BF533 system (courtesy of Analog Devices, Inc.)

many high-end microcontrollers. These processors have high-speed buses and advanced computational engines that support variable-length arithmetic operations in hardware. These features make the Blackfin processors suitable to replace other high-end DSP processors and microcontrollers. In the following sections, we further introduce the core architecture and its system peripherals.

5.1.3 Architecture: Hardware Processing Units and Register Files

Figure 5.2 shows that the core architecture consists of three main units: the address arithmetic unit, the data arithmetic unit, and the control unit.

5.1.3.1 Data Arithmetic Unit

The data arithmetic unit contains the following hardware blocks:

1. Two 16-bit multipliers represented as ⨝ in Figure 5.2.
2. Two 40-bit accumulators (ACC0 and ACC1). The 40-bit accumulator can be partitioned as 16-bit lower-half (A0.L, A1.L), 16-bit upper-half (A0.H, A1.H), and 8-bit extension (A0.X, A1.X), where L and H denote lower and higher 16-bit, respectively.

166 Chapter 5 Introduction to the Blackfin Processor

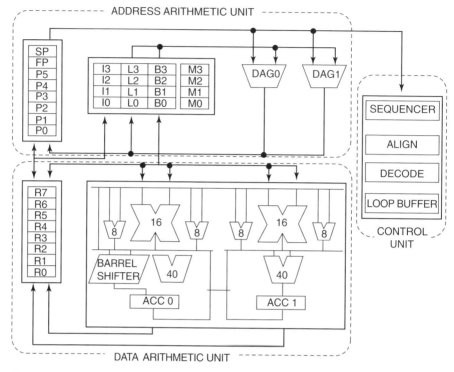

Figure 5.2 Core architecture of the Blackfin processor (courtesy of Analog Devices, Inc.)

3. Two 40-bit arithmetic logic units (ALUs) represented as ⟨40⟩ in Figure 5.2.
4. Four 8-bit video ALUs represented as ⟨8⟩ in Figure 5.2.
5. A 40-bit barrel shifter.
6. Eight 32-bit data registers (R0 to R7) or 16 independent 16-bit registers (R0.L to R7.L and R0.H to R7.H).

Computational units get data from data registers and perform fixed-point operations. The data registers receive data from the data buses and transfer the data to the computational units for processing. Similarly, computational results are moved to the data registers before transferring to the memory via data buses.

These hardware computational blocks are used extensively in performing DSP algorithms such as FIR filtering, FFT, etc. The multipliers are often combined with the adders inside the ALU and the 40-bit accumulators to form two 16- by 16-bit MAC units. Besides working with the multiplier, the ALU also performs common arithmetic (add, subtract) and logical (AND, OR, XOR, NOT) operations on 16-bit or 32-bit data. Many special instructions or options are included to perform saturation, rounding, sign/exponent detection, divide, field extraction, and other operations. In addition, a barrel shifter performs logical and arithmetic shifting, rotation, normalization and extraction in the accumulator. An illustrative experiment using the shift instructions is presented below in Hands-On Experiment 5.3. With the dual ALUs and multipliers, the Blackfin processor has the flexibility of operating two register pairs or four 16-bit registers simultaneously.

In this section, we use Blackfin assembly instructions to describe the arithmetic operations in several examples. The assembly instructions use algebraic syntax to simplify the development of the assembly code.

EXAMPLE 5.1 *Single 16-Bit Add/Subtract Operation*

Any two 16-bit registers (e.g., R1.L and R2.H) can be added or subtracted to form a 16-bit result, which is stored in another 16-bit register, for example, R3.H = R1.L + R2.H (ns), as shown in Figure 5.3. Note that for 16-bit arithmetic, either a saturation flag (s) or a no saturation (ns) flag must be placed at the end of the instruction. The symbol ";" specifies the end of the instruction. Saturation arithmetic is discussed in Chapter 6.

The Blackfin processor provides two ALU units to perform two 16-bit add/subtract operations in a single cycle. This dual 16-bit add/subtract operation doubles the arithmetic throughput over the single 16-bit add/subtract operation.

EXAMPLE 5.2 *Dual 16-Bit Add/Subtract Operations*

Any two 32-bit registers can be used to store four inputs for dual 16-bit add/subtract operations, and the two 16-bit results are saved in a single 32-bit register. As shown in Figure 5.4, the instruction R3 = R1+|-R2 performs addition in the upper halves of R1 and R2 and subtraction in the lower halves of R1 and R2, simultaneously. The results are stored in the high and low words of the R3 register, respectively.

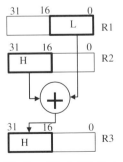

R3.H = R1.L+R2.H (ns);

Figure 5.3 Single 16-bit addition using three registers

168 Chapter 5 Introduction to the Blackfin Processor

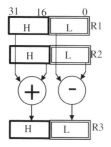

```
R3 = R1+|-R2;
```

Figure 5.4 Dual 16-bit add/subtract using three registers

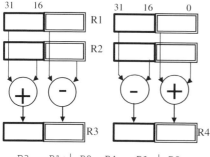

```
R3 = R1+|-R2,  R4 = R1-|+R2;
```

Figure 5.5 Quad 16-bit add/subtract using four registers

The Blackfin processor is also capable of performing four (or quad) 16-bit add/subtract operations in a single pass. These quad operations fully utilize the dual 40-bit ALU and thus quadruple the arithmetic throughput over the single add/subtract operation.

EXAMPLE 5.3 *Quad 16-Bit Add/Subtract Operations*

In quad 16-bit add/subtract operations, only the same two 32-bit registers can be used to house the four 16-bit inputs for these quad additions. In other words, two operations can be operated on the same pair of 16-bit registers. For example, the instructions `R3 = R1+|-R2, R4 = R1-|+R2` perform addition and subtraction on the halves of R1 and R2 as shown in Figure 5.5. Note that the symbol "," separates two instructions that are operated at the same cycle.

Besides the previous 16-bit operations, the Blackfin processor can also perform single 32-bit add/subtract using any two 32-bit registers as inputs.

5.1 An Architecture for Embedded Media Processing 169

EXAMPLE 5.4   *Single 32-Bit Operations*

The 32-bit result of the single 32-bit add/subtract operation is stored in another 32-bit register. For example, the instruction R3 = R1+R2 performs 32-bit addition of R1 and R2 and places the result in R3 as shown in Figure 5.6.

Similar to the dual 16-bit add/subtract, dual 32-bit add/subtract can also be carried out with the dual 40-bit ALUs.

EXAMPLE 5.5   *Dual 32-Bit Operations*

Example 5.4 can be extended to a dual 32-bit add/subtract. This operation is also similar to dual 16-bit operation, with the exception that the inputs and results are all 32 bits. For example, the instructions R3 = R1+R2, R4 = R1-R2 perform simultaneous addition and subtraction of R1 and R2 with the addition result saved in R3 and the subtraction result stored in R4, as depicted in Figure 5.7.

In the above examples of ALU operations, we add/subtract data in either 16- or 32-bit wordlength. The results may overflow because of the limited wordlength used

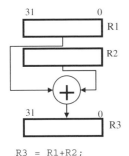

R3 = R1+R2;

Figure 5.6   Single 32-bit add/subtract using three registers

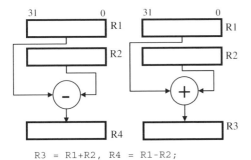

R3 = R1+R2, R4 = R1-R2;

Figure 5.7   Dual 32-bit add/subtract using four registers

Table 5.1 Arithmetic Modes and Options for the ALU Operations

Mode	Option	Example and Explanation
Dual and quad 16-bit operation (opt_mode_0)	S	Saturate the result at 16 bit R3 = R1+|-R2 (s);
	CO	Cross option that swaps the order of the results in the destination registers for use in complex math R3 = R1+|-R2 (co);
	SCO	Combination of S and CO options
Dual 32-bit and 40-bit operation (opt_mode_1)	S	Saturate result at 32 bit. R3 = R1+R2, R4 = R1-R2 (s);
Quad 16-bit operation (opt_mode_2)	ASR	Arithmetic shift right that halves the result before storing to the destination register R3 = R1+|-R2, R4 = R1-|+R2 (s,asr); Scaling is performed for the results before saturation.
	ASL	Arithmetic shift left that doubles the result before storing to the destination register

in storing the sums. A solution is provided in the arithmetic instructions to saturate the result at 16 or 32 bits, depending on the size of the operands. For example, in a 16-bit word, the result of add/subtract will not exceed the most positive or the most negative number of a 16-bit sign data on saturation. The modes and options available in the ALU operations are listed in Table 5.1. More details on different number formats and arithmetic overflow are provided in Chapter 6.

There are rounding options that can be used to perform biased rounding and saturation to a 16-bit result. The correct way to handle overflow, rounding, saturation, and other arithmetic issues is explained in Chapter 6. Besides performing arithmetic operations, ALU also allows 32-bit logical operations such as AND, OR, NOT, and XOR.

EXAMPLE 5.6 *32-Bit ALU Logical Operations*

This example shows 32-bit logical operations. For example, the instruction R3 = R1&R2 performs the bitwise AND operation on R1 and R2 registers, and the result is stored in R1. Other examples include R3 = R1|R2 (OR operation); R2 = ~R1 (NOT operation), and R3 = R1^R2 (XOR operation). Note that there is no 16-bit logical operation in the Blackfin processor.

So far, we have only performed addition and subtraction with the ALU. Figure 5.2 shows that two 16-bit multipliers are available in the Blackfin processor. These 16-bit multipliers perform single or dual 16-bit multiplications in a single cycle, and the result is stored in either 40-bit accumulators or 32-bit registers. The multipliers are also linked to the ALUs for implementing multiply-accumulate operations via the accumulators. The simplest multiplication is carried out with two 16-bit registers, and the result can be placed in a 16-bit or 32-bit register.

EXAMPLE 5.7 *Single 16-Bit Multiply Operations*

A 16-bit multiplication can easily be carried out on the Blackfin processor. Any two 16-bit register halves (e.g., R1.L and R2.H) can be multiplied together to form a 32-bit result, which is stored in the accumulator or data register (e.g., R3). For example, the operation of R3 = R1.L*R2.H is shown in Figure 5.8. In addition, a 16-bit result can also be stored in a half-register, for example, R3.H = R1.L*R2.H.

The 16-bit multiplication example can also be extended to a single 16-bit multiply/accumulate operation. The MAC operation is usually carried out iteratively, and the accumulator must be used to store the intermediate and final results.

EXAMPLE 5.8 *Single 16-Bit Multiply/Accumulate Operations*

This example is similar to Example 5.7 except that the 32-bit multiplication result is added to the previous result in the accumulator. For example, the instruction A0 += R1.L*R2.L multiplies the contents of R1.L with R2.L, and the result is added to the value in the accumulator A0 as shown in Figure 5.9. The final result is then stored back into the accumulator. This MAC operation is very useful in implementing FIR/IIR filters, which are described in Chapter 4. In addition, we can also transfer the final result to a half-register, for example, extending the previous example, R5.L = (A0 += R1.L*R2.L). This instruction truncates the accumulator result to 16-bit (i.e., ignore the lower 16 bits of A0) and stores the upper A0 (A0.H) in the lower half of the R5 register.

Similar to the dual add/subtract operations, dual multiply operations can also be carried out with two 16-bit multipliers, two 32-bit registers, and two accumulators.

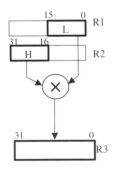

R3 = R1.L*R2.H;
Figure 5.8 Single 16-bit multiplication using three registers

172 Chapter 5 Introduction to the Blackfin Processor

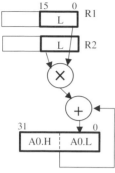

```
A0 += R1.L*R2.L;
```

Figure 5.9 Single 16-bit MAC using two registers and an accumulator

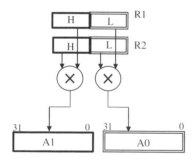

```
A1 = R1.H*R2.H, A0 = R1.L*R2.L;
```

Figure 5.10 Dual 16-bit multiplications using two registers and two accumulators

EXAMPLE 5.9 *Dual 16-Bit Multiply Operations Using Two Accumulators*

In this example, we perform A1 = R1.H*R2.H, A0 = R1.L*R2.L, where the two pairs of input are stored in the same registers R1 and R2 and the two 32-bit results are stored in the accumulators A0 and A1, as shown in Figure 5.10. Note that we use the symbol "," to separate the two parallel instructions.

Instead of using two accumulators as stated in Example 5.9, dual 16-bit multiplications can also be performed with two 32-bit registers to save the results.

EXAMPLE 5.10 *Dual 16-Bit Multiply Operations Using Two 32-Bit Registers*

When performing dual 16-bit multiplications using two 32-bit registers, the 32-bit destination registers must be used in pairs as R0:R1, R2:R3, R4:R5, or R6:R7. For example, the instructions R0 = R2.H*R3.H, R1 = R2.L*R3.L state that dual 32-bit results are stored in R0 and R1, respectively. Therefore, R0 and R1 must always be used as paired registers to store the results of high-word and low-word multiplications, respectively. Other pairs like R4:R5 and R6:R7 can also be used to replace R0:R1 in the above example.

Similar to Example 5.9, we can also double the throughput of single 16-bit multiply/accumulate by performing dual 16-bit multiply/accumulate operations but storing the results in two 16-bit destinations.

EXAMPLE 5.11 *Dual 16-Bit Multiply/Accumulate Operations with Two 16-Bit Destinations*

Example 5.9 can also be extended for dual-MAC operations to double its throughput, for example, A1 -= R1.H*R2.H, A0 += R1.L*R2.L. In addition, the dual MAC results can be stored into two 16-bit registers as shown in Figure 5.11, for example, R3.H = (A1 -= R1.H*R2.H), R3.L = (A0 += R1.L*R2.L). Note that the result in A1 must be stored to the high word of R3 and the result in A0 must be stored to the low word of R3.

The above example can also be extended to save in two 32-bit destination registers via accumulators A0 and A1.

EXAMPLE 5.12 *Dual 16-Bit Multiply/Accumulate Operations with Two 32-Bit Destinations*

In this case, the 32-bit destination registers in the dual 16-bit multiply/accumulate operations must be saved in pairs as R0:R1, R2:R3, R4:R5, or R6:R7, for example, R0=(A0+= R2.H*R3.H), R1=(A1+= R2.L*R3.L). Note that A1 is associated to the higher-numbered register (in this case, R1) and A0 is associated to the lower-numbered register (R0).

The above multiply and multiply/accumulate operations are executed as default with no option. The default option implies that the input data are of signed fractional number. However, the Blackfin processor is able to handle data of different formats. Possible options and descriptions are listed in Table 5.2. We discuss the different

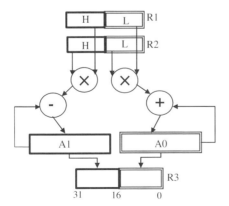

R3.H = (A1 -= R1.H*R2.H), R3.L = (A0 += R1.L*R2.L);

Figure 5.11 Dual 16-bit multiply/accumulate using three registers and two accumulators

Table 5.2 16-Bit Multiplier Options

Option	Description
Default (no option)	Input data operand is signed fraction.
(FU)	Input data operands are unsigned fraction. No shift correction.
(IS)	Input data operands are signed integer. No shift correction.
(IU)	Input data operands are unsigned integer. No shift correction.
(T)	Input data operands are signed fraction. When copying to the destination half-register, truncates the lower 16 bits of the accumulator contents.
(TFU)	Input data operands are unsigned fraction. When copying to the destination half-register, truncates the lower 16 bits of the accumulator contents.

number formats and the Blackfin arithmetic options (including why a shift correction is necessary) in Chapter 6.

Another unique feature of the Blackfin processor is the inclusion of four additional 8-bit video ALUs. A special set of video instructions is available for these video ALUs in image and video applications. Because the data registers are 32 bits, four 8-bit operations (for example, add, subtract, average, absolute) can be executed in a single instruction.

5.1.3.2 Address Arithmetic Unit

As shown in Figure 5.2, the address arithmetic unit consists of the following hardware units:

1. Two data address generators (DAG0 and DAG1) generate addresses for data moves to and from memory. The advantage of using two data address generators is to allow dual-data fetches in a single instruction.
2. Six 32-bit general-purpose address pointer registers (P0 to P5).
3. One 32-bit frame pointer (FP) pointing to the current procedure's activation record.
4. One 32-bit stack pointer (SP) pointing to the last location on the run time user stack.
5. A set of 32-bit data address generator registers:
   a. Indexing registers, I0 to I3, contain the effective addresses.
   b. Modifying registers, M0 to M3, contain offset values for add/subtract with the index registers.
   c. Base address registers, B0 to B3, contain the starting addresses of the circular buffers.
   d. Length value registers, L0 to L3, contain the lengths (in byte unit) of the circular buffers.

5.1 An Architecture for Embedded Media Processing

The main function of the address arithmetic unit is to generate addresses for accessing the memory in the Blackfin processor. The Blackfin processor is byte addressable. However, data can be accessed in 8-bit, 16-bit, or 32-bit via the pointer registers (P0–P5); 16-bit and 32-bit accesses via index registers (I0–I3); and 32-bit via the stack and frame pointer registers.

EXAMPLE 5.13

This example uses simple instructions to access data with different addressing modes. We use the register values and memory given in Figure 5.12 to illustrate the data movement.

1. Indirect addressing

    Square brackets "[ ]" within the instruction denote the use of index pointer and stack/frame registers as address pointers in data load/store operations. For example, the instruction R0 = [P0] implies that the pointer register, P0, is pointing to the address 0xFF80 0000, which contains a data 0x78; 0xFF80 0001 contains a data 0x56; 0xFF80 0002 contains a data 0x34; and 0xFF80 0003 contains a data 0x12. Note that "0x" denotes the number in a hexadecimal format. These data are loaded into R0 as 0x1234 5678 in a little-endian byte-ordering manner (i.e., lower-address byte loaded into lower bits of the register, and vice versa).

    In another example, [P1] = R0 specifies a store operation that stores the value in the data register to the pointed memory. Continuing from the preceding example, R0 contains 0x1234 5678 is stored into the memory pointed to by the P1 register (at address 0xFF80 1000). In other words, the memory 0xFF80 1000 now contains 0x78, 0xFF80 1001 contains 0x56, 0xFF80 1002 contains 0x34, and 0xFF80 1003 contains 0x12.

2. Indirect addressing that supports 16-bit and 8-bit data access

    The above example performed 32-bit data access. A similar example using 16-bit load operation can be specified as R0 = W[P0](z), where W represents word access and (z)

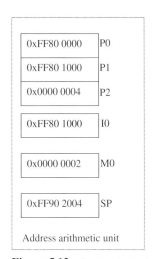
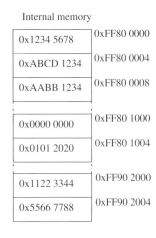

Figure 5.12 Current data values in memories and registers

is the option that states that the high bits must be zero filled. Here, R0 becomes 0x0000 5678. Similarly, an 8-bit load can be specified as R0 = B[P0](x), where B represents byte access and the (x) option specifies sign extension. Therefore, R0 becomes 0x0000 0078 because the most significant bit of 0x78 is 0 (a positive number).

Instead of using the pointer registers (P0–P5) as the address pointers, we can also use the index registers (I0–I3). However, the index registers only support 16-bit and 32-bit data access.

3. Post-modify operations

The address pointers (P0–P5) and index registers (I0–I3) can be modified after the load/store operations. This postmodification is useful in updating the pointers automatically to the next memory location for loading or storing the next data. Postmodification can be either ++ or –, which means postincrement or postdecrement, respectively. For example, R0 = [P0++] increments the value of P0 by 4 after the load operation. This means R0 = 0x1234 5678 and P0 = 0xFF80 0004 after the load operation. However, if R0 = W[P0++](z), R0 = 0x0000 5678 and P0 = 0xFF80 0002. The increment is now by 2 locations because it is a word load. Similarly, if R0 = B[P0++](z), P0 increases to 0xFF80 0001.

In a similar manner, a 32-bit store operation with postdecrement can be specified as [P1-] = R0. In this case, P1 is decremented to 0xFF80 0FFC after the store operation.

4. Pre-modify operations

The Blackfin processor also supports pre-modify instructions, but only at pointing to the stack pointer. For example, in the instruction [-SP] = R0, the stack pointer, SP, is first decremented from 0xFF90 2004 to 0xFF90 2000 before performing the store operation from R0 into the stack memory. If R0 = 0x1234 5678, this value is stored into the stack, starting from address 0xFF90 2000. It is important to note that the stack operates in 32-bit mode, and the push instruction is always predecremented before loading the register to the stack. In contrast, the pop instruction loads the content of the stack into a specified register and performs postincrement of the stack pointer, for example, R0 = [SP++].

5. Modify pointer registers

In some situations, we need to modify the pointer by more than one word increment/decrement. We can simply modify the pointer registers by an immediate value. For example, the instruction R1 = [P0 + 0x08] loads the value at location P0 + 0x08 = 0xFF80 0008 into the R1 register (i.e., R1 = 0xAABB 1234 after the load operation). Note that this is a pre-modify operation without updating to the pointer register. In this case, the P0 register still remains at 0xFF80 0000 after the operation.

If a postmodification of the pointer register is desired, the pointer register can be modified by another pointer register (e.g., R0 = [P0++P1], where P1 contains the offset value). Similarly, the index registers can be modified by using the M register as modifier. Note that both modify-increment and modify-decrement are supported in this addressing mode. For example, the instruction R1 = [I0 ++ M0] loads the value specified by I0 = 0xFF80 0000 (which is 0x0000 0000) into R1. The I0 register is then postincremented to 0xFF80 0002 because M0 = 0x2.

In addition to the above linear addressing mode, the Blackfin processor also offers circular addressing for accessing data in circular buffers. Circular buffering

5.1 An Architecture for Embedded Media Processing   177

is very useful for signal processing tasks such as FIR filtering. The application of circular buffering in DSP tasks is explained in Chapter 8. The circular buffer contains data that the address generator steps through repeatedly and wraps around when the pointer reaches the end of the circular buffer. The addressing of the circular buffer is governed by the length (L), base (B), modify (M), and index (I) registers. The L register sets the size of the circular buffer, and its value is always positive, with a maximum length of $2^{32} - 1$. When L = 0, the circular buffer is disabled. The B register specifies the base address of the circular buffer. The I register points to the address within the circular buffer, and it is controlled within B + L by the DAG. The index register is postmodified by the value specified in the M register after every access to the circular buffer. The value in the M register can be a positive or negative value, but its absolute value must be less than or equal to the length of the circular buffer, L.

EXAMPLE 5.14

Figure 5.13 shows that the memory range 0xFF80 0000–0xFF80 002B is set up as a circular buffer. In this example, 32-bit access is implemented. Therefore, the base address of the circular buffer must be 32-bit aligned (or the least significant two bits of starting address must always be "0"). The base register is chosen as B0 = 0xFF80 0000, and the index register is initialized to the base address as I0 = 0xFF80 0000. The length of the circular buffer is L0 = 44 (or 0x2C) bytes, and the modifier M0 = 16 (or 0x10) bytes.

With reference to Figure 5.13, the data access of the circular buffer can be explained as follows:

1st Access: The index register, I0, is pointing to the base address of the circular buffer at 0xFF80 0000. After the data 0x0000 0001 is loaded into a data register, I0 is modified to 0xFF80 0010 (i.e., 0xFF80 0000 + 0x10).

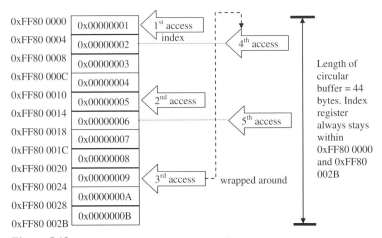

Figure 5.13   Example of a circular buffer with five data accesses

2nd Access: The data 0x0000 0005 is loaded into a data register, and the index is again modified to 0xFF80 0020 (i.e., 0xFF80 0010 + 0x10).

3rd Access: The data 0x0000 0009 is loaded into a data register and the index register, I0 = 0xFF80 0020 + 0x10 = 0xFF80 0030, which is outside the circular buffer range of 0xFF80 002B and is not a valid address. Therefore, I0 should be modified as 0xFF80 0020 + 0x10 − 0x2C = 0xFF80 0004, where the length of the circular buffer (0x2C) is subtracted after the postmodification.

4th Access: The data 0x0000 0002 is loaded into a data register and the index register, I0 = 0xFF80 0004 + 0x10 = 0xFF80 0014.

5th Access: The data 0x0000 0006 is loaded into a data register, and the index is modified to 0xFF80 0024 after the 5th access. This process continues, and the same update formula as in the 3rd access is used whenever the index register crosses the boundary.

QUIZ 5.1

1. The index pointer will return to 0xF800 0000 again at which access?
2. If the modifier (M) register is set at 0xFFFF FFF0, how do we perform data access and wrapping?
3. Change the above circular buffer to a 16-bit data access and determine how data can be 16 bit aligned.

5.1.3.3 Control Unit

The control unit shown in Figure 5.2 consists of the following blocks:

1. A program sequencer controls the instruction execution flow, which includes instruction alignment and instruction decoding. The address generated by the program sequencer is a 32-bit memory instruction address. A 32-bit program counter (PC) is used to indicate the current instruction being fetched.
2. The loop buffer is used to support zero-overhead looping. The Blackfin processor supports two loops with two sets of loop counters (LC0, LC1), loop top (LT0, LT1) and loop bottom (LB0, LB1) registers to handle looping. Hardware counters are used to evaluate the loop condition. Loop unit 1 has a higher priority than loop unit 0. Therefore, loop unit 1 is used for the inner loop and loop unit 0 is used for the outer loop.

The program sequencer controls all program flow, which includes maintaining loops, subroutines, jumps, idles, interrupts, and exceptions. Figure 5.14 illustrates the different program flows.

In the linear flow as shown in Figure 5.14(a), the PC moves from one instruction to the next sequentially. In the loop flow shown in Figure 5.14(b), the instruction within the loop block (i.e., immediately after the loop instruction and the end of the loop) is repeated N times. Zero-overhead loop registers are used to control the PC,

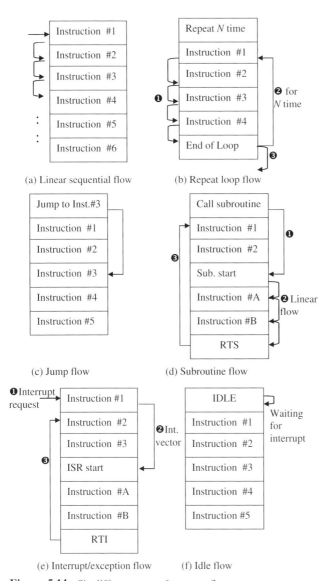

Figure 5.14 Six different types of program flow

and the PC jumps out of the loop once N repeats have been completed. In Figure 5.14(c), an unconditional jump instruction alters the program flow and sets the PC to point at another part of the memory. Similarly, a conditional branch instruction can also be used to direct the PC to another program section.

Figure 5.14(d) shows the program flow of the subroutine call. The CALL instruction temporarily interrupts sequential flow to execute instructions from a subroutine. Once the subroutine has completed, a return from subroutine (RTS) instruction resets the PC back to the instruction immediately after the CALL instruction. The return address is found in the RETS register, which is automatically loaded by the CALL instruction.

Interrupt can occur when a run time event (asynchronous to program flow) or an instruction that triggers an exceptional error (synchronous to program flow) occurs. The processor completes the current instruction and sets the PC to execute the interrupt service routine (ISR). Once the ISR has completed, a return from interrupt instruction RTI obtains the return address from the RETI register and returns to the instruction immediately after interrupted instruction, as shown in Figure 5.14(e).

Finally, in the idle flow shown in Figure 5.14(f), the IDLE instruction causes the processor to stop operating and hold its current state until an interrupt occurs. Subsequently, the processor services the interrupt and resumes normal operation. This idle program flow is frequently used in waiting for the incoming data sample and processing the data sample in the ISR.

A common feature of most DSP processors is the pipeline architecture. The pipeline is extensively used to maximize the distribution of workload among the processor's functional units, which results in efficient parallel processing among the processor's hardware. The Blackfin processor has a 10-stage instruction pipeline as shown in Table 5.3.

The sequencer ensures that the pipeline is fully interlocked, and this feature eases the task of the programmer in managing the pipeline. Figure 5.15 is a diagram showing the pipeline.

Figure 5.15 shows that the 1st instruction at instruction clock cycle #1 is in the IF1 stage. At the next clock cycle, the 2nd instruction is in the IF1 stage and, at the same time, the 1st instruction is in the IF2 stage. This process of overlapping different stages of instructions allows different functional units in the Blackfin processor to work simultaneously in the same clock cycle. As shown in Figure 5.15, the

Table 5.3 Stages of Instruction Pipeline

Pipeline Stage	Description
Instruction fetch 1 (IF1)	Start instruction memory access.
Instruction fetch 2 (IF2)	Intermediate memory pipeline.
Instruction fetch 3 (IF3)	Finish L1 instruction memory access.
Instruction decode (DEC)	Align instruction, start instruction decode, and access pointer register file.
Execute 1 (EX1)	Start access of data memory (program sequencer).
Execute 2 (EX2)	Register file read (data registers).
Execute 3 (EX3)	Finish access of data memory and start execution of dual-cycle instructions (multiplier and video unit).
Execute 4 (EX4)	Execute single-cycle instruction (ALU, shifter, accumulator).
Write back (WB)	Write states to data and pointer register files and process event.

Extracted from Blackfin Processor Hardware Reference [23].

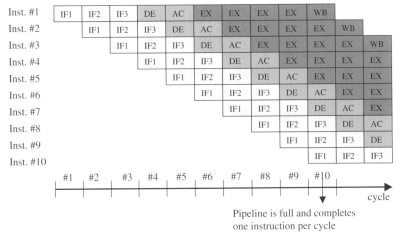

Figure 5.15 Ten pipeline stages of the Blackfin processor

10 pipeline stages are filled up by different instructions at the 10th clock cycle. This implies that each instruction can be executed in a single clock cycle when the pipeline is full, giving a throughput of one instruction per clock cycle. However, any nonsequential program flow (depicted in Fig. 5.14) can potentially decrease the processor's throughput. For example, a stall condition can occur when two instructions require extra cycles to complete because they are close to each other in the assembly program. In another example, a branch instruction causes the instruction after the branch to be invalid in the pipeline and these instructions must be terminated.

Another important feature of the Blackfin processor is zero-overhead looping. A program is given in Example 5.15 to explain the key features and characteristics of the hardware loop.

EXAMPLE 5.15

A simple assembly program to illustrate the setup of loop in the Blackfin processor is listed as follows:

```
P5 = 0x20;
LSETUP (loop_start, loop_end) LC0 = P5;
loop_start:
R2 = R0 + R1 || R3 = [P1++] || R4 = [I1++];
loop_end: R2 = R3 + R4;
```

In this example, the LSETUP instruction is used to load the three loop registers LC0, LB0, and LT0. The loop top address register takes on the address of loop_start, and the loop bottom address register takes on the address of loop_end. The starting address of the loop must be less than 30 bytes away from the LSETUP instruction. In the above program, loop_start follows immediately after the LSETUP instruction with zero overhead. The bottom address of the loop must be between 0 and 2,046 bytes from the LSETUP instruction. In other words, the instructions within the loop are at most 2,046 bytes long. In this example, the loop count register LC0 is set to 0x20 (loop for 32 times). The maximum count for the loop counter register is $2^{32} - 1$, and the minimum count is 2.

In addition, the Blackfin processor supports a four-location instruction loop buffer (similar to cache) that reduces instruction fetches during looping. Therefore, if the loop code is four or fewer instructions, no fetch from the instruction memory is necessary. However, if more than four instructions are present in the loop, only the first four instructions are stored in the buffer and the rest of the instructions must be fetched from instruction memory.

5.1.4 Bus Architecture and Memory

The Blackfin processor uses a modified Harvard architecture, which allows multiple memory accesses per clock cycle. However, the Blackfin processor has a single memory map that is shared between data and instruction memory. Instead of using a single large memory for supporting this single memory map, the Blackfin processor supports a hierarchical memory model as shown in Figure 5.16. The L1 data and instruction memory are located on the chip and are generally smaller in size but faster than the L2 external memory, which has a larger capacity. Therefore, transfer data from memory to registers in the Blackfin processor is arranged in a hierarchy from the slowest (L2 memory) to the fastest (L1 memory). The rationale behind hierarchy memory is based on three principles: (1) the principle of making the common case fast, where code and data that need to be accessed frequently are stored in the fastest memory; (2) the principle of locality, where the program tends to reuse instructions and data that have been used recently; and (3) the principle of smaller is faster, where smaller memory speeds up the access time.

For example, the memory map for the 4G (or 2^{32})-byte address space of the Blackfin BF533 processor is shown in Figure 5.17. There are 80K bytes of instruction memory from address 0xFFA0 0000 to 0xFFA1 3FFF and 64K bytes of data memory from address 0xFF80 0000 to 0xFF80 8000 (data bank A) and 0xFF90 0000 to 0xFF90 8000 (data bank B). In addition, there are 4K bytes of scratchpad memory for general data storage, such as the stack. Therefore, a total of 148K bytes of internal memory are available in the BF533 processor. These internal memories are all classified as internal L1 memory, and some of these memories also have the option of configuring as static random access memory (SRAM) or cache as shown in Figure 5.17. SRAM provides deterministic access time and very fast throughput;

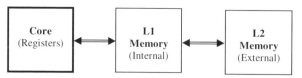

Figure 5.16 Hierarchical memory model

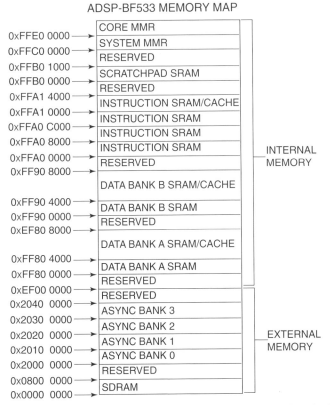

Figure 5.17 Memory map of the BF533 processor (courtesy of Analog Devices, Inc.)

thus it is suitable for DSP-based code. In contrast, cache provides both high performance and a simple programming model and is suitable for control-related tasks. The BF537 processor has the same size of internal data memory as the BF533, but only has 64K bytes of internal instruction memory.

The BF533 processor has 1K bytes of on-chip boot read-only memory (ROM) (BF537 has 2K bytes). The boot ROM includes a small boot kernel that can be either bypassed or used to load user's code from external memory devices (like flash memory or EEPROM) at address 0x2000 0000. The boot kernel completes the boot process and jumps to the start of the L1 instruction memory to begin execution of code from this address.

Figure 5.18 shows the memory architecture of the Blackfin processor. It uses four buses to link the L1 memory with the core processor: one 64-bit instruction bus, two 32-bit data-load buses, and one 32-bit data-store bus. Therefore, the Blackfin processor is capable of performing two data loads, or one data load and one data store, per cycle. It is important to note that the L1 memory is operating at the core clock frequency (CCLK).

184 Chapter 5 Introduction to the Blackfin Processor

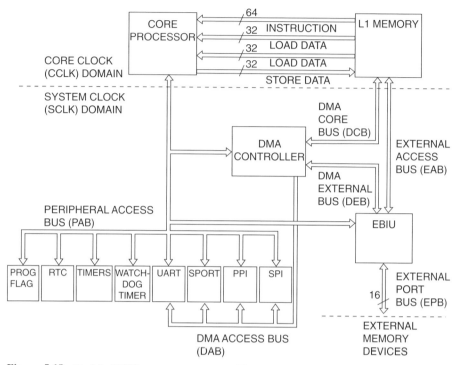

Figure 5.18 Blackfin BF533 processor memory architecture (courtesy of Analog Devices, Inc.)

The larger L2 memory of the BF533 processor can be located at address 0x0000 0000–0x0800 0000 (128M bytes) and 0x2000 0000–0x2040 0000 (four asynchronous banks with 1M bytes each). The former address is dedicated to the synchronous dynamic random access memory (SDRAM), and the latter address is dedicated to support asynchronous memories such as flash memory, ROM, erasable programmable ROM (EPROM), and memory-mapped I/O devices. In the BF537 processor, the addressable SDRAM memory has been increased to 512M bytes while maintaining the same addressable asynchronous memory as the BF533 processor. These off-chip L2 memories are used to hold large program and data. To access the slower external L2 memory, an external bus interface unit (EBIU) links the L1 memory with a wide variety of external memory devices. However, the EBIU is clocked by the slower system clock (SCLK), and the external data bus and address bus are limited to 16-bit and 20-bit width, respectively. In addition, the EBIU can work with the DMA controller to transfer data in and out of the processor's memory without the intervention of the processor core.

In the following sections, we take a closer look at the L1 instruction memory and the L1 data memory. We also introduce some of the features and characteristics of the Blackfin processor in the cache mode.

5.1.4.1 L1 Instruction Memory

The L1 instruction memory bank is illustrated in Figure 5.19. For the BF533 processor, there are three banks: (1) 32K bytes of SRAM in bank A, (2) 32K bytes (16K bytes for the BF537) of SRAM in bank B, and (3) 16K bytes of SRAM or cache in bank C. When bank C is configured as SRAM, the instruction memory is implemented as four single-ported subbanks with 4K bytes each. Simultaneous accesses to different banks can be carried out to speed up memory transfer. The processor core reads the instruction memory through the 64-bit-wide instruction bus.

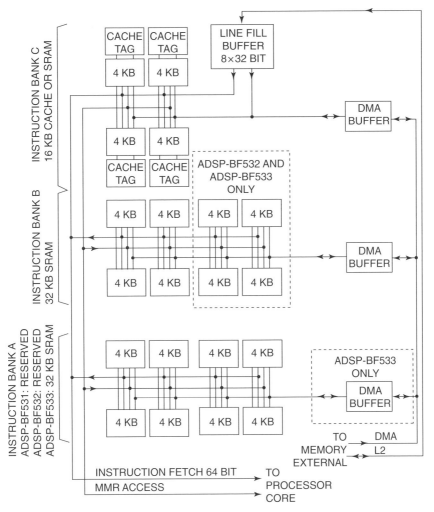

Figure 5.19 L1 instruction memory bank architecture (courtesy of Analog Devices, Inc.)

When the 16K bytes in bank C are configured as cache, a line-fill buffer (8×32-bit) transfers four 64-bit word bursts from external memory to cache when a cache miss occurs. The instruction remains in the cache whenever there is a cache hit. A cacheability and protection look-aside buffer (CPLB) provides control and protection to the cache during instruction memory access. However, DMA is not allowed to access bank C. More explanation on instruction cache configuration and usage is given in Chapter 7.

5.1.4.2 L1 Data Memory

As explained in Section 5.1.4, the BF533 contains 64 K bytes of L1 data memory. The L1 data memory is further divided into 8 subbanks as summarized in Table 5.4.

In the BF533 processor, the lower 16 K bytes for data bank A (0xFF80 0000–0xFF80 3FFF) and B (0xFF90 0000–0xFF90 3FFF) memories are always implemented as SRAM. In these SRAM memories, the L1 data memories can be accessed simultaneously with dual 32-bit DAGs and DMA as shown in Figure 5.20. When the data cache is enabled, either 16 K bytes of data bank A or 16 K bytes of both data banks A and B can be implemented as cache. Each bank (A and B) is a two-way set associative cache that can be independently mapped into the BF533 address space. However, no DMA access is allowed in cache mode. Similar to the instruction cache, the processor provides victim buffers and line-fill buffers for use when a cache load miss occurs. Again, more explanation of data cache configuration and usage is given in Chapter 7.

Table 5.4 BF533 L1 Data Memory

Subbank	Data bank A	Data bank B	Configured as
1	0xFF80 0000–0xFF80 0FFF	0xFF90 0000–0xFF90 0FFF	SRAM
2	0xFF80 1000–0xFF80 1FFF	0xFF90 1000–0xFF90 1FFF	
3	0xFF80 2000–0xFF80 2FFF	0xFF90 2000–0xFF90 2FFF	
4	0xFF80 3000–0xFF80 3FFF	0xFF90 3000–0xFF90 3FFF	
5	0xFF80 4000–0xFF80 4FFF	0xFF90 4000–0xFF90 4FFF	SRAM or cache options: (1) Both banks A and B as SRAM (2) Bank A as cache, bank B as SRAM (3) Both as cache
6	0xFF80 5000–0xFF80 5FFF	0xFF90 5000–0xFF90 5FFF	
7	0xFF80 6000–0xFF80 6FFF	0xFF90 6000–0xFF90 6FFF	
8	0xFF80 7000–0xFF80 7FFF	0xFF90 7000–0xFF90 7FFF	

5.1 An Architecture for Embedded Media Processing

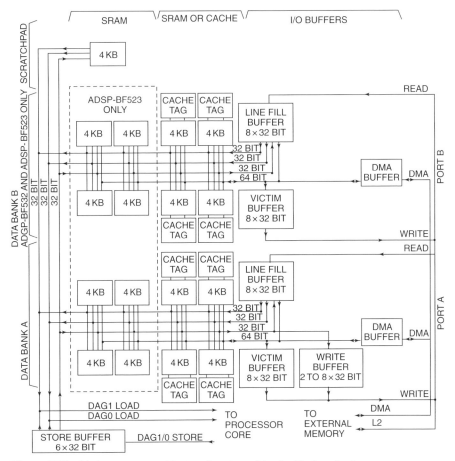

Figure 5.20 L1 data memory architecture (courtesy of Analog Devices, Inc.)

Note that in Figure 5.20 a dedicated 4K bytes of L1 scratchpad memory is available. However, this scratchpad memory cannot be configured as cache and accessed by DMA. It is typically used as stack for fast context switching during interrupt handling.

5.1.5 Basic Peripherals

As shown in Figure 5.18, the Blackfin BF533 processor has the following peripherals:

1. One watchdog timer is clocked by the system clock (SCLK). It generates an event when the timer expires before being updated by software.
2. One real-time clock provides a digital watch to the processor. It provides stopwatch countdown and alarm and maintains time of day.

3. Three general-purpose timers are configured as pulse-width modulation, width and period capture, and external event counter. These timers generate periodic waveform, pulse-width modulation waveform, etc. The BF537 processor has eight pulse-width modulation timers.

4. 16 Bidirectional general-purpose programmable flags (PF0–PF15). Each flag pin can be configured as an input, output, or interrupt pin. In the BF537 processor the programmable flags are named general-purpose inputs/outputs (GPIOs), and there are 48 GPIO pins.

5. One universal asynchronous receiver/transmitter (UART) has a maximum baud rate of up to SCLK/16. It interfaces with slow serial peripherals and serves as a maintenance port. The BF537 processor has two UART modules.

6. Two synchronous serial ports (SPORT0 and SPORT1) provide high-speed serial communication with a maximum speed of SCLK/2. This provides efficient interface with CODEC (coder-decoder).

7. One SPI provides high-speed serial communication of up to SCLK/4. It interfaces with another processor, data converters, and display.

8. One PPI provides a programmable parallel bus with a maximum communication rate of SCLK/2. It is used for high-speed data converters and video CODECs.

9. EBIU provides a glueless interface with external memories. Three internal 16-bit buses are connected to the EBIU: (a) the external access bus (EAB) is controlled by the core memory to access external memory; (b) the peripheral access bus (PAB) is used to access EBIU memory-mapped registers; and (c) the DMA external bus is controlled by the DMA controller to access external memory.

10. The DMA controller allows data transfer operations without processor intervention. There are three DMA buses: (a) the DMA access bus (DAB) allows peripherals to access the DMA channels; (b) the DMA external bus (DEB) links off-chip memory to the DMA channels; and (c) the DMA core bus (DCB) allows the DMA channels to gain access to the on-chip L1 memory.

These peripherals are connected to the system via PAB, DAB, DCB, DEB, and EAB. These buses and peripherals are operating at the SCLK, and the core processor and L1 memory run at the CCLK. The peripheral access bus accesses all peripheral resources that are mapped to the system memory-mapped register (MMR) space (in the memory map shown in Fig. 5.17). The three DMA buses (DAB, DCB, and DEB) provide access to on-chip and off-chip memory with little or no intervention from the core processor. There are six DMA-capable peripherals (PPI, SPORT0, SPORT1, SPI, UART, and memory) in the Blackfin BF533 processor, and 12 DMA channels and bus master support these devices. In the BF537 processor, there are four more DMA channels to support additional peripherals. There are 16 DMA channels in the BF537, of which 12 DMA channels support the peripherals (PPI, SPORT0, SPORT1, SPI, UART0, UART1, and Ethernet media access control) and four DMA channels support memory and handshaking memory. The DAB supports transfer of 16- or 32-bit data in and out of the L1 memory. If there is a conflict with the core access to the same memory, the DMA will always gain access because it has a higher priority over the core. The DEB and EAB support single word accesses of either 8-bit or 16-bit data types. A detailed examination of the DMA and its configuration is presented in Chapter 7.

Besides these system peripherals, the BF537 processor [24] also has the controller area network (CAN) 2.0B module, an I^2C-compatible two-wire interface (TWI) port, and a 10/100 Mbps Ethernet media access controller (MAC).

5.2 SOFTWARE TOOLS FOR THE BLACKFIN PROCESSOR

This section studies topics of programming the Blackfin processor. We have introduced the steps of loading the project file into the VisualDSP++ and performed some debugging and benchmarking of C programs. This section further examines the software development flow and tools. We illustrate the Blackfin data arithmetic and addressing arithmetic units with low-level programming and debugging.

5.2.1 Software Development Flow and Tools

C and assembly programs are the most commonly used in programming today's embedded signal processors including the Blackfin processor. With the advancement of the C compiler for embedded processors, it becomes more common and equally efficient to program in C instead of using assembly code. In addition, Analog Devices provides many optimized DSP functions that are available in the DSP run time library to ease the programming of DSP algorithms. However, assembly code is still useful for programmers who wish to optimize programs in terms of speed, memory resources, and power consumption.

A typical software development flow for code generation is illustrated in Figure 5.21. Users can use C/C++ and/or assembly source codes to program their applications. If the code is written in C, it needs to compile the C code to generate assembly code first, and passes this compiler-generated assembly file (.s) to the assembler. The assembly code can be fed directly to the assembler to generate the object (.doj) file. The linker maps the object file to an executable (.dxe) file using the information from the linker description file (.ldf). The .ldf file contains information on the memory mapping to the physical memory. More information on the linker and the .ldf file is provided in Section 5.2.3. The final .dxe file (which contains application code and debugging information) can be loaded into the VisualDSP++ simulator, the EZ-KIT, or other target boards to verify the correctness of the program. If not, debugging of the source code must be carried out and the process of compile-assemble-link-load is repeated. Once the software is verified, system

190 Chapter 5 Introduction to the Blackfin Processor

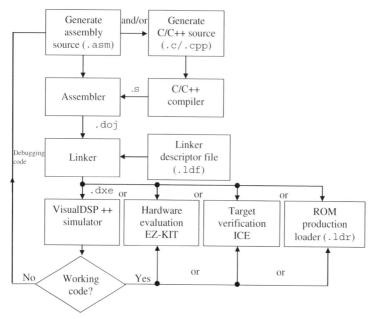

Figure 5.21 Software development flow

verification must be carried out. The programmer has the option of downloading the .dxe file into different hardware systems such as the Blackfin EZ-KIT or a custom Blackfin target board, or to load into an external flash memory. The last option is discussed at the end of this chapter.

VisualDSP++ can perform system verification and profile the performance of the code running on the actual processor. There is a need at this stage to make sure that the code will meet the real-time processing requirements, resource availability, and power consumption demand of the system. Therefore, VisualDSP++ is an integrated development and debugging environment (IDDE) that delivers efficient project management. VisualDSP++ allows programmers to edit, compile and/or assemble, and link the code to generate executable code; perform simulation of the Blackfin processor; and debug the code to correct errors and exceptions. VisualDSP++ also includes many advanced plotting and profiling capabilities for performing software and system verification. A VisualDSP++ kernel (VDK) allows users to perform task scheduling and resource allocation to address memory and timing constraints of programming. The VDK also includes standard library and framework, which ease the process of writing complex programs. In addition, there are several advanced features unique within the VisualDSP++ IDDE. They include profile-guided optimization (PGO), cache visualization, pipeline viewer, background telemetry channel (BTC) support, multiple processor support, integrated source code control, automation application program interface, and aware scripting engine. We discuss some of these advanced features in subsequent chapters.

5.2.2 Assembly Programming in VisualDSP++

This section uses several hands-on experiments to explain the writing and building of assembly code [28] and executing and debugging of the code in the VisualDSP++ environment.

HANDS-ON EXPERIMENT 5.1

Activate VisualDSP++ using the BF533 (or BF537) simulator, and perform the following steps:

1. Load a project file into VisualDSP++ by clicking on **File → Open → Project** and look for the project file `exp5_1.dpj` in directory `c:\adsp\chap5\exp5_1`.
2. Double-click on the `exp5_1.asm` file to view the content of the assembly code as shown in Figure 5.22. This assembly file is written based on Example 5.13.
3. Build the project by clicking on the **Build Project** icon located on the toolbar (or press **F7**), or click on the **Rebuild All** icon. Alternatively, click on **Project → Build Project** or **Project → Rebuild Project**. Several messages will be displayed in the **Output Window** during the building process. The executable code generated from the build operation will be automatically loaded into the simulator.
4. Note that an arrow on the left-hand side is pointing at the first instruction, `P0.L = buffa`.
5. Open the following register windows to view the register's status:
   a. **Register → Core → DAG Registers** to display all I, M, L, and B registers in the address arithmetic unit.

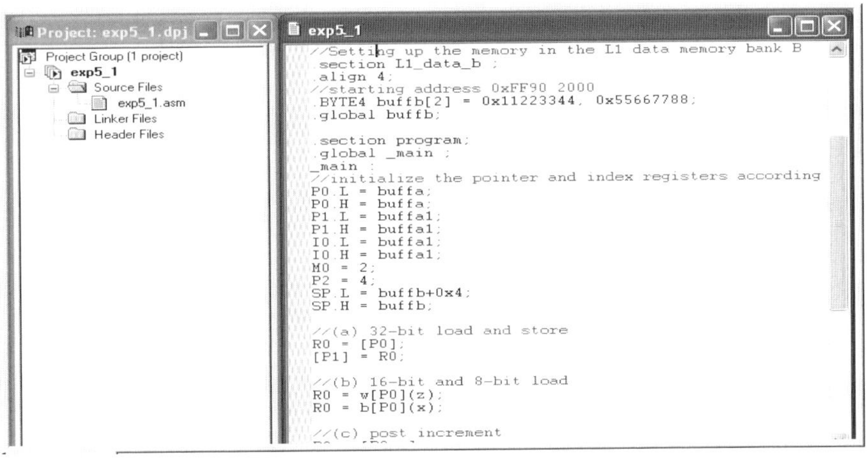

Figure 5.22 Snapshot of the project file `exp5_1.dpj`

192 Chapter 5 Introduction to the Blackfin Processor

b. **Register → Core → Data Register File** to display all the R registers in the data arithmetic unit.
c. **Register → Core → P Registers** to display all the P registers in the address arithmetic unit.
d. **Register → Core → Stack/Frame Registers** to display the stack pointer.

Right-click on these register windows and select `Hex32` as the display format. Adjust the window size with the mouse. The display windows can also be undocked from the rightmost column by right-clicking on the window and deselecting **Allow Docking**.

6. Display the memory by clicking on **Memory → Blackfin Memory**. Next, type in `buffa` in the **Blackfin Memory** window to view the data contained in address 0xFF80 0000. Similarly, users can also view memory at `buffa1` and `buffb`. Right-click on the memory window and click on **Select Format → Hex32**.

7. View the changes in registers and memory as we step through the program. Click on the **Step Into** icon on the toolbar to execute the program line by line, or press **F11** to do the same.

8. Compare the results with those obtained in Example 5.13. Comment on any disparity in results. Make the following changes (one at a time) to the existing program and rebuild the project. What happens with the changes?
   a. Change the `buffa1` address to start at 0xFF80 4000.
   b. Perform 8-bit and 16-bit store operations.
   c. Perform 8-bit and 16-bit accesses using the instructions `R1 = b[I0++M0]` and `R1 = w[I0++M0]`, respectively.
   d. Combine a P-register with an M-register as in the instruction `R0 = [P0+M0]`.

9. Complete this hands-on experiment by closing the project. Click on **File → Close → Project** `<filename>`. VisualDSP++ will prompt the user to close the project and save files.

HANDS-ON EXPERIMENT 5.2

This hands-on experiment implements circular buffers using assembly code in the VisualDSP++ BF533 (or BF537) simulator. The project file `exp5_2.dpj` for this experiment is located in directory `c:\adsp\chap5\exp5_2`. This experiment is based on Example 5.14 and can be loaded into the simulator by using the steps described in Hands-On Experiment 5.1. Build the project and perform the following steps:

1. Set a breakpoint (indicated by a red dot in the left column of source file window) by clicking on the **Toggle Breakpoint** icon, or press **Control + T** at the instruction line, `here: jump here;`. A red dot will appear at this instruction. Run the program from the first instruction to this breakpoint by clicking on the **Run** icon, or press **F5**. Observe the changes in the **Data Register File** window. The breakpoint can be cleared by clicking on the **Clear All Breakpoints** icon .

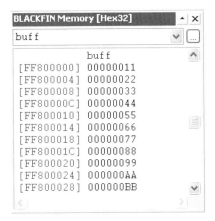

Figure 5.23 A newly edited data for memory buff

2. Click on the **Reload** icon to reload the existing program again. This will bring the PC to the beginning of the program. In this experiment, we change the content of the Blackfin memory by first selecting the memory buff, right-clicking on the specific memory data, and clicking on **Edit** to change the value for that memory location. Try changing the data in buff as shown in Figure 5.23. Rerun the program by clicking on the **Run** icon and observe the changed results.

3. Extend the program to perform 12 accesses of the circular buffer. Verify that the data are correct.

4. Change the data in buff from 4 bytes to 2 bytes, and modify the program for a 16-bit word access. Save the new file as exp5_2a.asm in the new directory c:\adsp\chap5\exp5_2a. Close the current project and create a new project by clicking on **File → New → Project**. Type in the new directory and specify the project name as exp5_2a.dpj. Add in the newly created assembly file exp5_2a.asm to the project by selecting **Project → Add to Project → File** and selecting exp5_2a.asm. Build the project and run the loaded executable file to verify the functionality of the new file.

HANDS-ON EXPERIMENT 5.3

This experiment investigates the shift (arithmetic and logic) and rotation operations. The arithmetic shift considers the most significant bit as the sign bit, and the logical shift treats the data to be shifted as an unsigned number. Open the project file exp5_3.dpj located in directory c:\adsp\chap5\exp5_3. Build the project and single-step through the code. There are four sections in the exp5_3.asm program:

1. Right shift by 4 bits.

    In this section, the R0 register is loaded with a value 0xFFFF A3C6. A logical right shift and an arithmetic right shift by 4 bits are written as R1 = R0>>0x04 and

R2 = R0>>>0x04, respectively. Note the difference in the right shift symbols for logical and arithmetic right shifts. The results are R1 = 0x0FFF FA3C and R2 = 0xFFFF FA3C. This shows that the arithmetic shift preserves the sign bit of the data and the logical shift simply fills in zeros in the most significant bits.

2. Left shift by 4 bits.

In this section, a value of 0x7FFF A3C6 is loaded into the R0 register. The logical and arithmetic left shifts by 4 bits are given as R1 = R0<<0x04 and R2 = R0<<0x04(s), respectively. Note that the symbol "<<" is applicable to both types of left shift. However, the arithmetic shift always ends with the saturation option(s). In this case, R1 = 0xFFFA 3C60 and R2 = 0x7FFF FFFF. The results show that the logical left shift treats the number as an unsigned number, but the arithmetic shift saturates the result to the maximum positive (or negative) values. A detailed explanation of the signed number format is given in Chapter 6.

3. Another shifting instruction.

Besides using a constant shift magnitude, another type of shifting instructions uses the lower 16 bits of the register to contain the shift magnitude. In this example, we use R1.L and R2.L to contain the shift magnitude. A logical right shift by 4 bits is stated as R5 = LSHIFT R0 by R2.L, and an arithmetic right shift by 4 bits is stated as R4 = ASHIFT R0 by R2.L. In this case, the R2.L register contains the 2's complement binary pattern (−4). Examine and verify the results of both shifts. A negative (or right) and a positive (or left) shift magnitude correspond to multiplications by $2^{(-shift)}$ and $2^{(shift)}$, respectively. The ASHIFT instruction can shift a 32-bit data register or 40 bit accumulators from −32 to 31 bits.

4. Rotate instructions.

In the final section of this experiment, the rotate instructions are written to perform bit rotation of the binary word. For example, the instruction R1 = ROT R0 by 1 rotates the contents in R0 leftward by one bit, via the CC bit. In other words, the CC bit is in the rotate chain. Therefore, the first bit rotated into the register is the initial value of the CC bit. Subsequently, the CC bit is replaced by the bit rotated out of the register. Similar to the shift instruction, a positive rotate magnitude specifies a left rotate, and a negative rotate magnitude specifies a right rotate. Single-step through the instructions in this section and verify the results. The CC bit can be viewed by clicking on **Register → Core → Status → Arithmetic Status**. It is also easy to see the rotate results by using the **binary** display format.

EXERCISE 5.1

In this hands-on exercise, some arithmetic operations are programmed in the VisualDSP++ environment. We examine the arithmetic instructions by revisiting the examples given in Examples 5.1–5.6. A simple template project file exercise5_1.dpj located in directory c:\adsp\chap5\exercise5_1 can be modified to implement the following simple arithmetic instructions:

    a. Single 16-bit add
    b. Dual 16-bit add

c. Quad 16-bit add
   d. Single 32-bit add
   e. Dual 32-bit add
   f. 16-bit and 32-bit add with options
   g. 32-bit logical operations

   Verify the results obtained from VisualDSP++ with a calculator.

EXERCISE 5.2

In this hands-on exercise, we revisit the multiply and multiply-accumulate operations given in Examples 5.7–5.12 using VisualDSP++. To simplify the explanation, only a signed integer number is considered here. Therefore, the (IS) option must be inserted at the end of all multiply and MAC instructions in this exercise. A simple template project file exercise5_2.dpj located in directory c:\adsp\chap5\exercise5_2 can be used to implement the following simple arithmetic instructions:

   a. Single 16-bit multiply operation
   b. Single 16-bit multiply/accumulate operation
   c. Dual 16-bit multiply operations using two accumulators
   d. Dual 16-bit multiply operations using two 32-bit registers
   e. Dual 16-bit multiply/accumulate operations with two 16-bit destinations
   f. Dual 16-bit multiply/accumulate operations with two 32-bit destinations

Verify the results obtained from VisualDSP++ with a calculator. What happens to the multiplication results when the (IS) option is removed from the instruction? Explain the difference in the multiplication results. When loading a 32-bit integer to a 16-bit register, which part of the 32-bit number is loaded (the most or least significant 16 bits)?

5.2.3 More Explanation of Linker

The linker [36] is important in the software build process. As shown in Figure 5.21, the linker generates a complete executable program (.dxe). The linker also resolves all external references, assigns an address to relocatable code and data spaces, and generates optional memory map. Its output can be read by loader, simulator, and debugger. The linker is controlled by the linker description file (.ldf) [39], which describes the target system and provides a complete specification for mapping the linker's input files into the physical memory.

The VisualDSP++ IDDE provides an Expert Linker, which uses the LDF wizard to create the .ldf file. The Expert Linker defines the target memory map and allows object sections to be placed in different memory sections by simple

196   Chapter 5   Introduction to the Blackfin Processor

drag-and-drop placement. The following hands-on experiments will launch the LDF wizard, view the memory map, and map a user-defined section into the memory.

HANDS-ON EXPERIMENT 5.4

This hands-on experiment shows a quick and easy way to create the .ldf file. We use the previous project file, exp5_1.dpj, to illustrate the creation of the .ldf file. In the previous case, there is no specific .ldf file under the Linker Files section, and VisualDSP++ uses the default .ldf file adsp-BF533.ldf for the BF533 processor (or adsp-BF537.ldf for the BF537 processor), which is located in directory C:\Program Files\Analog Devices\VisualDSP 4.0\Blackfin\ldf.

1. Open the project file exp5_1.dpj. Click on **Tools → Expert Linker → Create LDF...**; a **Create LDF** window appears. To continue, click on **Next**. A new window appears as shown in Figure 5.24. Change the .ldf filename and choose the programming language type (**Assembly** in this case). Click on **Next** to proceed. Note that when working with a mixed C and assembly program, **Project type C** is chosen.

    A new window (**Step 2 of 3**) appears. This window allows users to select processor type and other properties. Click on **Next**, and the last window confirms the selected choice; click on the **Finish** button to generate the .ldf file. A memory map view of the generated .ldf file is shown in Figure 5.25. It is also observed that the .ldf file appears under the Linker Files folder in the **Project Window**.

2. Explore the memory map. The **Input Section** pane in the left column shows the input sections defined in the source code. Right-click on the object file exp5_1.doj under

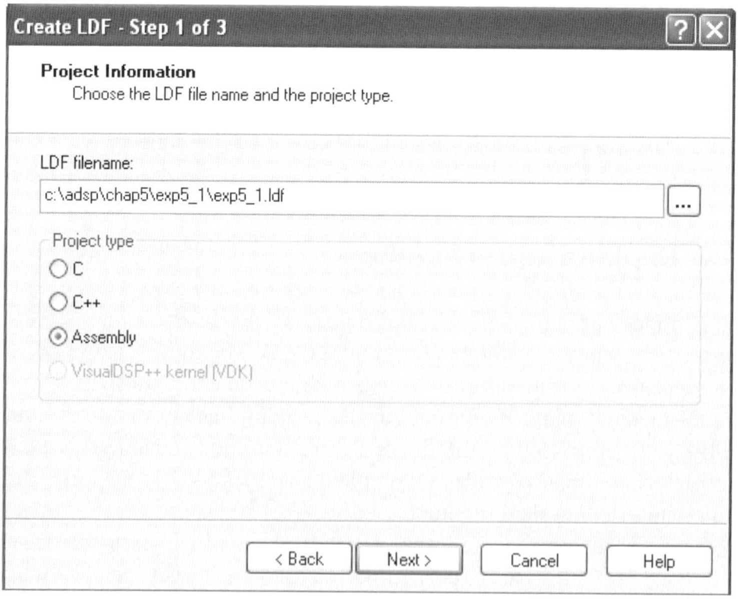

Figure 5.24   LDF wizard window

5.2 Software Tools for the Blackfin Processor   197

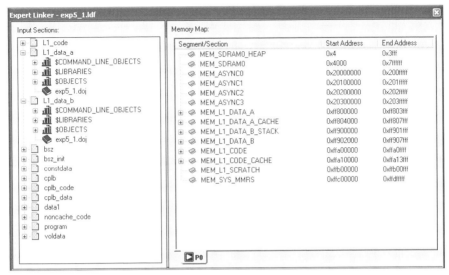

Figure 5.25 Memory map view of the `exp5_1.ldf` file

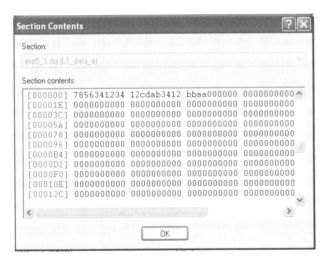

Figure 5.26 **Section Contents** window

the `L1_data_a` section. Click on **View Section Contents** to display the section contents as shown in Figure 5.26. This shows that the data are arranged in a little-endian manner (i.e., lower-byte data in the lower memory location).

The **Memory Map** pane in the right column of the **Expert Linker Window** defines the target system's memory, memory types, and their address range. Right-click on the **Memory Map** pane and select **View Mode** → **Graphical Memory Map** to display a memory map. Zoom in and out of the memory map to have a better view. Close the **Expert Linker Window** when finished.

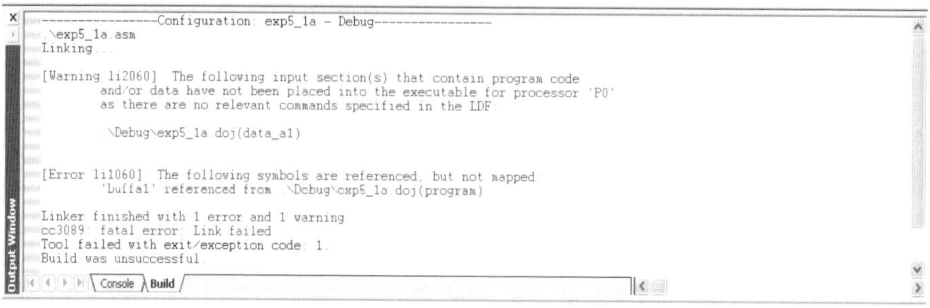

```
----------------Configuration: exp5_1a - Debug----------------
.\exp5_1a.asm
Linking...

[Warning li2060] The following input section(s) that contain program code
    and/or data have not been placed into the executable for processor 'P0'
    as there are no relevant commands specified in the LDF
    
    .\Debug\exp5_1a.doj(data_a1)

[Error li1060] The following symbols are referenced, but not mapped
    'buffa1' referenced from .\Debug\exp5_1a.doj(program)

Linker finished with 1 error and 1 warning
cc3089: fatal error: Link failed
Tool failed with exit/exception code: 1.
Build was unsuccessful.
```

Figure 5.27 Output window with linker error

3. Modify the existing source file (`exp5_1.asm`) by adding a new user-defined section as follows before the data `buffa1`:

   ```
   .section data_a1;              // newly inserted section
   .BYTE4 buffa1[2] = 0x00000000, 0x01012020;
   ```

   In this case, the data `buffa1` is no longer specified under the `L1_data_a` memory. Rebuild the program again. A linker error appears in the **Output Window** as shown in Figure 5.27.

   The error arises because the section "`data_a1`" is not linked to any of the memory sections. Double-click on the `exp5_1.ldf` file. A red cross (X) appears on the `data_a1` icon. Click on the (+) sign of the icon, and it is expanded into the linker macros `$COMMAND_LINE_OBJECTS` and `$OBJECTS`. Map the object `exp5_1.obj` under `$OBJECTS` onto the `MEM_L1_SCRATCH` output section by drag and drop. Alternatively, mapping to another valid output section like `data_L1_data_a` or `data_L1_data_b` is also possible. What is the starting address of `buffa1` now? Single-step through the code and check the changes in the data movement.

4. Rebuild the whole program with the option of creating a `.map` file. Before building the project, click on **Project → Project Options → Link**. Select **Generate Symbol Map** to produce a `.map` file under the **debug** directory. Open the `exp5_1.map` file, and an HTML window is displayed in the web browser. Observe the word size (in bytes) of the data and instructions used in this program.

5.2.4 More Debugging Features

This section introduces additional debugging features. Besides using the register windows to view the register values, users can also customize the display window by clicking on **View → Debug Windows → Expressions**. An **Expressions** window appears, and users can click on it and type in the variable names and register names (register names must be preceded with a "$" sign) as shown in Figure 5.28.

Another feature of the debugger is that it supports a linear profiling window when running the program in simulation mode. This window displays the percentage of execution time or cycle counts needed to run every line of the code. This feature

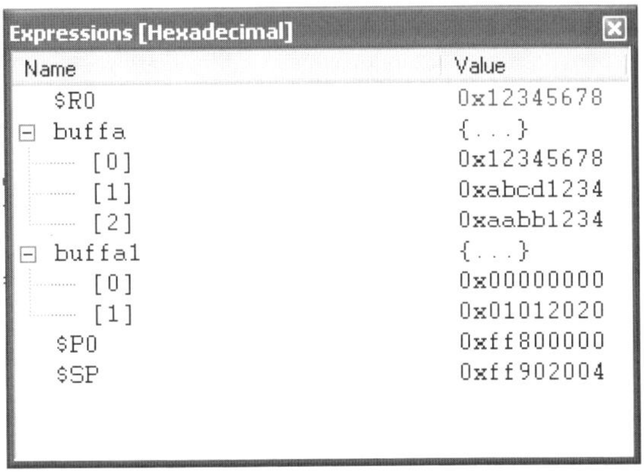

Figure 5.28 Expressions window

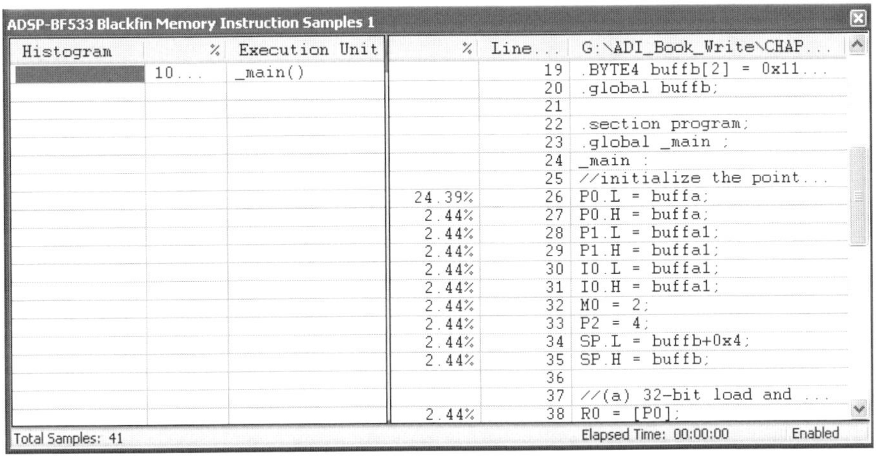

Figure 5.29 Linear Profile window

can be displayed by clicking on **Tools → Linear Profiling → New profile**; a window is displayed as shown in Figure 5.29. Single-step through the code to update the results in the **Linear Profile** window. The number of instruction cycle count can be viewed by right click, followed by **View Sample Count**. This linear profiling feature is a useful and quick tool in determining bottlenecks in the code. However, to compute an accurate cycle count for real-time signal processing, another profiling method is recommended in subsequent chapters.

So far, we have introduced the architecture of the Blackfin processor using the assembly code and the VisualDSP++ simulator. This low-level programming language provides an in-depth appreciation of the use of some of the arithmetic and addressing modes. However, when we program any embedded signal processing task on the Blackfin processor, it is often wise to start from the high-level programming language of C/C++ and check whether its real-time performance can be fulfilled. Otherwise, we can optimize the C program and rewrite some time-critical segments of code in low-level assembly language. The optimization methods are presented in Chapter 8. In the following sections, we examine the design and real-time implementation (in C) of an eight-band graphic equalizer using the Blackfin simulator and the EZ-KIT. The VisualDSP++ IDDE can be configured for both simulation and real-time implementation.

5.3 INTRODUCTION TO THE FIR FILTER-BASED GRAPHIC EQUALIZER

The eight-band FIR-based graphic equalizer is an extension of the digital FIR filter design in Section 2.3.2. We design eight separate FIR filters with eight different frequency specifications instead of designing a single FIR filter. The output of each FIR filter has a user-adjustable gain that is used to amplify or attenuate the specific frequency contents of an audio signal. Thus it can compensate for signal components that are distorted by recording devices, boost some frequency contents of the signal to make it sound better, or remove undesired band-limited noise. The block diagram of a stereo eight-band graphic equalizer is shown in Figure 5.30. The input signals, $x_L(n)$ and $x_R(n)$, may be connected directly to the output in applications that allow the original signal to pass through when all of the gains at the outputs of bandpass filters are set to 0. In other applications, such as the removal of band-limited noise, only bandpass filter outputs with attenuation are added to form the overall outputs, $y_L(n)$ and $y_R(n)$. Usually, the gains attached to the filters are adjusted in tandem for the left and right channels.

In the following experiments, we investigate the design and implementation of the eight-band graphic equalizer that covers the frequency range from 0 to 24 kHz. The sampling rate is 48 kHz. The frequency specifications for the eight-band graphic

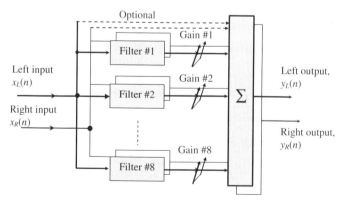

Figure 5.30 Block diagram of a stereo 8-band graphic equalizer

5.3 Introduction to the Fir Filter-Based Graphic Equalizer

Table 5.5 Frequency Specifications for the 8-Band Graphic Equalizer

Band Number	Passband Frequency (Hz)
#1	0–200
#2	200–400
#3	400–800
#4	800–1,600
#5	1,600–3,200
#6	3,200–6,400
#7	6,400–12,800
#8	12,800–24,000

equalizer are summarized in Table 5.5. Basically, Filter #1 is a low-pass filter from 0 to 200 Hz, and Filter #8 is a high-pass filter from 12,800 to 24,000 Hz. The rest of the filters are bandpass filters. The passband and stopband ripples of these filters are specified as 1 and 40 dB, respectively, and the window (Hanning) design method is used to derive the coefficients of these filters.

HANDS-ON EXPERIMENT 5.5

This experiment uses the MATLAB FDATool to design the eight FIR filters. Open FDATool by typing the following command in the MATLAB **Command Window**:

```
fdatool
```

A window is opened as shown in Figure 5.31. Users can key in the parameters for different filter specifications. The filter order of 255 is specified, and the Window (Hann) FIR filter **Design Method** is used for all filters. Note that the default **Filter arithmetic** is set to double precision floating-point (which is 64-bit floating-point format). The filter arithmetic can be viewed by clicking on the **Set quantization parameter** icon. Click on the **Design Filter** tab at the bottom of the window to design the filter. Users can check the magnitude response, phase response, impulse response, pole-zero plot, coefficients, etc. by clicking on the respective icons. A list of icons and their definitions can be found in Hands-On Experiment 3.1. Observe the characteristics of the FIR filters by filling in Table 5.6. The implementation cost depends on the number of multiplications and additions and the memory used to implement the FIR filter.

Once each filter has been designed, it can be saved in a file band-i.fda, where i is the filter number in the graphic equalizer.

The gain of the ith FIR filter, Gain(i), can be set in a range of ±12 dB with steps of 3 dB, as shown in Table 5.7. For example, a gain of 3.9811 applied to the output of any filter corresponds to an amplification of 12 dB at that frequency band.

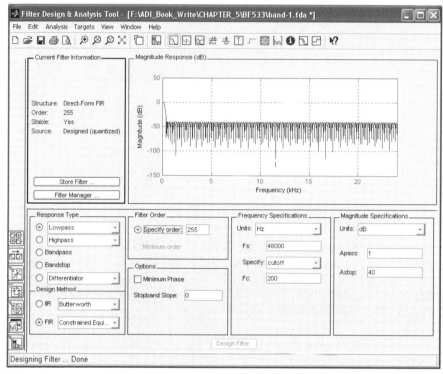

Figure 5.31 FDATool window for designing Filter #1

Table 5.6 FIR Filter Characteristics

FIR Filter Parameters and Responses	Characteristics
Average order of the IIR filter	
Magnitude response	
Phase response	
Group-delay response	
Impulse response	
Pole-zero plot	
Implementation cost	

5.4 DESIGN OF GRAPHIC EQUALIZER USING BLACKFIN SIMULATOR

This section introduces the steps required to complete a fixed-point DSP system design. We use an eight-band FIR based graphic equalizer to illustrate the steps needed to convert the filter design from double-precision floating-point format to 16-bit fixed-point format. This conversion is needed because many fixed-point

Table 5.7 Gain Table Applied to the FIR Filter

Gain Setting in dB ($20\log A$)	Gain Setting in Linear Scale ($A = 10^{(dB/20)}$)
+12 dB	3.9811
+9 dB	2.8184
+6 dB	1.9952
+3 dB	1.4125
+0 dB	1.0000
−3 dB	0.7079
−6 dB	0.5012
−9 dB	0.3548
−12 dB	0.2512
−∞ dB (mute)	0.0000

embedded processors (including Blackfin) are optimized for 16-bit fixed-point arithmetic. The floating-point coefficients must be quantized to 16-bit fraction before porting these coefficients to the fixed-point processors. FDATool can quantize the floating-point coefficients to a 16-bit fixed-point format. A more detailed explanation of the concept behind this conversion is provided in Chapter 6. In addition, FDATool supports a comprehensive simulation on the fixed-point analysis of the digital filter. Therefore, this tool provides a mean to detect and correct any fixed-point implementation problem before actual implementation on the Blackfin processor.

HANDS-ON EXPERIMENT 5.6

We extend the previous FDATool experiment by converting the double-precision floating-point coefficients into fixed-point coefficients. Activate FDATool and open the previously saved session for the individual filter. Click on the **Set quantization parameter** icon again and change the **Filter arithmetic** parameters to those shown in Figure 5.32. The **Numerator word length** is set to 16 bits (for implementation on 16-bit processors), and the **Numerator fractional length** is set to 15 bits to form a word with a single sign bit and 15 fractional bits. This 1.15 format is commonly used in most 16-bit embedded processors as it produces an effective result for fixed-point multiplication. More details on the number range of this format and its arithmetic precision are provided in Chapter 6.

Click on **Apply** to perform the quantization. Examine the magnitude response and enable **View → Show Reference Filter(s)**. The reference filter refers to the filter designed in Hands-On Experiment 5.5 using the 64-bit double-precision floating-point format. We can check whether the filter characteristics (magnitude and phase) with 16-bit fixed-point precision are close to those with the 64-bit floating-point precision. If the quantized filter degrades slightly with 16-bit precision, we can use the 16-bit quantized coefficients for implementation. Unfortunately, we are not able to directly port the fractional coefficients (with 1 sign

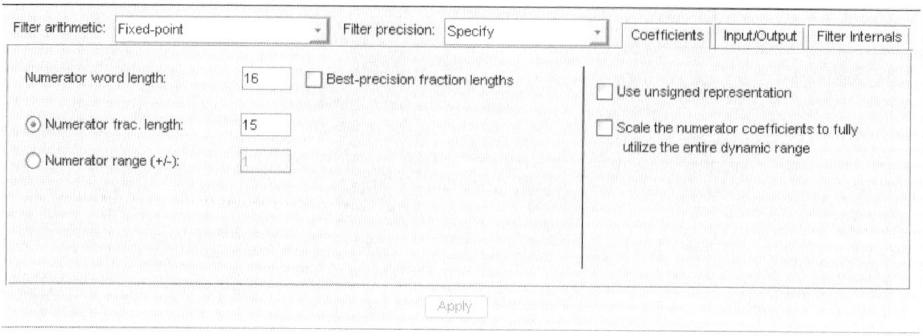

204 Chapter 5 Introduction to the Blackfin Processor

Figure 5.32 Set quantization parameter

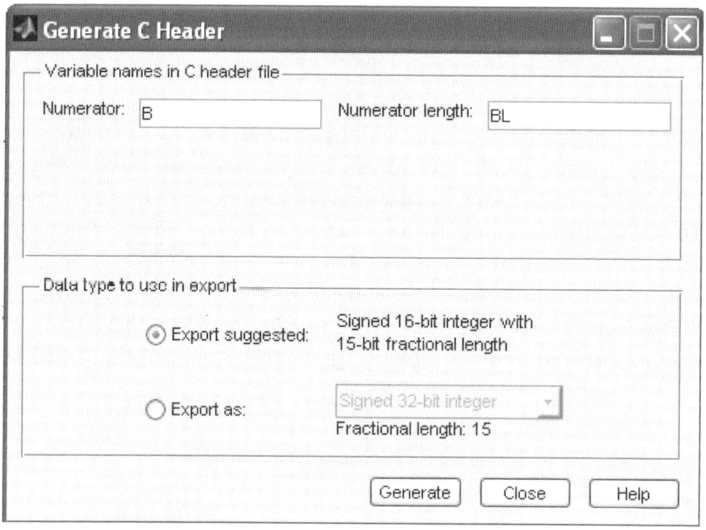

Figure 5.33 **Generate C Header** window

bit and 15 fractional bits) into the C program because there is no data type available to represent fractional number in the Blackfin C compiler. Instead, we perform another conversion from fractional fixed-point number into 16-bit integer (fract16). This conversion can be easily performed by using the following conversion equation:

fixed-point integer (or fract16) = round [fixed-point fractional number × 32,768], (5.4.1)

where the round operation is used to round the result to the nearest integer.

The FDATool automatically performs this conversion when we click on **Targets → Generate C header...**, and a new window shown in Figure 5.33 is generated. Use the suggested settings and click on **Generate**. Save the header file. Because the header file generated from MATLAB contains a specially defined integer constant type that is not defined in the VisualDSP++ C compiler, we cannot use this header file directly. Instead, we simply extract the coefficients from the header file and save them in a data (.dat) file. We will create eight data files from "band0a.dat" to "band7a.dat" for the eight filters, which correspond to bands #1 to #8 of the equalizer. These files are ported into the Blackfin memory in the following experiment.

HANDS-ON EXPERIMENT 5.7

In this experiment, we port the fixed-point coefficients generated from Hands-On Experiment 5.6 into the Blackfin memory. The C program, mainEQ.c in directory c:\adsp\chap5\exp5_7, implements the eight-band graphic equalizer with eight parallel FIR filters. This C file is added to the project file, exp5_7.dpj, and run on the BF533 (or BF537) VisualDSP++ compiled simulator. The filter coefficients for eight frequency bands are included in the C file. A predefined gain table is set in the array variable fract16 band_gain[9] to specify the nine gain settings in the first column of Table 5.8. Conversion from dB scale to linear scale must be carried out for a 16-bit gain value, represented in (1.15) format. Because this arithmetic representation limits the number range to within ±1, the gain value must be scaled down by 4 to limit the gain value to the number range of the (1.15) format. This step is shown in the third column of Table 5.8. The last column of Table 5.8 shows that the gain is converted into integer format (fract16) based on Equation 5.4.1 for use in the C program.

Because we are connecting the eight FIR filters of equal length in parallel for each channel, we can sum these eight parallel FIR filters (including band gain) into one combined FIR filter. This combined FIR filter can greatly reduce the workload of the graphic equalizer. The combined FIR filter can be derived by multiplying each set of filter coefficients with its respective band gain before summing the coefficients of the eight filters. Two C callable functions, mult_fr1x32 and add_fr1x32, are used in the main program, mainEQ.c, to perform multiplication and addition, respectively, of two 16-bit numbers and save the result in a 32-bit variable. The combined coefficients allCoeff [i] are rounded to 16 bits.

Build the project file exp5_7.dpj and open the display window by clicking on View → Debug Windows → Plot → Restore. Select ImpulseResponse.vps and

Table 5.8 Gain Settings of the Graphic Equalizer

Gain Setting in dB ($20\log_{10} A$)	Gain Setting in Linear Scale ($A = 10^{(dB/20)}$)	Gain Setting Scaled to (1.15) Format (A/4)	(1.15) Format in Hexadecimal (fract16) Hex[(A/4)×32,768]
+12 dB	3.9811	0.9953	0x7F66
+9 dB	2.8184	0.7046	0x5A30
+6 dB	1.9952	0.4988	0x3FD9
+3 dB	1.4125	0.3531	0x2D32
+0 dB	1.0000	0.2500	0x2000
−3 dB	0.7079	0.1769	0x16A5
−6 dB	0.5012	0.1253	0x100A
−9 dB	0.3548	0.0887	0x0B5B
−12 dB	0.2512	0.0628	0x080A

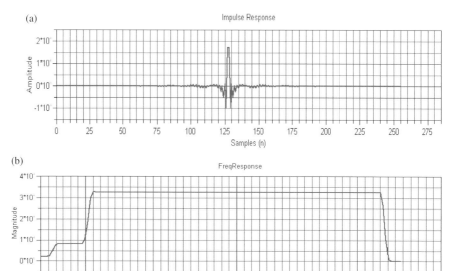

Figure 5.34 Impulse response (a) and frequency response (b) of combined FIR filter

FreqResponse.vps to display the impulse and frequency responses of the combined filter as shown in Figure 5.34 (a) and (b), respectively.

EXERCISE 5.3

1. Modify the C code to enhance the low-frequency bands (band #1 to band #3) to +12 dB, middle-frequency bands (band #4 to band #6) to 0 dB, and high-frequency bands (band #7 to band #8) to −12 dB.

2. Use the equiripple filter design method instead of the window filter design method. Obtain the frequency response and impulse response of the combined filter with the new filter design method. Observe any difference with the plots obtained in Hands-On Experiment 5.7.

3. Modify the C code to input a noisy signal from the file (sineNoise-3sec_48k.dat) and attenuate the noise by −12 dB using the 8-band graphic equalizer. Hint: Refer to Hands-On Experiment 2.3.

5.5 IMPLEMENTATION OF GRAPHIC EQUALIZER USING BF533/BF537 EZ-KIT

This section explores the real-time implementation of the eight-band FIR filter-based graphic equalizer using the Blackfin BF533 or BF537 EZ-KIT. In addition, we explore how to program an application as a stand-alone system without needing the computer to download the code to the EZ-KIT via the USB cable. In other words, the graphic equalizer code can be downloaded to the Blackfin memory on power up. This feature turns the EZ-KIT into a portable device (battery operated) that can be used directly in many real-life applications.

HANDS-ON EXPERIMENT 5.8

In this experiment, we convert the C file in Hands-On Experiment 5.7 into a real-time graphic equalizer that can input musical signal and perform graphic equalization on a pair of stereo signals. The processed signals are sent to headphones (or loudspeakers) for playback.

Activate VisualDSP++ for the BF533 (or BF537) EZ-KIT and open exp5_8_533.dpj (or exp5_8_537.dpj) in directory c:\adsp\chap5\exp5_8_533 (or c:\adsp\chap5\exp5_8_537). Build the project, and the executable file (exp5_8_533.dxe or exp5_8_537.dxe) is automatically loaded into the memories of the EZ-KIT. In this program, we have programmed four push buttons and six LEDs on the BF533 and BF537 EZ-KITs as shown in Table 5.9. Input any musical signal (sampled at 48 kHz) to the input port (ADC1) of the BF533 EZ-KIT or the stereo line-in port of the BF537 EZ-KIT. Run the project and connect the processed signal at the output port (DAC1) of the BF533 EZ-KIT or the stereo line-out port of the BF537 EZ-KIT to a pair of loudspeakers. Users can adjust the gains of the filters with the push button settings listed in Table 5.9 and listen to the equalized music.

The combined left and right FIR filters of the graphic equalizer are implemented by using the following FIR filter functions (available in the DSP run time library) to process the left (sCh0LeftIn) and right (sCh0RightIn) channels separately:

```
fir_fr16(sCh0LeftIn, sCh0LeftOut, INPUT_SIZE, &stateFIRL);
fir_fr16(sCh0RightIn, sCh0RightOut, INPUT_SIZE, &stateFIRR);
```

The processed left (Ch0LeftOut) and right (Ch0RightOut) signals are sent to the DAC of the EZ-KIT. The combined left and right FIR filters must be initialized by the init.c file, using the following functions:

```
fir_init(stateFIRL, allCoeff, firDelayL, NUM_COEFF, 0);
fir_init(stateFIRR, allCoeff, firDelayR, NUM_COEFF, 0);
```

The combined FIR filter coefficients are defined as allCoeff, which are passed to the fir_fr16 function as a structured variable. Note that the array variables firDelayL and firDelayR represent the delay line for the left and right filters, respectively. These array variables must be initialized to zero and should not be modified by the user program.

EXERCISE 5.4

1. Instead of user-adjusted gain control, we can predefine the equalizer gain settings to perform different audio enhancement modes. The gain settings are listed below from low- to high-frequency band for the eight-band equalizer:

Table 5.9 Functions of Push Buttons and LEDs on the BF533/BF537 EZ-KITs

Functions	BF533 EZ-KIT	BF537 EZ-KIT
Switch from higher to lower filter band (from #7 to #0)	SW4	SW10
Switch from lower to higher filter band (from #0 to #7)	SW5	SW11
Decrease the gain level of the active filter band in steps of −3 dB per press	SW6	SW12
Increase the gain level of the active filter band in steps of +3 dB per press	SW7	SW13
Indicate the filter band number (in binary)	LED6|LED5|LED4	LED3|LED2|LED1
Band #1	0 | 0 | 0 (all off)	0 | 0 | 0 (all off)
Band #2	0 | 0 | 1	0 | 0 | 1
⋮	⋮ ⋮ ⋮	⋮ ⋮ ⋮
Band #8	1 | 1 | 1 (all on)	1 | 1 | 1 (all on)
Indicate the gain level of the active band	LED7|LED8|LED9 The blink rate of these LEDs indicates the gain level. When the blink rate is fast, higher gain amplification; when the blink rate is slow, more attenuation	LED4|LED5|LED6 The blink rate of these LEDs indicates the gain level. When the blink rate is fast, higher gain amplification; when the blink rate is slow, more attenuation

    a. Bass enhanced: [+6 +6 +6 +3 +3 +3 0 0] dB
    b. Treble enhanced: [0 0 +3 +3 +3 +6 +6 +6] dB
    c. Rock: [+6 +3 −3 −3 0 +3 +6 +6] dB
    d. Pop: [0 +3 +3 +6 +6 −3 −3 −3] dB

Simplify the programs in Hands-On Experiment 5.8 to implement the above fixed settings.

2. Implement the BTC feature of the VisualDSP++ to view the input and output frequency responses for Hands-On Experiment 5.8. Verify that the output frequency response matches with the gain settings selected by the user. Hint: Use the BTC example given in Hands-On Experiment 3.7.

3. Determine the cycle counts required to run (1) the overall interrupt service routine and (2) FIR filters, and (3) compute the combined FIR filter coefficients in the BF533/BF537 EZ-KIT. Hint: Use the CYCLES and CYCLES2 registers.

5.5 Implementation of Graphic Equalizer Using BF533/BF537 EZ-KIT   209

4. Turn on the statistical profiler (**Tools** → **Statistical Profiling** → **New Profile**) and examine the execution percentage of running different tasks in the graphic equalizer.

HANDS-ON EXPERIMENT 5.9

So far, we have loaded our compiled code from the computer to the EZ-KIT via the USB cable. VisualDSP++ also provides a feature to create and download a loader file to the flash memory of the EZ-KIT. In this experiment, we go through the steps in creating a stand-alone system that runs the above-described eight-band graphic equalizer on power up of the EZ-KIT. The steps are stated as follows:

1. Open the VisualDSP++ and load the project file exp5_8_533.dpj for the BF533 EZ-KIT or exp5_8_537.dpj for BF537. These projects are located at directories c:\adsp\chap5\exp5_8_533 and c:\adsp\chap5\exp5_8_537, respectively. Because we have finished debugging the eight-band equalizer in the preceding experiment, we can rebuild the project using the release build by changing the toolbar menu from Debug to Release. This step will remove all debug information and reduce the memory usage in the Blackfin processor. Release build can also optimize real-time performance.

2. Click on **Project** → **Project Options** and change the **Target Type** to **Loader file** as shown in Figure 5.35. The loader file (.ldr) is essentially the same as the

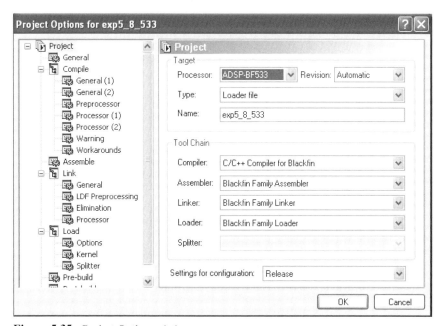

Figure 5.35 **Project Options** window

210 Chapter 5 Introduction to the Blackfin Processor

executable file (.dxe), with the exception that the loader file does not contain any debug information and symbols.

3. Click on **Options** under the **Load** menu in the **Project Options** window. A new window will appear as shown in Figure 5.36. Select the output format as shown. Make sure that **Output Width** is set to **16-bit**. Users can also specify a different loader file (.ldr) name (which uses the same name as the project file by default) by typing in the **Output file** field.

4. Click on **OK** and rebuild the project. A loader file exp5_8_533.ldr for the BF533 EZ-KIT or exp5_8_537.ldr for the BF537 EZ-KIT is created in the respective Debug folder.

5. Click on **Tools → Flash Programmer...** to turn on the **Flash Programmer** window as shown in Figure 5.37. Click on the ⋯ icon next to the **Driver file** and search for the default driver for the BF533 or BF537 EZ-KIT. For the BF533 EZ-KIT, use

    ...\VisualDSP 4.0\Blackfin\Flash Programmer Drivers\ADSP-BF533 EZ-kit Lite\BF533EzFlash.dxe.
    For the BF537 EZ-KIT, use ...\VisualDSP 4.0\Blackfin\Flash Programmer Drivers\ADSP-BF537 EZ-kit Lite\BF537EzFlash.dxe.

6. Click on the ⋯ icon next to the **Data file** to search for the loader file created in Step 4. Click on **Load File** to complete the loading.

7. Click on **Settings → Boot load...** to verify that the eight-band graphic equalizer program has been loaded into the flash memory of the EZ-KIT. The program is now booting from the flash memory. Alternatively, turn off the EZ-KIT, unplug the

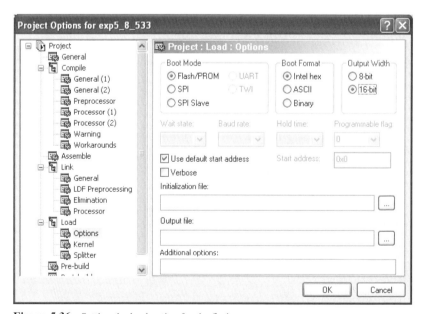

Figure 5.36 Setting the load option for the flash programmer

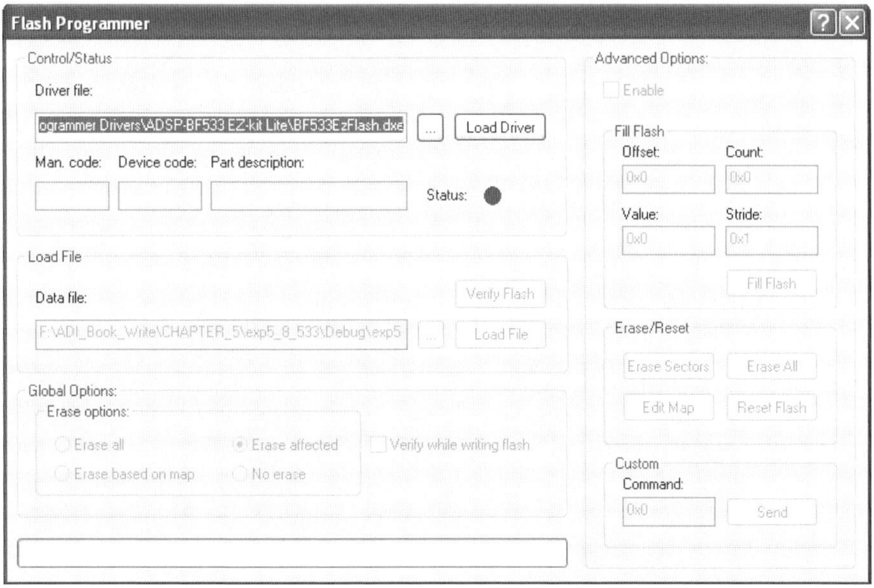

5.6 Implementation of Graphic Equalizer Using LabVIEW Embedded Module 211

Figure 5.37 **Flash Programmer** window

USB, and turn on the EZ-KIT again. The program should run immediately on power up.

8. Test the functionality of the graphic equalizer to confirm that the program is working properly in stand-alone mode.

5.6 IMPLEMENTATION OF GRAPHIC EQUALIZER USING LABVIEW EMBEDDED MODULE FOR BLACKFIN PROCESSORS

Graphic equalization is a very scalable technology allowing a designated number of audio frequency bands to be extracted, amplified, and reassembled to improve or process audio signals. Commercial graphic equalizers use a wide variety of analog and digital technology to achieve similar functionality. Most have a standard user interface using slider bars allowing each band to be individually amplified or attenuated. The algorithm for equalizing audio is computationally intensive in its theoretical form but can be simplified and implemented in LabVIEW with the same principles discussed in the moving-average filter application in Chapter 2.

In the following exercises, you will simulate, prototype, and deploy a graphic equalizer using the FIR filter coefficients derived in previous examples. The eight-band graphic equalizer simulation allows you to load custom coefficients and an audio signal. This gives you the ability to modify and listen to the effects of different equalizer gains. The equalizer is then run on the Blackfin EZ-KIT, demonstrating a real-time filtering application created with graphical programming.

HANDS-ON EXPERIMENT 5.10

In this experiment, an eight-band graphic equalizer is implemented, using FIR filtering to modify the frequency content of audio signals. The simulation allows you to hear the effects of the equalizer and see its results in both time and frequency domains. Custom filter coefficients can also be loaded to test the results of your own equalizer designs. You can easily modify the gain of each frequency band and observe the differences in filter characteristics.

Navigate to the program called FIR_EQ_Sim.exe located in directory c:\adsp\chap5\exp5_10. In Figure 5.38, we see the user interface for FIR_EQ_Sim.exe. Separate tabs show various plots of the audio signal and allow you to customize the gain applied to each band. The **Time Signal** tab shows the time-domain input signal. Click the **Enable Filter** button to turn on equalization. The **Frequency Signal** tab shows both the input signal and its frequency content, and the graph can be viewed with either a linear or a logarithmic (dB) scale. The **Frequency Bands** tab shows the magnitude response of each of the eight filters

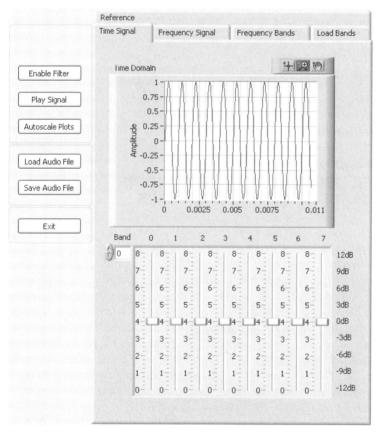

Figure 5.38 Eight-band graphic equalizer (user interface for FIR_EQ_Sim.exe)

individually. Note that although each of the individual filters is not ideal, the overall passband is relatively flat when the filter are combined. The **Load Bands** tab gives us the ability to open new sets of coefficient data files and change the gains applied by the equalizer slider positions. The set of eight slider bars at the bottom of the user interface corresponds to the gain applied to each of eight frequency bands.

The default input signal is a 1-kHz sine wave sampled at 48 kHz with a buffer length of 512 samples, as seen in Figure 5.38. Different wave files can be loaded for additional filter testing. Custom coefficients can be loaded from data files in `fract16`, a common delimited format for custom equalizer simulations.

Click the **Frequency Bands** tab and adjust the values of the Band 1 slider bar. As you change the selected gain for that frequency band, note how the magnitude response for that band changes in the graph as well.

Now load and experiment with the audio file `speech_tone_48k.wav` that we used in Hands-On Experiment 2.5. Recall that this audio file contains tonal noise that degrades the overall quality of the audio. Can you identify the frequency band that contains the tonal noise? Experiment with the equalizer gain settings to attenuate the noise as much as possible while still retaining the rest of the signal. How would these settings need to change if the tonal noise had a different frequency? How would you redesign the filter if you needed finer resolution control over high frequencies and less resolution control over low frequencies while keeping just eight bands?

HANDS-ON EXPERIMENT 5.11

This experiment implements an eight-band FIR filter on the Blackfin processor with the LabVIEW Embedded Module for Blackfin Processors. This project is run in **Release** mode to take advantage of the excellent speed optimization when the various debugging features are not necessary. There are two key processing steps in this experiment. First, the single FIR filter Blackfin library function is the only processing VI that modifies the original signal. The computationally intensive portion of this VI is the calculation of the FIR filter coefficients. This filter coefficient calculation is packaged into a subVI for code modularity and is used as a single processing block.

Open the `FIR Equalizer-BF5xx.lep` project appropriate for your Blackfin hardware in directory `c:\adsp\chap5\exp5_11`. You will find that the project is composed of three files, one for each of the main processing blocks in the project. Double-click on the top-level VI, `FIR Equalizer-BF5xx.vi`, and open the block diagram to study how the equalizer is implemented. Note that the processing algorithm is divided into distinct sections for buffer initialization, coefficient calculation, and filtering, where each of these operations has its own subVI. This application streams a real-time audio signal from the Blackfin audio input, processes it, and generates an output signal from the Blackfin EZ-KIT audio out port. When **SW4** of BF533 (or **SW10/PB4** of BF537) is pressed, the selected FIR filter is enabled and applied to the signal. Pay special attention to the block diagram of the **Init Eq Coeffs subVI** (shown in Fig. 5.39), which implements the calculation of the filter coefficients.

This block diagram shows how to calculate the coefficients of the FIR filter. Graphical representation of the algorithm illustrates the parallelism of the code to be easily seen. Parallelism is valuable in understanding complex algorithms with recurring symmetry. For each of the eight bands, the input coefficients are multiplied by the gain selected for that band.

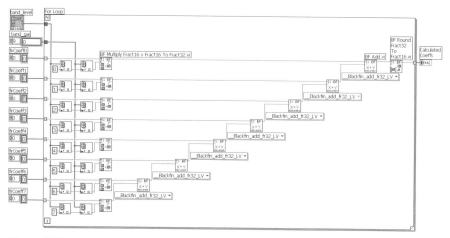

Figure 5.39 Block diagram for coefficient calculation

We then sum these products to calculate a single set of coefficients. Once the coefficients are calculated by the **Init Eq Coeffs subVI**, they are passed to the **FIR filter** function block along with the audio buffer for real-time processing.

Connect the computer audio output to the input of the Blackfin processor and the output of the Blackfin processor to loudspeakers or headphones. On the computer, play the audio signal that you would like to equalize. Compile and run the project on the Blackfin EZ-KIT. To enable the equalizer, press and hold **SW4** of BF533 (or **SW13/PB1** of BF537).

The application can be customized with different graphic equalizer gain settings, or by changing the filter coefficients for each band. Customize the gains for different bands by changing the selected preset. Use the same settings simulated in Hands-On Experiment 5.10 to attenuate the 1-kHz sine wave from the speech_tone_48k.wav audio file and evaluate its performance with the Blackfin processor implementation. Does it behave the same? Recompile the project and run it. Can you hear the difference in the audio? Experiment with the various gain presets and try creating your own.

5.7 MORE EXERCISE PROBLEMS

1. State whether the following Blackfin instructions are correct or not. If not, state the reasons.
    (a) R6 = R2.L+|-R1.H;
    (b) R6.H = R2.L+R1.H;
        R6.L = R2.L-R1.H;
    (c) R3 = R0+|+R1, R4 = R2+|-R1;
    (d) R3 = R0+R1, R4 = R2-R1;
    (e) R4.H = R4.H&R2.H;
    (f) R0.L = R2.H*R1.L;
    (g) A0.H += R2.H*R3.H;

(h) R0.H = (A0 += R2.H*R3.H);
(i) R1 = (A0 += R2.H*R3.H);
(j) A1 = R2.H*R3.L, A0 = R2.L*R3.H;
(k) R0.H = (A1 += R5.H*R6.H), R0.L = (A0 += R5.L*R6.L);
(l) R1 = (A1 + = R5.H*R6.H), R2 = (A0 += R5.L*R6.L);

2. The Blackfin memory starting from 0xFF80 0100 contains the data shown in the second column of Table 5.10. The pointer P1 = 0xFF80 0100, and P2 = 0x0000 0002. Fill in all the update registers in the last column of Table 5.10 after executing every instruction.

3. Load the project in Hands-On Experiment 5.1 and use the pipeline viewer (**View → Debug Windows → Pipeline Viewer**) to examine any stall in the program when single-stepping through the program.

4. A 10-band graphic equalizer based on the ISO octave-band center frequency is shown in Table 5.11. Design this 10-band graphic equalizer with FDATool and verify its fixed-point implementation using the VisualDSP++ compiled simulator for the BF processor.

Table 5.10 Blackfin Memory and Instructions

Address	Content	Instruction (arrange in sequential top-down manner)	Update Registers
0xFF80 0100	0x12	R1 = [P1++]	
0xFF80 0101	0x34	R1 = W[P1--](x)	
0xFF80 0102	0x56	R1 = B[P1++](z)	
0xFF80 0103	0x78	P0 = [P1 + 0x03]	
0xFF80 0104	0x9A	[P0++] = R1	
0xFF80 0105	0xBC	R1.H = W[P1++]	
0xFF80 0106	0xDE	R1.L = W[P1++P2]	
0xFF80 0107	0xF0	P1 - = P2	
0xFF80 0108	0xAA	W[P1++] = R1	
0xFF80 0109	0xBB	B[P1++] = R1	

Table 5.11 ISO Octave Center Frequencies with Band Limits

Octave Band Center Frequency	Band Limits
31.5 Hz	22 Hz–44 Hz
63 Hz	44 Hz–88 Hz
125 Hz	88 Hz–176 Hz
250 Hz	176 Hz–353 Hz
500 Hz	353 Hz–707 Hz
1,000 Hz	707 Hz–1,414 Hz
2,000 Hz	1,414 Hz–2,825 Hz
4,000 Hz	2,825 Hz–5,650 Hz
8,000 Hz	5,650 Hz–11,250 Hz
16,000 Hz	11,250 Hz–22,500 Hz

5. Implement the graphic equalizer in Problem 4 with the Blackfin BF533/BF537 EZ-KIT. Benchmark the cycle counts needed to complete the graphic equalizer. Profile the percentage of processing time for the tasks in the graphic equalizer. Feed a stereo musical signal (sampled at 48 kHz) to the input of the EZ-KIT and perform the various gain settings of the graphic equalizer to hear the processed signal from the output of the EZ-KIT.

6. A noisy speech signal, noisy.wav, is recorded offline. This signal contains band-limited noise that can be removed with the graphic equalizer. Analyze the band-limited noise in order to design a suitable graphic equalizer to remove the band-limited noise with the Blackfin BF533/BF537 EZ-KIT. Users can implement the graphic equalizer either in C or the LabVIEW Embedded Module.

7. Instead of using switch to control the gain of the graphic equalizer, the gain control can also be performed by using the BTC. Modify the C files in Hands-On Experiment 5.8 to perform the gain control via software.

8. The project file in Hands-On Experiment 5.8 computes the combined FIR filter coefficients at every CODEC interrupt, even though there is no adjustment of the gain setting by the user. Rewrite the C files such that the combined filter coefficients are computed only when the push buttons are pressed.

9. Table 5.8 shows that the user-adjustable gain of the graphic equalizer is divided by 4 in order to scale all the gain values to less than 1. Examine the project file in Hands-On Experiment 5.8 and state how this gain reduction has been compensated at the output of the graphic equalizer.

10. In Hands-On Experiment 5.8, a single combined filter is used to implement FIR filtering in the eight-band graphic equalizer. A more direct approach is to implement eight FIR filters, one for each band, and add up the gain-adjusted outputs of these eight filters. Implement this direct approach and benchmark its implementation cost (cycle count and data memory usage) with reference to the single combined FIR filter.

Chapter 6

Real-Time DSP Fundamentals and Implementation Considerations

In Chapter 5, we introduced the architecture of the Blackfin processor and moved data between memories and registers. This chapter introduces important topics on representing data in the fixed-point Blackfin processors, processing and handling digital signals with finite wordlength, segmenting signal samples into blocks for processing, and evaluating the resources (speed, memory, peripherals, and power management) for real-time tasks. We use many examples and hands-on experiments to explain these topics and show a step-by-step approach in designing an embedded signal processing application. This effective embedded system design moves from floating-point simulations to fixed-point representations using MATLAB and then to porting the fixed-point code and data into the Blackfin processor. We also introduce more advanced debugging and profiling features in the VisualDSP++ IDDE to meet real-time requirements.

6.1 NUMBER FORMATS USED IN THE BLACKFIN PROCESSOR

This section defines number representations in digital systems with different formats and performs different arithmetic operations based on these formats. In particular, we use the Blackfin processor that supports 8-, 16-, 32-, and 40-bit fixed-point data to explain different number formats.

6.1.1 Fixed-Point Format

In fixed-point processors, a number is represented with a series of binary digits of 1s and 0s. An unsigned number without sign information is always positive. For

---

Embedded Signal Processing with the Micro Signal Architecture. By Woon-Seng Gan and Sen M. Kuo
Copyright © 2007 John Wiley & Sons, Inc.

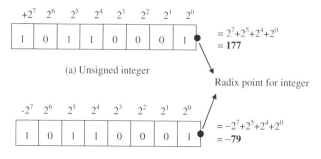

(b) Signed integer (2's complement)

Figure 6.1 Examples of 8-bit binary data format for integer numbers of 177 (a) and –79 (b)

example, an 8-bit unsigned binary number of 1011 0001 represents 177 in base 10 integer. However, in a signed number representation, the same 8-bit binary digits represent –79 in 2's complement format. The 2's complement format is the most popular signed number in DSP processors, including the Blackfin processor.

In addition, most DSP processors support both integer and fractional data formats. In an integer format, the radix point is located to the right of the least significant bit (LSB). For example, the integer number 79 has the binary pattern shown in Fig. 6.1(a). The smallest magnitude of an integer number is 1 (the weighting of the LSB). The negative number –79 in 2's complement format is shown in Figure 6.1(b). Note that the sign bit in 2's complement format has a negative weight.

In the fractional number format, the radix point is located within the binary number. For example, a radix point can be positioned to the left of the four LSBs with the weights indicated in Figure 6.2(a) for the unsigned fractional number and in Figure 6.2(b) for the signed fractional number. Note that the number to the right of the radix point assumes a fractional binary bit, with a weighting of 2^{-p} where $p = 1, 2, 3,$ and 4. In this case, the lowest fractional increment is 2^{-4} (or 0.0625). For the number to the left of the radix point, the weighting increases from 2^q where $q = 0, 1, 2,$ and 3. The weighting of the MSB (or sign bit) depends on whether the number is signed or unsigned.

An $(N.M)$ notation describes any fractional number, where N is the number of bits to the left of the radix point (integer part) and M is the number of bits to the right of the radix point (fractional part). The symbol "." represents the radix point. The total number of bits in the data word is $B = N + M$. In the example shown in Figure 6.2, this is called the (4.4) format. An integer number can be named in the $(B.0)$ format. For example, the number given in Figure 6.1 is called the (8.0) format or 8-bit integer.

6.1 Number Formats Used in the Blackfin Processor

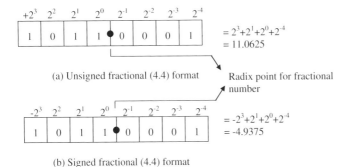

Figure 6.2 Example of 8-bit binary data formats for a fractional number

Table 6.1 Examples of Representing Numbers in Different Formats

Hexadecimal Number	(16.0) Format or Integer	(4.12) Format	(1.15) Format
0x7FFF			
0x8000			
0x1234			
0xABCD			
0x5566			

QUIZ 6.1

Interpret the 16-bit hexadecimal numbers (with prefix 0x) given in Table 6.1 as

1. (16.0) format or integer
2. (4.12) format
3. (1.15) format

The (1.15) format displayed in the last column of Table 6.1 is commonly used in 16-bit fractional number representation. Figure 6.3 shows the weightings of bits for the (1.15) format. The radix point is positioned one bit to the right of the MSB. Therefore, the maximum positive number in (1.15) format is $1 - 2^{-15}$ (= 0.999969482421875), which has a hexadecimal representation of 0x7FFF because all bits are "1" except the MSB. The minimum negative number in (1.15) format is −1 (0x8000). Therefore, the (1.15) format has a dynamic range of [+0.999969482421875 to −1], and numbers exceeding this range cannot be represented in (1.15) format. For example, +1.0 cannot be represented in (1.15) format. The smallest increment (or precision) within the (1.15) format is 2^{-15}. In the Blackfin processor, the data type `fract16` represents the (1.15) format.

Table 6.2 lists all 16 possible $(N.M)$ formats for 16-bit numbers. Different formats give different dynamic ranges and precisions. There is a trade-off between the

-2^0	2^{-1}	2^{-2}	2^{-3}	2^{-4}	2^{-5}	2^{-6}	2^{-7}	2^{-8}	2^{-9}	2^{-10}	2^{-11}	2^{-12}	2^{-13}	2^{-14}	2^{-15}
1	0	1	1	0	0	0	1	1	0	1	1	0	0	0	1

$$x = -1 + 2^{-2} + 2^{-3} + 2^{-7} + 2^{-8} + 2^{-10} + 2^{-11} + 2^{-15} = -0.61178588867$$

Figure 6.3 An example of a number represented in (1.15) format

Table 6.2 Dynamic Ranges and Precisions of 16-Bit Numbers Using Different ($N.M$) Formats

Format ($N.M$)	Largest Positive Value (0x7FFF)	Least Negative Value (0x8000)	Precision (0x0001)
(1.15)	0.999969482421875	−1	0.00003051757813
(2.14)	1.99993896484375	−2	0.00006103515625
(3.13)	3.9998779296875	−4	0.00012207031250
(4.12)	7.999755859375	−8	0.00024414062500
(5.11)	15.99951171875	−16	0.00048828125000
(6.10)	31.9990234375	−32	0.00097656250000
(7.9)	63.998046875	−64	0.00195312500000
(8.8)	127.99609375	−128	0.00390625000000
(9.7)	255.9921875	−256	0.00781250000000
(10.6)	511.984375	−512	0.01562500000000
(11.5)	1,023.96875	−1,024	0.03125000000000
(12.4)	2,047.9375	−2,048	0.06250000000000
(13.3)	4,095.875	−4,096	0.12500000000000
(14.2)	8,191.75	−8,192	0.25000000000000
(15.1)	16,383.5	−16,384	0.50000000000000
(16.0)	32,767	−32,768	1.00000000000000

dynamic range and precision. As the dynamic range increases, precision becomes coarser. For example, when we change the format from (1.15) to (2.14), we get a larger dynamic range from [+0.999969482421875 to −1] to [1.99993896484375 to −2]; however, we reduce the precision from 2^{-15} to 2^{-14}. The largest dynamic range of a 16-bit number can be obtained by using the (16.0) format or integer; however, this format has the worst precision of 1. Therefore, the selection of the right format depends on the dynamic range and precision required by the given DSP application.

A number in ($N.M$) format cannot be represented in the programs because most compilers and assemblers only recognize numbers in integer or (16.0) format. Therefore we must convert the fractional number in ($N.M$) format into its integer equivalent, and its radix point must be accounted for by the programmers. For example, to convert a number 0.6 in (1.15) format to its integer representation, we multiply it by 2^{15} (or 32,768) and round the product to its nearest integer to become 19,661 (or 0x4CCD). As shown in Table 6.3, a scaling factor 2^M is needed for converting the ($N.M$) format to the integer. Note that the number range in

6.1 Number Formats Used in the Blackfin Processor

Table 6.3 Scaling Factors and Dynamic Ranges for 16-Bit Numbers Using Different (N.M) Formats

Format	Scaling Factor (2^M)	Range in Hex (fractional value)
(1.15)	$2^{15} = 32,768$	0x7FFF (0.99) → 0x8000 (−1)
(2.14)	$2^{14} = 16,384$	0x7FFF (1.99) → 0x8000 (−2)
(3.13)	$2^{13} = 8,192$	0x7FFF (3.99) → 0x8000 (−4)
(4.12)	$2^{12} = 4,096$	0x7FFF (7.99) → 0x8000 (−8)
(5.11)	$2^{11} = 2,048$	0x7FFF (15.99) → 0x8000 (−16)
(6.10)	$2^{10} = 1,024$	0x7FFF (31.99) → 0x8000 (−32)
(7.9)	$2^9 = 512$	0x7FFF (63.99) → 0x8000 (−64)
(8.8)	$2^8 = 256$	0x7FFF (127.99) → 0x8000 (−128)
(9.7)	$2^7 = 128$	0x7FFF (511.99) → 0x8000 (−512)
(10.6)	$2^6 = 64$	0x7FFF (1,023.99) → 0x8000 (−1,024)
(11.5)	$2^5 = 32$	0x7FFF (2,047.99) → 0x8000 (−2,048)
(12.4)	$2^4 = 16$	0x7FFF (4,095.99) → 0x8000 (−4,096)
(13.3)	$2^3 = 8$	0x7FFF (4,095.99) → 0x8000 (−4,096)
(14.2)	$2^2 = 4$	0x7FFF (8,191.99) → 0x8000 (−8,192)
(15.1)	$2^1 = 2$	0x7FFF (16,383.99) → 0x8000 (−16,384)
(16.0)	$2^0 = 1$ (integer)	0x7FFF (32,767) → 0x8000h (−32,768)

Table 6.4 Example of Converting Numbers in (N.M) Format to Hexadecimal Representations

Number	(1.15) Format	(2.14) Format	(8.8) Format	(16.0) Format
0.5				
1.55				
−1				
−2.0345				

hexadecimal, [0x7FFF to 0x8000], remains the same for all 16-bit number formats; the only difference is the fractional value that is represented by the hexadecimal integer.

QUIZ 6.2

Determine the hexadecimal integer representation of the numbers in different (N.M) formats as shown in Table 6.4.

Conversely, we can divide an integer number by the associate scaling factor to obtain the fractional number for the specific format. In Quiz 6.1, the answers can be easily obtained by using the scaling factors listed in Table 6.3.

The major advantage of using fixed-point fractional representation is that this format adheres to the basic arithmetic operations of most fixed-point signal proces-

222 Chapter 6 Real-Time DSP Fundamentals and Implementation Considerations

sors and does not require additional libraries or hardware logic. In the following sections, we examine fixed-point addition and multiplication in the Blackfin processor.

6.1.1.1 Binary Addition

In Section 5.1.3.1, we introduced addition and multiplication in the Blackfin processor without considering the data format. Overflow occurs when an arithmetic operation produces a number that exceeds the number range. For example, when two numbers (0.5 and 0.7) in (1.15) format are added together, the result is 1.2, which exceeds the valid (1.15) number range of [+0.999969482421875 to −1]. The result 1.2 is overflowed into the negative number range and becomes −0.8. As shown in Table 5.1, a saturation mode is available in the Blackfin processor that forces the overflowed number to be fixed at the maximum positive or negative value. Therefore, in the above example, the result 1.2 will be saturated to the maximum positive number of +0.999969482421875 instead of overflowing into a negative number of −0.8. In the saturation mode, the error between the actual number and the saturated result is smaller compared to the error without saturation.

HANDS-ON EXPERIMENT 6.1

This VisualDSP++ experiment performs add/subtract operations on four sets of numbers with the Blackfin instructions given in Example 5.1. The arithmetic operations are (a) 0.5 + 0.7, (b) −0.3 − 0.7, (c) 0.1 + 0.8, and (d) −0.5 − 0.7. These add/subtract operations are carried out separately, using (1.15) and (2.14) formats. This experiment shows how to load these numbers into the Blackfin memory and add (or subtract) with saturated (s) and nonsaturated (ns) modes. The steps of the experiment are as follows:

1. Load the project file into the VisualDSP++ BF533/BF537 simulator by clicking on **File → Open → Project** and look for the project file `exp6_1.dpj` in directory `c:\adsp\chap6\exp6_1`.
2. Examine the file `exp6_1.asm` and fill in the missing hexadecimal numbers stated in the program. Build the project and single-step through the code.
3. Fill in Table 6.5 after executing addition and subtraction for the four sets of numbers presented in (1.15) and (2.14) formats.
4. Change the mode to nonsaturation (or overflow) option and repeat Steps 2 and 3.

Table 6.5 Examples of Add/Subtract with Saturation and Nonsaturation Modes

Arithmetic	Result in (1.15) Format		Result in (2.14) Format	
	Saturate (s)	Overflow	Saturate (s)	Overflow
0.5 + 0.7				
−0.3 − 0.7				
0.1 + 0.8				
−0.5 − 0.7				

Explain why there is no difference between the results of saturation and nonsaturation (overflow) options using the (2.14) format for add/subtract operations.

Modify the above program to add a series of 10 numbers {0.5, 0.7, −0.3, −0.4, 0.21, −0.12, 0.8, −0.6, −0.18, 0.11} in (1.15) format using data registers. Compare the results with both saturated and nonsaturated options. Click on **Registers → Core → Status → Arithmetic Status** to observe any register overflow (V flag) in this series of additions. Which option gives the right result and why?

6.1.1.2 Binary Multiplication

In binary multiplication, the input numbers can be represented by different data formats, and the result depends on the input data formats. If two input numbers have formats of $(N.M)$ and $(P.Q)$, then the format of the product is $(N + P).(M + Q)$. For example, the product of multiplying two 16-bit numbers in (1.15) format results in a 32-bit result in (2.30) format. In another example, if the inputs are in (1.15) and (2.14) formats, the multiplication result is in (3.29) format. When both inputs are signed numbers, there is an extra sign bit, as shown in Figure 6.4 for the case of multiplying two fractional numbers in (1.15) format.

The example shown in Figure 6.4 can be extended to all multiplications with signed fractional data format. For example, if a number in (2.14) format is multiplied by a number in (3.13) format, the multiplication result is in (5.27) format, but the result must be shifted left by one bit to get a correct (4.28) format with only one sign bit. The LSB of the multiplication result is zero filled after the shift operation. The Blackfin processor can automatically shift the multiplication results left by one bit before writing to the register. This is the default option as stated in Table 5.2. However, the left shift is not necessary when multiplying a signed number by an unsigned number, or multiplying two unsigned numbers. This is the case when the (FU) option in Table 5.2 is used.

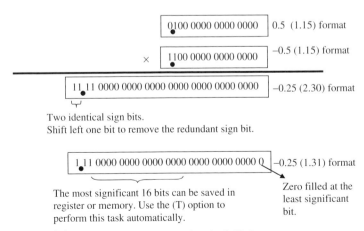

Figure 6.4 Multiplication of two numbers in (1.15) format

224 Chapter 6 Real-Time DSP Fundamentals and Implementation Considerations

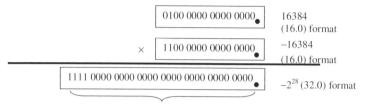

Figure 6.5 Multiplication of two signed integers in (16.0) format

In the case of integer (signed or unsigned) multiplication, there is no need to perform any shifting after the multiplication. For example, multiplying two numbers in (16.0) format results in a number in (32.0) format. Figure 6.5 illustrates the concept behind the multiplication of signed integers. The 32-bit integer multiplication result must be saved in two 16-bit memory locations. This is because every bit in the integer multiplication has a significant weighting, and huge error can be generated when even a few least significant bits are omitted. In the Blackfin processor, the (IS) and (IU) options stated in Table 5.2 are used for signed and unsigned integer multiplication, respectively.

HANDS-ON EXPERIMENT 6.2

Perform the following multiplication operations with the VisualDSP++ BF533/BF537 simulator:

1. 0.5×-0.5 (both numbers are in (1.15) format)
2. $16,384 \times -16,384$ (both numbers are in (16.0) format)
3. 0.25 (in (1.15) format) $\times 1.25$ (in (2.14) format)
4. -1.5 (in (2.14) format) $\times 1.5$ (in (2.14) format)
5. $2^{-15} \times 2^{-15}$ (both numbers are in (1.15) format)

Load the project file exp6_2.dpj in directory c:\adsp\chap6\exp6_2 and fill in the missing numbers in the program exp6_2.asm. Open the following register windows:

- Arithmetic status window: **Registers → Core → Status → Arithmetic Status**. Observe the AV0 and AV1 bits, which indicate saturation in the A0 and A1 accumulators, respectively. The V bit indicates data register overflows when written from ALU to data register. A value of "1" indicates overflow.
- Accumulators window: **Registers → Core → Accumulators**. Observe the A0 and A1 accumulators. From the **Accumulators Window**, it is observed that the accumulators A0 and A1 are 40-bit accumulators that consist of three fields, A0.X, A0.H, A0.L and A1.X, A1.H, A1.L. The symbol "X" denotes the 8 extended bits (or guard bits), and the symbols "H" and "L" denote the upper and lower 16 bits of the remaining 32-bit register, respectively.
- Data register file window: **Registers → Core → Data Register File**. Observe the data registers used.

Interpret and verify the results. In addition, answer the following questions:

1. Will the multiplication of two numbers in (1.15) format result in overflow if the result is stored in (1.31) format? Any advantage of using (1.15) format for multiplication?
2. Observe and explain the values of A0.X and A1.X after every multiplication from the five multiplication operations.
3. Will the multiplication of two numbers in (2.14) format result in overflow if the result is stored in (2.30) format?
4. Can you perform -1×-1 in (1.15) format?

6.1.1.3 Binary Multiply-Add

The multiply-add operation is one of the most commonly used operations in DSP algorithms. As introduced in Chapters 2 to 4, DSP algorithms such as digital filtering and transform require extensive multiply-add operations. Therefore, it is important to examine the data formats that may cause overflow and use the register extension in the accumulator when performing multiply-add in the Blackfin processor.

When multiplying two numbers in (1.15) format, a result in (1.31) format (after shift) is obtained. The extended registers A0.X and A1.X allow the result in the accumulator to sign extend to (9.31) format. This (9.31) format implies that there are 1 sign bit, 8 integer bits, and 31 fractional bits. Therefore, the result of the accumulator can range from 0x7F FFFF FFFF (most positive number of $256 - 2^{-31}$) to 0x80 0000 0000 (most negative number of -256).

When the Blackfin processor performs multiply-add operations such as in FIR filtering, the extended registers support the sequence of additions without overflow. These extended registers with 8 guard bits provide the headroom for accumulated products to temporarily overflow without setting the AV0/AV1 flag. In the worst case, a number 0x7FFF is multiplied by another number 0x7FFF, and the product is accumulated in the accumulator. If the same multiply-add operation continues for more than 2^8 (or 256) times, the 40-bit accumulator is saturated to 0x7F FFFF FFFF without overflow. This worst-case scenario is shown in the following experiment.

HANDS-ON EXPERIMENT 6.3

Load the project file `exp6_3.dpj` in directory `c:\adsp\chap6\exp6_3`. Set the loop counter LC0 to 0x100 (256 times). Build and run the project. Increase the loop counter and observe the AV0 flag in the **Arithmetic Status Window**. It is easy to run this program by setting a breakpoint on the instruction "`idle`" and running the program from `_main` to `idle`. It is important to note that this is a very special case, which may never happen in carefully designed DSP applications. The objective of this experiment is to demonstrate how the result in the accumulator can be overflowed into the guard bits.

HANDS-ON EXPERIMENT 6.4

In this experiment, we use the BF533/BF537 simulator to perform multiply-add operations in FIR filtering with the 16-bit multiply-add instruction. The input data and filter coefficients can be arranged in the memory as shown in Figure 6.6. The multiply-add operation is repeated 10 times to complete a single pass of 10-tap FIR filtering.

Load the project file exp6_4.dpj in directory c:\adsp\chap6\exp6_4 to perform the single-MAC FIR filtering. Modify the program exp6_4.asm to take full advantage of the dual MACs within the Blackfin processor. We can save almost 50% of the instruction cycles for the same 10-tap FIR filtering. To profile the instruction cycles needed to run the loop with a single MAC versus dual MACs, we set a breakpoint on the following line:

    LSETUP (begin_loop, end_loop) LC0=P1;

We also set another breakpoint after the instruction where the accumulator has been stored to the data register R4 as follows:

    W[I2++] = R4.H;

The cycle registers can be viewed by clicking on **Registers → Core → Cycles**. Right-click on the **Cycles Window** and select Unsigned Integer. Run the program (**F5**) from the start until the first breakpoint of the single-MAC program, then reset the counter by double-clicking on the number in the CYCLES register and typing 0, followed by **Enter**. Run

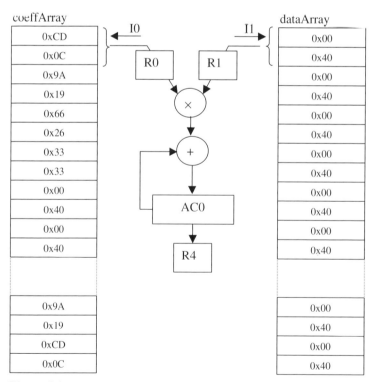

Figure 6.6 Program setup for performing 16-bit multiply-accumulate in the Blackfin processor

the program (**F5**) to the next breakpoint and record the instruction cycle. Repeat the same step for the dual-MAC program and note the difference in the instruction cycle count. For checking, the cycle counts for single and dual MACs are 42 and 23 cycles, respectively. Obviously, the dual-MAC program is restricted to an FIR filter with an even number of coefficients. Extend the above program to an odd-length filter.

6.1.1.4 Truncation and Rounding of Multiplication Results

It was observed in preceding sections that the multiplication of two 16-bit numbers results in a 32-bit product. This result may be stored to memory or register with 16-bit (8-bit or lesser precision). In other words, a 32-bit result with a precision of 2^{-31} in (1.31) format can be either truncated or rounded to 16-bit with a precision of 2^{-15} in (1.15) format, or to 8-bit with a precision of 2^{-7} in (1.7) format as shown in Figure 6.7.

Truncation reduces the wordlength by simply masking out the lower 16 bits or 24 bits. In Table 5.2, there are two options, (T) and (TFU), to perform truncation of a 32-bit multiplication result into 16-bit so it can be stored in half-register. However, these options are applicable to fractional multiplication only.

The Blackfin processor supports rounding of multiplication results with the RND option. Similar to truncation, rounding also reduces the precision of the number by removing the least significant bits. Instead of simply throwing away the P least significant bits out of the N-bit result (where $N > P$), we can modify the remaining $(N - P)$ bits to more accurately represent its former value. Two types of rounding are used in the Blackfin processor: unbiased (or convergent) rounding and bias (round-to-nearest) rounding.

Unbiased rounding returns the number closest to the original number. When the original number lies exactly halfway between two numbers, unbiased rounding returns the nearest even number (the rounded result has an LSB of 0). For example, a number 0.25 (0.01) in 3-bit (1.2) format lies midway between 0.0 and 0.5. When we perform unbiased rounding to two bits in (1.1) format, we get 0.0 because the LSB must be 0.

Biased rounding returns the closest number to the original number. However, the original number that lies exactly halfway between two numbers always rounds up to the larger of the two. Using the same example as above, the number 0.25 is

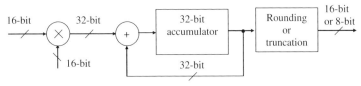

Figure 6.7 Block diagram of the multiply-accumulate functional block in typical digital signal processors

228   Chapter 6   Real-Time DSP Fundamentals and Implementation Considerations

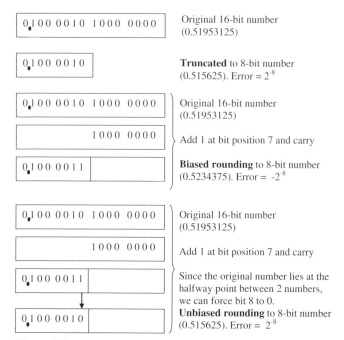

Figure 6.8   Steps in performing truncation and rounding (biased and unbiased) from 16-bit number to 8-bit number

rounded up to 0.5. Because this method always rounds up, it produces bias in the rounding. It is also important to note that biased rounding and unbiased rounding result in the same number when the number is not lying halfway between the rounded numbers. In addition, when the truncated value is an odd number, both biased rounding and unbiased rounding result in the same number, even when the original number is lying halfway between the rounded numbers. For example, round an 8-bit number 0.6875 in (1.7) format to 4 bits in (1.3) format with both biased and unbiased rounding methods.

EXAMPLE 6.1

Figure 6.8 shows the difference in result of reducing a 16-bit number to 8 bits with truncation and unbiased and biased rounding. We also show the steps in performing both unbiased and biased rounding operations.

QUIZ 6.3

Perform 32-bit to 16-bit truncation and rounding (biased and unbiased) for the following numbers. These numbers are all in (1.31) format and should be truncated and rounded to (1.15) format.

1. 0x0000 8001.
2. 0x0001 8000.
3. 0x0001 7FFF.

Can we round or truncate an integer multiplication result?

Blackfin processors provide several options for biased rounding. The options RND12, RND, and RND20 extract 16-bit values from bit 12, bit 16, and bit 20, respectively. The RND12 option prescales the input operands by left shifting 4 (16 − 12) bits. The RND20 option prescales the input operands by right shifting 4 (16 − 20) bits. The RND option has no prescaling. By default (with no option specified), the Blackfin processor performs unbiased rounding. The RND_MOD bit of the arithmetic status register ASTAT specifies the rounding mode. When RND_MOD = 1 the processor uses biased rounding, whereas when RND_MOD = 0 the processor uses unbiased rounding.

HANDS-ON EXPERIMENT 6.5

Load the project file exp6_5.dpj in directory c:\adsp\chap6\exp6_5 into the BF533/BF537 VisualDSP++ simulator. Build and run the project. Fill in Table 6.6 for different truncation and rounding modes. Comment on the differences between truncation and unbiased rounding.

Table 6.6 Example of Truncating a 32-Bit Multiplication Result into 16 Bits

Multiplication Result	Truncation	Unbiased Rounding	Biased Rounding
0x6675 × 0x5547			
0xA000 × 0x0002			
0x6000 × 0x0002			

6.1.2 Fixed-Point Extended Format

In some applications like high-end audio signal processing, extended precision (32 bit) is normally preferred to handle the large dynamic range (greater than 100 dB) of audio signals. The Blackfin processor is suited for extended-precision arithmetic because the register file is based on 32-bit registers. These registers can be treated as either a single 32-bit word or two 16-bit halves.

The Blackfin processor's instruction set supports a single-cycle 32-bit addition of the form Rn = Rm + Rp, where n, m, or p is the number index of the register. Similarly, 32-bit subtraction can be carried out.

32-Bit multiplication involves more operations and memories to derive the final 64-bit result. Usually, a fractional (1.31) format is used for storing the data. For example, we can perform 32-bit multiplication with several single-cycle 16-bit

230 Chapter 6 Real-Time DSP Fundamentals and Implementation Considerations

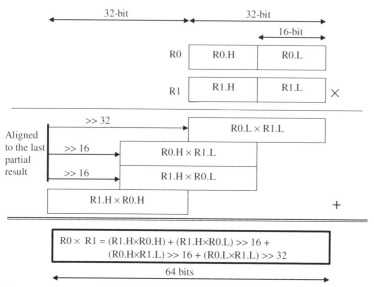

Figure 6.9 A 32-bit multiplication using several 16-bit multiplications

multiplications. As shown in Figure 6.9, two 32-bit operands are loaded into R0 and R1 registers and can be partitioned into two 16-bit halves (R0.H, R0.L, R1.H, R1.L) for four 16-bit multiplications. The partial results must be correctly aligned with reference to the MSB before being added together. Because multiplication is carried out in (1.31) format, the final result is in (1.63) format with the redundant sign bit removed. In most applications, we only need to save the most significant 32-bit result. Therefore, the (1.63) format is either rounded or truncated to (1.31) format to fit into a 32-bit data register or a 4-byte memory.

To reduce the computational load, the lowest 16-bit multiplication (R0.L × R1.L) can be removed. This is possible because the contribution from this lowest 16-bit multiplication is negligible if the final result is truncated to (1.31) format (see Fig. 6.9). Therefore, the reduced 32-bit multiplication can be stated as:

```
R0×R1 = (R1.H×R0.H)+(R1.H×R0.L)>>16+(R0.H×R1.L)>>16
```

HANDS-ON EXPERIMENT 6.6

This experiment demonstrates how to perform 32-bit multiplication. We use the same data as in Hands-On Experiment 6.4, but extend the 16-bit data and coefficients into 32-bit data sets. Therefore, the 10-tap FIR filter is now operating with a 32-bit precision. Load the project file `exp6_6.dpj` in directory `c:\adsp\chap6\exp6_6` into the BF533/BF537 simulator to perform the single-MAC FIR filtering using 32-bit precision. Note that reduced 32-bit multiplication is being used in this program.

It is noted that there are two types of multiplication in the program. The first type uses the mixed-mode (M) option and is applied to the (R1.H × R0.L) and

(R0.H × R1.L) multiplications. The mixed-mode option treats the first operand as signed and the second operand as unsigned to produce a 32-bit result. In fractional multiplication, the multiplication of two numbers in (1.15) and (0.16) formats produces a result in (1.31) format. There is no automatic left shift by 1 bit. The second type of multiplication, (R0.H × R1.H), is performed in the default sign fractional option. Extend this experiment to perform a full 32-bit multiplication. Compare the filter outputs (in 32 bits) with reduced and full 32-bit multiplications.

6.1.3 Fixed-Point Data Types

The Blackfin C compiler supports eight scalar data types and two fractional data types as shown in Table 6.7. Fractional data types can be represented in either `fract16` (1.15) format or `fract32` (1.31) format. However, these fractional data types are reserved only for fractional value built-in functions. In Chapter 8, we give more examples using these data types, explain different programming styles, and use built-in functions and a DSP library to ease the task of writing code.

6.1.4 Emulation of Floating-Point Format

In general, fixed-point processors perform fixed-point arithmetic. Fixed-point processors have higher processing speed, lower power consumption, and lower cost compared to floating-point processors, such as the SHARC processor [47] from Analog Devices. Floating-point processors usually consist of 32-bit registers and thus offer higher precision and wider dynamic range compared to fixed-point processors.

Although Blackfin processors are designed for native fixed-point computations, they can also emulate floating-point operations in software. To standardize the

Table 6.7 Fixed-Point Data Types

Type	Number Representation
char	8-bit signed integer
unsigned char	8-bit unsigned integer
short	16-bit signed integer
unsigned short	16-bit unsigned integer
int	32-bit signed integer
unsigned int	32-bit unsigned integer
long	32-bit signed integer
unsigned long	32-bit unsigned integer
fract16	16-bit signed (1.15) fractional number
fract32	32-bit signed (1.31) fractional number

floating-point formats, the Institute of Electrical and Electronics Engineers (IEEE) introduced a standard for representing floating-point numbers in 32-bit (single precision), ≥43-bit (extended single precision), 64-bit (double precision), and ≥79-bit (extended double precision) formats. These formats are stated in the IEEE-754 standard [7].

The floating-point number consists of three fields: sign bit, exponential bits, and mantissa bits. The sign bit represents the sign of the number (1 for negative and 0 for positive). The exponential bits contain the value to be raised by a power of two, and the mantissa bits are similar to the fractional bits in the fixed-point number format. As shown in Figure 6.10(a), the IEEE floating-point (single precision) format is expressed as

$$x = -1^s \times 2^{(\exp-127)} \times 1.\text{man}, \quad (6.1.1)$$

where the sign bit is b_{31}, the exponential bits (exp) are 8 bits from b_{30} to b_{23}, and the mantissa bits (man) are 23 bits from b_{22} to b_0. The double-precision format shown in Figure 6.10(b) can be expressed as

$$x = -1^s \times 2^{(\exp-1023)} \times 1.\text{man}, \quad (6.1.2)$$

where the sign bit is b_{63}, the exponential bits are 11 bits from bits b_{62} to b_{52}, and the mantissa bits are 52 bits from b_{51} to b_0.

The exponential value is used to offset the location of the binary point left or right. In the IEEE standard, the exponential value is biased by a value of 127 (single precision) or 1,023 (double precision) to obtain positive and negative offsets. A set of rules for representing special floating-point data types is stated in the IEEE-754 standard.

An important difference can be observed between fixed-point and floating-point numbers. In fixed-point representation, the radix point is always at the same location. In contrast, the floating-point number has a movable radix point that can be positioned to represent very large or very small numbers. Therefore, floating-point DSP processors automatically scale the number to obtain the full-range representation of the mantissa, which is done by increasing or decreasing the exponential value for small or large numbers, respectively. In other words, floating-point processors track the number and adjust the value of the exponent.

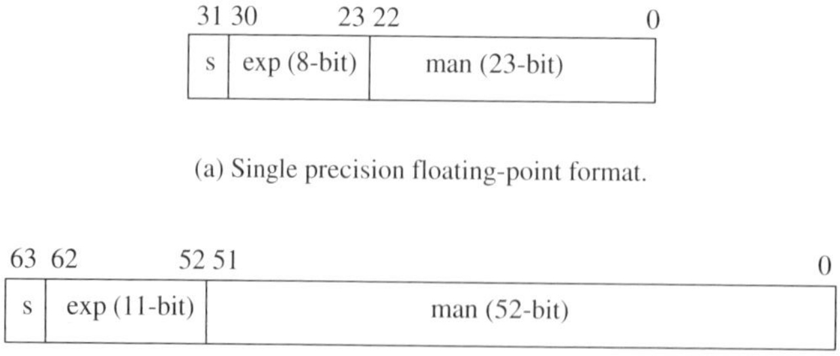

```
  31 30      23 22                    0
 | s | exp (8-bit) |    man (23-bit)    |
```

(a) Single precision floating-point format.

```
  63 62         52 51                              0
 | s | exp (11-bit) |         man (52-bit)          |
```

(b) Double precision floating-point format.

Figure 6.10 IEEE-754 floating-point formats

Table 6.8 Floating-Point Data Types

Type	Number Representation
`float`	32-bit IEEE single-precision floating point
`double`	32-bit IEEE single-precision floating point

6.1.4.1 Floating-Point Data Types

The Blackfin C compiler supports two floating-point data types as shown in Table 6.8. Note that the data type `double` is equivalent to `float` on Blackfin processors because 64-bit values are not supported directly by the hardware. Therefore, 64-bit double-precision floating-point numbers must be implemented with software emulation. In general, it is not recommended to perform floating-point operations on the Blackfin processors because of higher cycle count compared to the fixed-point operations.

6.1.4.2 Floating-Point Addition and Multiplication

To perform floating-point addition and subtraction, the smaller of the two numbers must be adjusted such that they have the same exponential value.

EXAMPLE 6.2

Add two floating-point numbers 0.5×2^2 and 0.6×2^1. The smaller of these numbers is 0.6×2^1 and must be adjusted to $0.3 - 2^2$ by increasing the exponent to match 0.5×2^2. The two numbers now can be added to become 0.8×2^2. The final answer has a mantissa of 0.8 and an exponent of 2.

QUIZ 6.4

The floating-point numbers given in Example 6.2 are not defined in IEEE-754 format. Express the numbers in single-precision IEEE-754 floating-point format and perform the floating-point addition.

Floating-point multiplication can be carried out in a more straightforward manner. The mantissas of the two numbers are multiplied, whereas the exponential terms are added without the need to align them.

EXAMPLE 6.3

Multiply 0.5×2^2 by 0.6×2^1. In this case, the mantissa 0.5 is multiplied by 0.6 to get 0.3, and the exponents are added together $(2 + 1)$ to become 3. The result is 0.3×2^3. How do we perform the multiplication with IEEE-754 format?

6.1.4.3 Normalization

Normalization of floating-point numbers is an important step of floating-point representation. A floating-point number must be normalized if it contains redundant sign bits in the mantissa. In other words, all bits in the mantissa are significant and provide the highest precision for the number of available mantissa bits. Normalization is required when comparing two floating-point numbers. After the two numbers are normalized, the exponents of these numbers are compared. If the exponents are equal, the mantissas are examined to determine the bigger of the two numbers.

EXAMPLE 6.4

This example demonstrates how to normalize a floating-point number. A floating-point number 16.25 is represented in binary as $10,000.01 \times 2^0$, which can be normalized to 1.000001×2^4 by left-shifting the binary point by 4 bits. Normalization is used to maximize the precision of a number. Therefore, the normalized binary number always has a significant bit of 1 and follows the binary point.

QUIZ 6.5

Express the following floating-point numbers in normalized single-precision IEEE-754 format:

1. 0.125
2. 23.125
3. Smallest positive number 2^{-126}
4. Largest positive number $1.99999988079 \times 2^{127}$

6.1.4.4 Fast Floating-Point Emulation

To reduce the computational complexity of using the IEEE-754 standard, we can relax the rules by using a fast floating-point format [40]. This two-word format is employed by the Blackfin processor to represent short (32 bit) and long (48 bit) fast floating-point data types. The two-word format consists of an exponent that is a 16-bit signed integer and a mantissa that is either a 16-bit (short) or a 32-bit (long) signed fraction as shown in Figure 6.11.

Floating-point emulation (both IEEE-754 and fast formats) on fixed-point DSP processors is very cycle intensive because the emulator routine needs to take care of both the exponent and the mantissa. It is only worthwhile to perform emulation on small sections of data that require high dynamic range and precision. A better approach is to use the block floating-point format, which is described next.

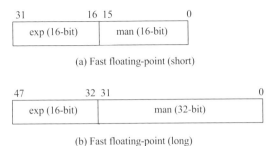

(a) Fast floating-point (short)

(b) Fast floating-point (long)

Figure 6.11 Fast floating-point formats

6.1.5 Block Floating-Point Format

In block floating-point format, a set (or block) of data samples share a common exponent. A block of fixed-point numbers can be converted to block floating-point numbers by shifting each number left by the same amount and storing the shift value as the block exponent. Therefore, the data samples are still stored in fixed-point format, for example, (1.15) format, and the common exponent is stored as a separate data word. This arrangement results in a minimum memory requirement compared to the conventional floating-point format. In general, the common exponent for the group of numbers is derived from the number with the largest absolute value.

A block floating-point format combines the advantages of both fixed-point and floating-point formats. The block floating-point format allows a fixed-point processor to increase its dynamic range as a fixed-point computation without the extensive overhead and memory needed to emulate floating-point arithmetic. It has a common exponent for all data values within the same data block and preserves the precision of a fixed-point processor.

EXAMPLE 6.5

This example groups and represents 10 numbers with block floating-point numbers. These numbers are 0.012, 0.05, −0.03, 0.06, −0.009, −0.01, 0.07, 0.007, −0.09, and 0.01. These numbers can be represented in fixed-point (1.15) format hexadecimal as 0x0189, 0x0666, 0xFC29, 0x07AE, 0xFED9, 0xFEB8, 0x08F6, 0x00E5, 0xF476, and 0x0148. They are associated to a common exponent of 3. It is noted that these fixed-point numbers in (1.15) format have at most three nonsignificant, redundant sign bits. Therefore, each data value within this block can be normalized by left-shifting of three bits to become 0x0C48, 0x3330, 0xE148, 0x3D70, 0xF6C8, 0xF5C0, 0x47B0, 0x0728, 0xA3D8, and 0x0A40 with a common exponent of 0.

The Blackfin processor provides exponent detection and sign bit instructions to adjust the exponential value for the block of data samples. The sign bit instruction SIGNBITS returns the number of sign bits in a number minus 1 (also the reference exponent). It can be used in conjunction with the ASHIFT instruction to normalize the number. The exponent detection instruction EXPADJ is used to identify the largest magnitude of the numbers based on their reference exponent derived from the sign bit instruction. If block floating-point numbers have an exponent that is less than the reference exponent, right-shift these numbers by P bits (where P is the reference exponent minus the detected exponent) to maintain the reference exponent before the next process.

For example, if the reference exponent of the block floating-point numbers given in Example 6.5 is 3, and the exponent is changed to 1 (or 1 redundant sign bit) after processing, the numbers within the block can be shifted by 2 bits to maintain the reference exponent of 3. Therefore, the block floating-point format can be used to eliminate the possibility of overflowing.

6.2 DYNAMIC RANGE, PRECISION, AND QUANTIZATION ERRORS

This section discusses several practical issues on the dynamic range and precision of digital signals. We also study a common error occurring in fixed-point implementation called the quantization error. Many factors contribute to this error, including nonperfect digitization of the input signal, finite wordlength of the memory (or register), and rounding and truncation of fixed-point arithmetic. MATLAB experiments are developed to illustrate these important concepts.

6.2.1 Incoming Analog Signal and Quantization

As introduced in Chapters 1 and 2, the incoming analog signal, $x(t)$, is sampled at a regular interval of T_s seconds. These samples must be quantized into discrete values within a specific range for a given number of bits. Therefore, the analog-to-digital converter (ADC) performs both sampling and quantization of the analog signal to form a discrete-time, discrete-amplitude data sample called the digital signal as shown in Figure 6.12.

For a B-bit ADC with full-scale (peak to peak) voltage of V_{FS}, the physical quantization interval (step) or LSB is defined as

$$\Delta = \frac{V_{FS}}{2^B}. \tag{6.2.1}$$

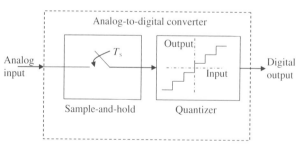

Figure 6.12 Sampling and quantization of the ADC

6.2 Dynamic Range, Precision, and Quantization Errors

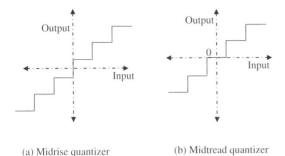

(a) Midrise quantizer (b) Midtread quantizer

Figure 6.13 Two different quantizers: midrise (a) and midtread (b)

For example, the AD1836A CODEC used in the BF533 EZ-KIT has four ADC channels. The maximum ADC resolution is 24 bits, and the full-scale voltage is 6.16 V. Therefore, the number of quantization levels is $2^{24} = 16{,}777{,}216$ and the quantization step $\Delta = 0.367\,\mu V$.

Equation 6.2.1 is applicable for a linear quantizer that has a midrise characteristic as shown in Figure 6.13(a). In the midrise quantizer, there are 2^B quantization levels. There is however, no zero output level in the midrise quantizer. An alternate linear quantizer has a midtread characteristic as shown in Figure 6.13(b). In contrast to the midrise quantizer, the midtread quantizer always has an odd number ($2^B - 1$) of quantization levels. The midtread quantizer has a zero output level, and it is often preferred for audio signals.

A quantization error (or noise) exists between the analog-valued discrete-time signal and its corresponding discrete-valued digital signal. The quantization errors depend on the type of quantizer used. For a simple linear quantizer, quantization errors are distributed uniformly between $\pm\Delta/2$. When an analog-valued discrete-time sample is halfway between two discrete levels, the quantizer quantizes the analog sample into the upper discrete level. In this situation, a maximum absolute quantization error of $\Delta/2$ occurs. In other words, the quantization error of an ideal ADC can never be greater than $\pm 1/2$ of LSB. The mean and variance of the quantization error $e(n)$ are:

$$m_e = 0 \qquad (6.2.2)$$

$$\sigma_e^2 = \Delta^2/12. \qquad (6.2.3)$$

HANDS-ON EXPERIMENT 6.7

This experiment quantizes an incoming analog signal $x(t)$ with the MATLAB program exp6_7.m in directory c:\adsp\chap6\MATLAB_ex6. It generates a 200-Hz sine wave sampled at 4 kHz.

```
fs = 4000;          % sampling rate is 4 kHz
f  = 200;           % frequency of sinewave is 200 Hz
```

```
n = 0:1/fs:3;           % time index n to cover 3 seconds
xn = sin(2*pi*f*n);     % generate sinewave
figure(1),plot(n, xn);grid on;
soundsc(xn_, fs);       % listen to the quantized signal
```

This sine wave can be quantized to 16 bits with a scaling factor of 32,768 for (1.15) format with the following MATLAB code:

```
xn_fix_1_15 = fix((2^15)*sin(2*pi*f*n/fs));
index = (xn_fix_1_15 == 2^15);
xn_fix_1_15(index) = 2^15-1;
figure, plot(n, xn_fix_1_15); grid on;
soundsc(xn_fix_115, fs); % listen to the quantized signal
```

Repeat the same process of quantizing the sinewave to:

1. 8 bits, using (1.7) format with scaling factor of 128
2. 4 bits, using (1.3) format with scaling factor of 8
3. 2 bits, using (1.1) format with scaling factor of 2

Can you still hear the sine wave encoded with 2 bits? Perform FFT over a single period of these quantized signals and observe their magnitude spectrum plots. Explain the differences in what you hear. Replace the sine wave with a speech wave file, timit.wav, and repeat the quantization process with different wordlengths. Can you still perceive the speech signal using 2 bits?

6.2.2 Dynamic Range, Signal-to-Quantization Noise Ratio, and Precision

The dynamic range of a digital signal is defined as the difference between the largest and smallest signal values. If noise is present in the system, the dynamic range is the difference between the largest signal level and the noise floor. Dynamic range is also commonly defined by taking the logarithm of the ratio between the largest signal level and the noise floor (assuming signal is inaudible below the noise floor) as follows:

$$\text{dynamic range} = 20 \log_{10} \left(\frac{\text{largest signal level}}{\text{noise floor}} \right) (\text{dB}) \qquad (6.2.4)$$

The signal-to-quantization noise ratio (SQNR) has the same definition as dynamic range expressed in Equation 6.2.4 when noise is measured in the absence of any signal. In theory, there is an increase of SQNR and dynamic range by approximately 6 dB for every bit increased. Therefore,

$$\text{dynamic range} = \text{SQNR} \approx 6 \times B \text{ (dB)}, \qquad (6.2.5)$$

where B is the wordlength of the ADC. A more accurate expression is stated in Chapter 7. For example, if an ADC has a wordlength of 16 bits, the dynamic range is approximately 96 dB. In addition, the overall dynamic range of the system is always bounded by the ADC. When performing signal processing, a higher precision

6.2 Dynamic Range, Precision, and Quantization Errors

(>16 bits) is normally preferred to maintain this level of dynamic range. There are other sources of quantization errors, which are explained in Section 6.2.3.

QUIZ 6.6

A 16-bit ADC is used for "CD-quality" audio. The 16-bit digital sample is transferred to a processor that operates on 16-bit wordlength. However, three bits of truncation error occurs when a simple audio processing algorithm is performed in the processor. These data samples are sent to the 16-bit DAC.

1. What is the SQNR at the ADC?
2. What is the SQNR after processing?
3. How do we compensate for the loss of the SQNR?
4. Is there a way to increase the SQNR of the overall system?

QUIZ 6.7

A 12-bit converter is used to convert a temperature reading to a digital value once every second, that is, a sampling frequency of 1 Hz.

1. What is the temperature range to be displayed if the temperature resolution is 0.1°C?
2. If oversampling of the temperature readings is allowed, specify the oversampling factor that allows an increase of resolution to 16 bits. Hint: Every factor of 4 times oversampling results in an increase of 1-bit precision.
3. What is the maximum temperature range that can be achieved with 16-bit precision if the temperature resolution is now 0.01°C?

EXAMPLE 6.6

The AD1836 audio CODEC [41] offers 24-bit, 96-kHz multichannel audio capability. It has a dynamic range of 105 dB. A digital audio system has 20-dB headroom and uses either 16-bit or 32-bit wordlength in the Blackfin processor as shown in Figure 6.14.

The dynamic range of the 16-bit processor is not sufficient to preserve the data samples from the 24-bit ADC. Therefore, a 32-bit wordlength can be used to preserve the 24-bit precision. Using 32-bit wordlength and assuming that each bit contributes to 6 dB of SQNR gain, there are approximately 14 bits below the noise floor of the AD1836 CODEC. This extra floor room is useful in keeping the error in arithmetic computation below the noise floor. In conclusion, preserving the quality of 24-bit samples requires at least 32 bits (or extended precision) for signal processing. However, to keep the programming simple, we use the 16-bit mode of the AD1836 CODEC and perform 16-bit computations in the Blackfin processor in most of the experiments and exercises in this book.

240 Chapter 6 Real-Time DSP Fundamentals and Implementation Considerations

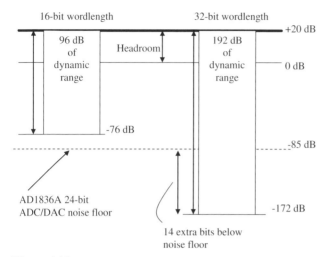

Figure 6.14 Comparison of AD1836 CODEC noise floor with different wordlengths

Precision defines the resolution of the digital signal representations. In fixed-point format, precision equals to the size of the LSB of the fraction (see Table 6.2). For example, the precision of (1.15) format is 2^{-15} and the precision of (16.0) format is 1. Therefore, the wordlength of the fixed-point format governs its precision. For the floating-point format, precision is derived as the minimum difference between two numbers with a common exponent.

QUIZ 6.8

Complete Table 6.9 by stating the precision and dynamic range of the number formats. Comment on the differences between fixed-point and floating-point representations.

By default, MATLAB is operating in IEEE-754 double-precision floating-point format (64 bits). Type in the following MATLAB commands to confirm:

```
eps        % precision used in MATLAB
realmax    % maximum number represented in MATLAB
realmin    % smallest number represented in MATLAB
```

6.2.3 Sources of Quantization Errors in Digital Systems

Besides analog-to-digital and digital-to-analog quantization noise, there are other sources of quantization errors in digital systems. These error sources include:

1. Coefficient quantization
2. Computational overflow
3. Quantization error due to truncation and rounding

6.2 Dynamic Range, Precision, and Quantization Errors

Table 6.9 Precision versus Dynamic Range

Number Format	Precision	Dynamic Range (for positive range)	Dynamic Range in dB
(0.16) fixed point			
(1.15) fixed point			
(2.14) fixed point			
(16.0) fixed point			
IEEE-754 single precision			
IEEE-754 double precision			

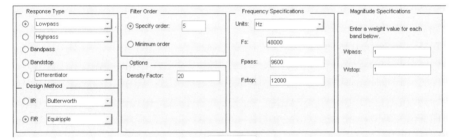

Figure 6.15 Parameters used for designing an FIR filter

6.2.3.1 Coefficient Quantization

When a digital system is designed for computer simulation, the coefficients are generally represented with floating-point format. However, these parameters are usually represented with a finite number of bits in fixed-point processors with a typical wordlength of 16 bits. Quantization of coefficients can alter the characteristics of the original digital system. For example, coefficient quantization of an IIR filter can affect pole/zero locations, thus altering the frequency response and even the stability of the digital filter.

HANDS-ON EXPERIMENT 6.8

This example uses FDATool (introduced in Chapter 4) to examine the filter coefficient quantization effects. FDATool allows quick examination of various quantization errors including coefficient quantization. Open FDATool and design a 6-tap FIR low-pass filter by entering the parameters shown in Figure 6.15. Click on the **Design Filter** button to start the filter design. Observe the magnitude response by clicking on the magnitude response icon .

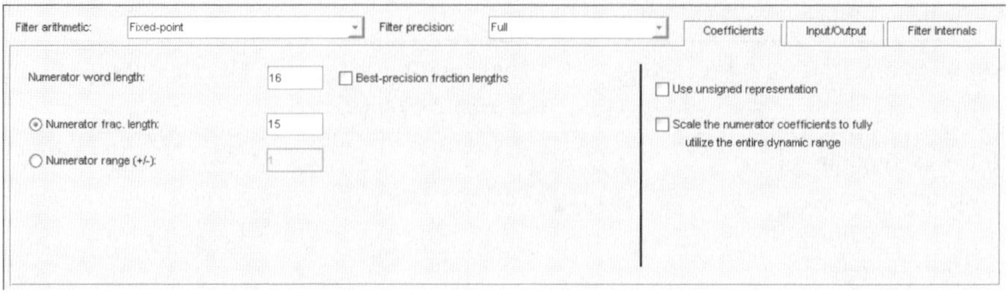

242   Chapter 6   Real-Time DSP Fundamentals and Implementation Considerations

Figure 6.16 Coefficient quantization settings

```
Quantized Numerator:
-0.0675048828125
 0.2841796875
 0.444244384765625
 0.444244384765625
 0.2841796875
-0.0675048828125

Reference Numerator:
-0.0674941238280376
 0.2841670814341582
 0.44424423787605283
 0.44424423787605283
 0.2841670814341582
-0.0674941238280376
```

Figure 6.17 Comparison of quantized and reference filter coefficients

We can quantize the filter coefficients by using the fixed-point (1.15) format in the Blackfin processor. Simply click on the **Set quantization parameters** icon to open a new window as shown in Figure 6.16 and set the wordlength and fractional length as 16 and 15, respectively. Click on **Apply** to start the quantization.

Display the filter coefficients by clicking on the icon at the top row of the window. A window displays the coefficients quantized in (1.15) format and reference floating-point double-precision format as shown in Figure 6.17. It is noted that there are variations between the reference and quantized coefficients. The difference is called the coefficient quantization error. Compute the sum of the squared quantization error for the 6-tap FIR filter. In this example, the quantization error has a marginal effect on the reference filter response based on double-precision floating-point format.

Repeat the experiment with (1.31) and (1.7) formats. Compare the quantization errors with different data formats. In particular, examine the frequency response and pole/zero plots for the (1.7) format and comment on the results.

As shown in the preceding experiments, coefficient quantization can affect the frequency response of the filter. In the case of an IIR filter, the coefficient quantization can even cause instability if the quantized poles lie outside the unit circle. Therefore, careful design of the quantized IIR filter must be performed to ensure stability before porting the coefficients to the fixed-point processor.

HANDS-ON EXPERIMENT 6.9

A simple 2nd-order IIR filter with a complex pole-pair at $0.9998\angle \pm 1.04$ radians (where $\angle$ is the angular frequency) is first designed with double precision, using FDATool. Click on the **Pole/Zero Editor** icon in FDATool. A new window that shows the pole/zero plot of the filter is displayed. Remove all the zeros and poles by right-clicking on the z-plane, choosing **Select All**, and then pressing **Delete**. A blank window appears. Click on the **Add pole** icon and select **Conjugate** in the menu. Enter the magnitude and phase as shown in Figure 6.18. Observe the impulse response and note that it will gradually decay to zero.

Click on the **Set quantization parameter** icon. A new window appears; select Fixed-Point under **Filter arithmetic**. We now implement the designed IIR filter with the fixed-point (1.7) format as shown in Figure 6.19.

Click on **Apply** and examine the impulse response of the quantized IIR filter. Note that the quantized filter using the (1.7) format produces an oscillatory impulse response, which is undesirable. Examine the quantized and reference (using double-precision floating-point format) coefficients of the IIR filter and explain the differences.

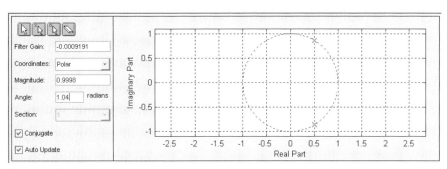

Figure 6.18 Pole-zero editor

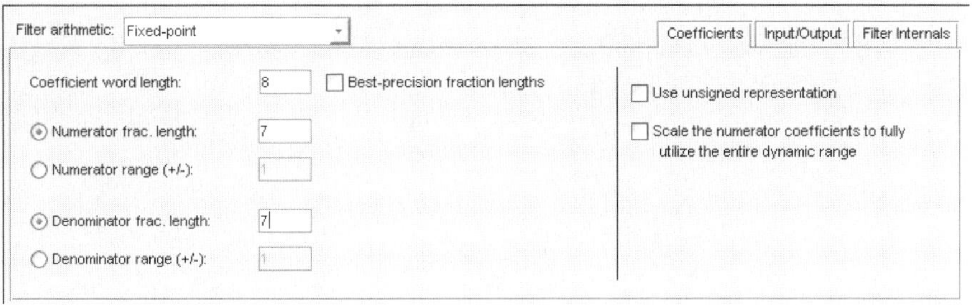

Figure 6.19 Quantizing filter coefficients

6.2.3.2 Computational Overflow

Because of the finite memory/register length, the results of arithmetic operations may have too many bits to be fitted into the wordlength of the memory or register, for example, when adding two 8-bit numbers in (1.7) format, 0111 0000b (0.875) and 0100 1111b (0.6171875), and saving the result into another 8-bit register as 1011 1111b (−0.5078125). This is the wrong result, because adding two positive number results in a negative number! Figure 6.20 shows an 8-bit number circle for adding the above two numbers shown in hexadecimal. The positive number starts from the 12 o'clock position (or 0x00) of the number circle and increments in a clockwise direction to the most positive number of 0x7F. The next position clockwise after the 0x7F is immediately the most negative number of 0x80. The negative number range is decremented (in clockwise direction) toward 0xFF, which is the smallest negative number. When adding the preceding numbers, we first locate 0x70 (0.875) and add 0x4F (0.6171875) in a clockwise manner to reach the overflowed answer at 0xBF. When performing subtraction, the direction is changed to counterclockwise. It is also noted that when adding two numbers of different signs overflow will never occur. Therefore, the number circle is a useful tool to illustrate the concept of number overflow in add/subtract operations.

The number circle in Figure 6.20 can be extended to illustrate any addition/subtraction overflow in any fixed-point and floating-point formats. A general rule in fixed point arithmetic is that the sum of M B-bit numbers requires $B + \log_2 M$ bits to represent the final result without overflow. For example, if 256 32-bit numbers are added, a total of $32 + \log_2 256$ (=40) bits are required to save the final result. In the case of the Blackfin processor, 40-bit accumulators are available to ensure that no overflow occurs when adding 256 numbers in an extreme case. This extreme case is demonstrated in Hands-On Experiment 6.3.

One method to avoid arithmetic overflow is to perform scaling, followed by truncation or rounding after every partial sum in order to fit into the given wordlength. Scaling is a process of reducing the value of a number with a certain constant so that the end result can fit into the wordlength of the processor. In

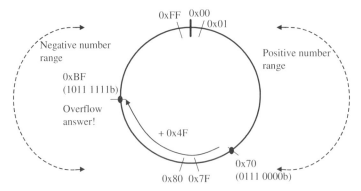

Figure 6.20 A 8-bit number circle and its addition

DSP systems, scaling can be applied to the signal or to the internal parameters of the system. For example, the coefficients of the FIR filter can be scaled down such that it has a gain of 1. Alternatively, we can also scale the input signal. However, scaling reduces the dynamic range (or SQNR) of the digital system. For example, right shift of the input signal by 1 bit (or scale by 0.5) results in a loss of 6 dB in the dynamic range of the input signal. Several methods of deriving scaling factors are:

1. Scaling by sum of magnitude of impulse response (L_1 norm). In this method, we constrain the magnitude of the digital system at any node to be less than 1 for a system using (1.15) format. If the maximum input signal x_{max} to the digital system is $(1 - 2^{-15})$, the output of the digital system is restricted to $|y(n)| < 1$ provided that the scaling factor is limited by

$$G < \frac{1}{x_{max} \sum_{k=0}^{N-1} |h_k|}, \qquad (6.2.6)$$

where h_k is the impulse response of the filter with length N. The summation term, $\sum_{k=0}^{N-1} |h_k|$, is also called the L_1 norm of the impulse response h_k, and it is denoted as $\|h_k\|_1$. The scaling factor G can be applied to either the input signal or the coefficients of the filter. This scaling factor is the most stringent and guarantees no overflow.

2. Scaling by square root of sum of squared magnitude of impulse response (L_2 norm). Besides using the magnitude of the impulse response, we can also relax the restriction by using a scaling factor as follows:

$$G < \frac{1}{x_{max} \left(\sum_{k=0}^{N-1} h_k^2 \right)^{1/2}}. \qquad (6.2.7)$$

The term $\left(\sum_{k=0}^{N-1} h_k^2 \right)^{1/2}$ is called the L_2 norm of impulse response, and it is denoted as $\|h_k\|_2$. The L_2 norm is always less than the L_1 norm of h_k.

3. Scaling by maximum of frequency response (Chebyshev norm). The preceding two scaling methods are suitable for wideband signals. The third method to determine the scaling factor is applicable when the input is a narrowband (or sinusoidal) signal. In this method, the magnitude response (or system gain) at the input frequency is first determined, which is multiplied by the maximum input signal x_{max} to determine the scaling gain as follows:

$$G < \frac{1}{x_{max} \max[H(\omega_k)]}. \qquad (6.2.8)$$

The term $\max[H(\omega_k)]$ is known as the Chebyshev norm of the frequency response $H(\omega)$. The Chebyshev norm guarantees that the steady-state response of the system to a sine wave input will never overflow. We will use an example to illustrate the effects of using different scaling methods.

EXAMPLE 6.7

A simple 2nd-order IIR filter with the transfer function

$$H(z) = \frac{1 + 0.72z^{-1} + z^{-2}}{1 + 0.052z^{-1} + 0.8z^{-2}}$$

is implemented in MATLAB file example6_7.m. Observe the impulse response and compute the L_1 and L_2 norms. Also, compute the frequency response of the transfer function and pick the highest magnitude to compute the Chebyshev norm. It is noted that L_1 norm $\geq$ Chebyshev norm $\geq L_2$ norm. The L_1 norm is the most conservative. We can observe the frequency responses after scaling by the three norms as shown in Figure 6.21. The most conservative scaling has all its magnitude below 1. It is observed in Figure 6.21 that the L_2 norm allows magnitude to exceed 1 within a normalized frequency from 0.2 to 0.55. The Chebyshev norm has its maximum frequency response at around 0.5.

In this example, we only compute the scaling factor based on the output of the IIR filter. However, overflow can occur in other addition nodes. Assuming that (1.15) format is used and multiplication nodes will not overflow, there is a need to compute the scaling factor for all addition nodes. The overall scaling factor for the filter is based on the largest norm. Fortunately, we can skip the scaling factor computation for all nodes, if the nonsaturating

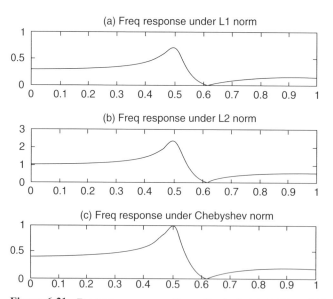

Figure 6.21 Frequency responses after scaling by L_1 (a), L_2 (b), and Chebyshev (c) norms

mode is used and internal nodes are allowed to overflow. The final output node must be computed to ensure that no overflow is allowed at the final result. This intermediate overflow is also demonstrated in Hands-On Experiment 6.1 on the Blackfin processor.

Another method to avoid overflow is use of the saturation mode of the processor. The saturation mode limits any positive or negative number from exceeding its most positive (0x7FFF) or most negative (0x8000) for a 16-bit number, respectively. In consecutive summations, the saturation mode can be applied to every summation, but this step will produce too much error. A better approach is to allow intermediate overflow during consecutive summation and only apply saturation in the final addition.

In Section 6.1.1.2, we discussed several aspects of overflow due to multiplication in different formats. In particular, (1.15) format is preferred because multiplying two numbers in (1.15) format cannot lead to overflow, with the exception of $-1 \times -1 = 1$. A method to overcome this exceptional case is to saturate the result to $1 - 2^{-15}$.

6.2.3.3 Truncation and Rounding

As discussed in Section 6.1.1.4, truncation or rounding is used after multiplication to store the result back into the memory. Truncation or rounding must also be used in other arithmetic operations such as addition, division, square root, trigonometric functions, and other operations. In general, the distribution of errors caused by rounding is uniform, resulting in quantization errors with zero mean and variance of $\Delta^2/12$. Truncation has a bias mean of $\Delta/2$ and a variance of $\Delta^2/12$.

EXAMPLE 6.8

There are several MATLAB functions that can be used to truncate and round numbers. These functions include `round`, which rounds a number to the nearest integer; `floor`, which rounds a number toward floor (or negative infinity); `ceil`, which rounds a number toward ceiling (or positive infinity); and `fix`, which rounds a number toward zero (or truncation). We can simply plot a 1-Hz sine wave, sampling at 20Hz, using different rounding and truncation methods to illustrate the behaviors of rounding and truncation. Run `example6_8.m` located in directory `c:\adsp\chap6\MATLAB_ex6` to obtain the graphs shown in Figure 6.22. This example shows that the `round` function has the smallest squared error and the `fix` function has the highest squared error. From Figure 6.22, the `floor` function has negative cumulative bias error, whereas the `ceil` function has positive cumulative bias error. We can also relate the rounding schemes in MATLAB to the rounding schemes in the Blackfin processor as stated in Section 6.1.1.4. Unbiased and bias rounding in Blackfin processors are equivalent to the `round` and `ceil` functions in MATLAB, respectively.

In an IIR filter, limit cycles can occur because of the truncation and rounding of multiplication results or addition overflow. The limit cycles cause periodic oscillations in the output of the IIR filter, even when no input signal is applied to the filter. We use a simple example to illustrate this phenomenon.

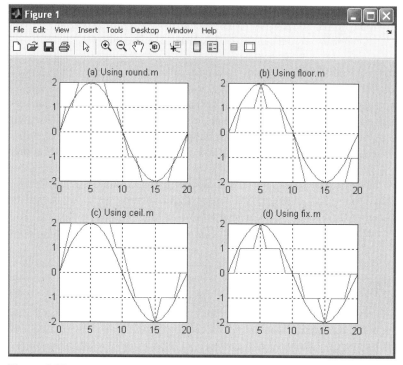

Figure 6.22 Performance of different rounding schemes used in MATLAB

HANDS-ON EXPERIMENT 6.10

A simple 2nd-order IIR filter is shown in Figure 6.23. The summation of the feedback paths is rounded before storing to the delay elements. Different rounding and truncation schemes can be examined to test their suitability. A MATLAB file, exp6_10.m, is written to implement the IIR filter with different rounding schemes. The initial internal state of the IIR filter is given as $y(-2) = 0$ and $y(-1) = 0.001$, and no input is applied to the filter. Examine the response (limit cycle oscillation) of this filter under different rounding schemes. Suggest a method to remove the limit cycle oscillation.

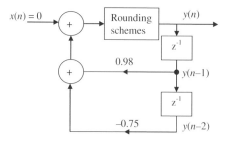

Figure 6.23 A 2nd-order IIR filter with no input applied

6.2.3.4 Overall Quantization Errors

In this section, we summarize quantization errors found in a typical digital system. We use the 4-tap FIR filter shown in Figure 6.24 to illustrate different quantization errors. The quantization error e_{in} first occurs when the analog input signal is converted into digital samples. The ADC wordlength and conversion error degrades the dynamic range of the digital signal as compared to the analog signal. The quantized signal is passed into the signal buffer for processing. The dynamic range of the signal processing path depends on the program definitions. These definitions include wordlength used to declare coefficients of filter and whether truncation or rounding is applied at the final multiply-accumulate of the FIR filter. The coefficient quantization error e_{coeff} is the error introduced when converting a coefficient from a reference floating-point to a fixed-point representation. There is saturation error e_{sat} if saturation mode is set during the multiply-accumulate operations and results exceed the dynamic range of the data format. However, saturation error in the accumulator will only occur in an extreme case. In addition, truncation error e_{trunc} or rounding error e_{round} will only be considered when transferring the accumulator result to memory location with shorter wordlength, before passing to the DAC for conversion into analog signal. The wordlength of DAC also restricts the final dynamic range and contributes to the DAC quantization error e_{out}. A fixed-point simulation tool like FDATool can be used to evaluate the performance of the digital system under different fixed-point quantization errors.

In general, using a double-precision (32-bit) arithmetic in a fixed-point processor ensures that the 16-bit or 24-bit signal samples from the ADC are not impaired by the quantization noise. Alternatively, a 32-bit fixed-point or floating-point processor is required to maintain the dynamic range of the input signal. The

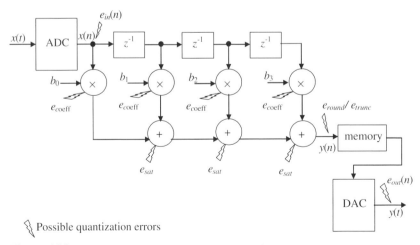

Figure 6.24 Quantization error sources in a 4-tap FIR filter

use of 32-bit arithmetic is particularly crucial for IIR filtering. We examine this case in Section 6.5.

6.3 OVERVIEW OF REAL-TIME PROCESSING

This section discusses important topics on real-time signal processing and defines the terms used in measuring real-time performance. We introduce concepts of real-time and non-real-time processing and explain the sample-by-sample and block processing modes. To benchmark real-time performance in embedded signal processing, we measure the cycle counts of running different tasks on the Blackfin processor. A brief introduction of the power measurement and its relation to the processing speed is also given, but a detailed discussion on power-saving features of the Blackfin processor is postponed to Chapter 8.

6.3.1 Real-Time Versus Offline Processing

A real-time system processes data at a regular and timely rate. As shown in Figure 6.25, the inputs of a real-time system are often associated with signal capturing devices like microphones, cameras, thermometers, etc. The inputs can also come from digital media streaming devices like audio and video players. The outputs can be devices like loudspeakers, video display, etc. that play back the processed signals.

To be more explicit, a real-time system must satisfy certain response time constraints. The response time is defined as the time between the arrival of input data sample(s) and the output of processed data sample(s). For example, the response time constraint for a typical speech processing system is to digitize the analog speech, process the digital speech, and output the processed signal within a given sampling period. If the response time exceeds the sampling period, the new speech sample cannot be retrieved on time and thus violates the real-time constraint. A more detailed explanation is provided in Sections 6.3.2 and 6.3.3.

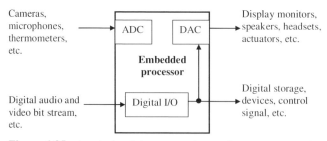

Figure 6.25 A typical real-time system that receives, processes, and transmits data

In contrast, an offline processing system is not required to complete the task within allocated time. For example, we can sample a noisy speech signal, save it in a data file, and run a program on a computer to read the speech samples from the file and perform noise reduction. After processing, we can store the clean signal to a data file and play it back with a loudspeaker. In this way, there is no timing constraint to complete the overall receive-process-transmit chain. Offline processing always involves extensive memory storage, and it is often used in film and music postproduction. The Blackfin simulator can be used to simulate DSP algorithms in an offline manner because the test data are stored in a data file and no time constraint is imposed on the processing. However, when the same algorithm is implemented on the Blackfin processor, real-time processing must be carried out on the incoming signal.

6.3.2 Sample-by-Sample Processing Mode and Its Real-Time Constraints

The sample-by-sample processing mode requires that all operations must be completed within the given sampling period. As shown in Figure 6.26, an audio signal can be sampled at a sampling period of every T_s seconds. Latency (or response time) of processing can be defined as the total time from the instant the data sample is read in to the time the digital output is written to the memory. This latency contains the three subintervals listed below:

1. T_{in} is the time needed for the processor to copy the current sample from the ADC into the processor memory. It also includes the program access time.
2. T_{sp} is the time needed for processing the current data samples. This duration depends on the complexity of the algorithm and the efficiency of the program.
3. T_{out} is the time needed to output the processed data to the DAC.

The overall overhead time for sample-by-sample processing is denoted as T_{os}, which includes both T_{in} and T_{out} and the response time to interrupt. At the beginning of every sampling interval, the ADC samples new data, and the DAC sends out the processed data.

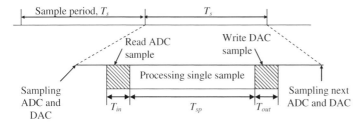

Figure 6.26 Timing details for sample-by-sample processing mode

The real-time constraint of the sample-by-sample processing is that the processing time T_{sp} must satisfy

$$T_{sp} \leq T_s - T_{os}. \qquad (6.3.1)$$

The advantages of using sample-by-sample processing are:

1. Delay between the input and the output is always kept within one sampling interval.
2. Single-sample storage for input and output samples. In some applications, multiple channels are acquired, processed, and output within the sampling interval. The memory requirement is increased to store the multichannel data samples.
3. Results are kept current within the sampling period.

The disadvantage of sample-by-sample processing is the overhead of program setup, program access, and latency in reading a new data sample and writing the processed sample in every sampling interval. The processor must be fast enough to complete all processing tasks before the arrival of the next input sample to avoid distortion caused by missing data. A possible method to reduce the processing speed requirement is to group a block of data samples and perform processing on this block of data as a batch. The block processing method is introduced in Section 6.3.3.

In addition, some DSP algorithms, such as the fast Fourier transform described in Chapter 3, require a block of data samples for processing. In this type of block processing algorithm, sample-by-sample processing cannot be implemented and the block processing mode must be applied.

6.3.3 Block Processing Mode and Its Real-Time Constraints

In the block mode, data samples are gathered into an input buffer of N samples and a whole block of samples are processed after the buffer is full. At the same time, a new block of N samples are acquired. Figure 6.27 shows block processing for a block of five samples. The block processing system starts by sampling the first five input samples from the ADC to form block i. A more detailed description of how data can be acquired into the processor is presented in Chapter 7. The system continues to sample another five data samples to form block $i + 1$. At the same time, the processor operates on data samples in block i and sends the five previously processed samples to the DAC. During the next block period, $i + 2$, another five newer samples are acquired. The processor operates on the data samples in block $i + 1$ and outputs the processed data samples in block i. Therefore, the data samples are output to the DAC after a delay of $2NT_s$ seconds.

To perform the block processing shown in Figure 6.27, we need to continuously save data samples in different memory buffers. Therefore, the memory requirement is increased. A memory buffering approach known as double (or ping pong) buffer-

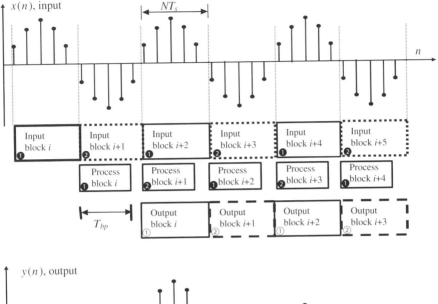

Figure 6.27 Block processing mode ($N = 5$).

ing is recommended. As shown in Figure 6.28, double buffering uses two memory buffers of length N for the input of data and another two buffers of the same length for the output. When the processor is operating on data in buffer(in) ❶, new input samples $x(n)$ are saved in buffer(in) ❷. The function of these two buffers is alternated every NT_s seconds. This "ping-pong" switching mechanism between data acquisition and processing is shown in Figure 6.28, with the labels identifying the buffer used in every block. In the same fashion, the output of data to the buffer and the sending of data out to the DAC are also alternated between two output buffers, buffer(out) ① and buffer(out) ②.

To meet real-time constraints for block processing shown in Figure 6.27, the computational time for block processing T_{bp} must satisfy

$$T_{bp} \leq NT_s - T_{ob}, \qquad (6.3.2)$$

where NT_s is the block acquisition time in seconds and T_{ob} is the overhead for block processing, which is mainly caused by program setup and the response time to get data in and out of the processor. In general, the overhead for block processing is lower than for sample-by-sample processing because the program access and setup time for block data transfer is lower than for single-sample transfer. That is,

254 Chapter 6 Real-Time DSP Fundamentals and Implementation Considerations

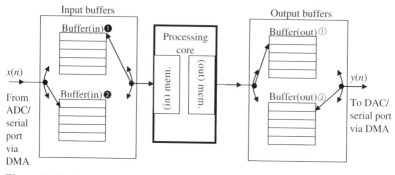

Figure 6.28 Implementation of double buffering

$$T_{ob} < NT_{os}. \qquad (6.3.3)$$

Therefore, more time is available for the processor to process signal. However, the disadvantages of using block processing are:

1. Four memory buffers of length N are required for holding input and output data samples with the double-buffering method. In addition, another two memory buffers (in and out) are needed for internal processing by the processor. A detailed explanation of the data acquisition program and how to reduce the memory buffer is given in Chapter 7.
2. A delay of $2NT_s$ is incurred in block processing.
3. More complicated programming is needed to manage the switching between buffers.

A detailed introduction on setting up ADC and DAC to transfer data samples into the internal memory of processor using the serial ports and the DMA is provide in Chapter 7.

QUIZ 6.9

An analog signal is sampling at 48 kHz. Frequency analysis using FFT is applied to a buffer of 5 ms.

1. How many data samples in the 5-ms buffer?
2. What is the order of the FFT that can be used if the FFT is based on the radix-2 algorithm?
3. What is the frequency resolution?
4. What is the memory requirement if double buffering is used?
5. What is the maximum time available for the processor to compute FFT?

The core clock frequency F_{core} used by the processor and the sampling frequency f_s of the real-time system determine the total cycle counts possible within the data acquisition period. The total time (or deadline) needed for sampling a block of N samples is NT_s (or N/f_s) seconds. The deadline cycle count is therefore given as

$$N_{\text{deadline\_cycle}} = \frac{NF_{\text{core}}}{f_s}. \qquad (6.3.4)$$

As a result, the three ways to increase the deadline cycle are to (1) increase the block size N at the expense of memory, (2) increase the core clock frequency F_{core} at the expense of higher power consumption, and (3) reduce sampling frequency f_s at the expense of coarser frequency resolution.

6.3.4 Performance Parameters for Real-Time Implementation

This section examines three important topics of real-time implementation for embedded systems:

1. Speed and clock frequency of different processors
2. Memory requirements for DSP algorithms
3. Power consumption of the Blackfin processors

These topics are crucial for selecting a suitable processor to meet the real-time demand of the given application. We use the Blackfin processor to examine these topics.

The current Blackfin processors operate at a clock speed ranging from 400 to 750 MHz. The instruction cycle time is the inverse of the clock speed. For example, if the CYCLE register in the Blackfin processor has recorded 2,000 cycles to complete a task under a clock speed of 600 MHz, the execution time for this task is computed as $2{,}000 \times (1/600\,\text{MHz}) = 3.33\,\mu s$.

The computational speed of the fixed-point processor is often specified as million instructions per second (MIPS), million multiply-accumulate computation (MMAC), or million operations per second (MOPS). In floating-point processors, million floating-point operations per second (MFLOPS) is commonly used. These numbers are only applicable to describe how fast an individual processor can perform a task, and cannot be used for comparison among different processors. The main reason for this is that the term "instruction" refers to different operations in different processors. For example, in a RISC-like processor, an instruction can be just a simple addition; in a complex instruction set computer (CISC)-like processor, a single instruction can perform multiply-accumulate and shift operations. Therefore, these benchmark numbers are only applicable as an approximation of whether a certain DSP algorithm can be implemented on a particular processor.

To compute the MIPS needed for the algorithm running within a sampling period, the user can simply profile the code, using the following formula:

$$\text{MIPS} = (\text{instruction cycle/sample}) \times (\text{sample/s}). \qquad (6.3.5)$$

For example, an interrupt service routine (ISR) is profiled to complete a task in 2,000 cycles. The sampling frequency used is 48 kHz, and the ISR is running on the Blackfin processor at CCLK = 600 MHz. Assume that the 600-MHz processor executes one instruction per cycle, which results in a total of 600 MIPS. From Equation 6.3.5, the algorithm requires 96 MIPS (or 16% of the total available MIPS) to complete the task.

If block processing is used, Equation 6.3.5 can be computed as:

$$\text{MIPS} = (\text{instruction cycle}/\text{samples per block}) \times (\text{sample}/\text{s}) \times (1/\text{block size}).$$

(6.3.6)

As explained in Section 6.3.3, the instruction cycle to compute a block of samples is shorter compared to sample-by-sample processing. This is mainly due to the reduced overhead in response to ISR and function call. For example, when processing the above ISR in a block size of 32, the cycle count profiled is 60,000. Therefore, using Equation 6.3.6, the block algorithm only requires 90 MIPS or 15% of the total processor MIPS.

EXAMPLE 6.9

A digital signal processor is used to filter two channels of audio signal at a sampling frequency of 48 kHz. If a 300-tap FIR filter is used and the processor can execute one MAC instruction in a single cycle, we need 300 MACs for each channel and a total of 600 MACs for two channels. The dual-channel filtering must be completed within $1/(48\,\text{kHz}) = 20.83\,\mu\text{s}$. The processor requires more than $600 \times 48\,\text{kHz} = 28.8\,\text{MIPS}$ plus other overhead. This approximation gives us an idea about selecting a processor with the right MIPS (more than 30 MIPS) to handle the task.

EXAMPLE 6.10

A Blackfin processor with an instruction cycle time of 2 ns has been chosen to implement an FIR filter. Based on a benchmark given by ADI, a block FIR filtering requires (number of samples/2) × (2 + number of taps) cycles. If the number of samples per block is 64 and the sampling frequency is 48 kHz, we can find the maximum number of taps available for processing as: $32 \times (2 + \text{maximum number of taps}) \times 2\,\text{ns} < 64/(48\,\text{kHz})$. Therefore, the maximum number of taps ≈ 20,831 taps.

HANDS-ON EXPERIMENT 6.11

This experiment uses the Blackfin BF533/BF537 EZ-KIT to run a simple FIR filter on stereo channels at a sampling frequency of 48 kHz. The CYCLE register is embedded in the main program (process_data.c) to benchmark the time needed to process two FIR filters. A background telemetry channel (BTC) is set up to display the cycle count. Load the project file `exp6_11_533.dpj` (or `exp6_11_537.dpj`) located in directory `c:\adsp\chap6\exp6_11_533` for the BF533 EZ-KIT (or `c:\adsp\chap6\exp6_11_537` for the BF537 EZ-KIT) and build the project. Run the program by clicking on the **Run** icon. To view the cycle count, click on **View → Debug Windows → BTC Memory**. A **BTC Memory** window appears. Right-click on this window to change the display format by clicking on **Select Format → Hex32**. Right-click again to turn on the **Auto Refresh** option and set a refresh rate of 1 se. Press the **SW6** (**SW11**) button on the BF533 (or the BF537) EZ-KIT to start the filter operation. Observe the cycle count and compute the execution time in seconds. The core clock frequency used in this program is 270 MHz based on the initial setup, and block filtering is used in this program. Modify the code to profile the cycle count to filter one channel of signal. Halt the program after running the program. Is the single-channel cycle count half that of the stereo channels?

HANDS-ON EXPERIMENT 6.12

Following the preceding experiment, we use the statistical profiling tool to evaluate program efficiency and identify the program segment that takes up most of the execution time. From the **Tools** menu, select **Statistical Profiling → New Profile**. Run the program and press the **SW6** button for the BF533 (or **SW11** for the BF537) to activate FIR filtering. The percentage of execution time is shown in the profile window illustrated in Figure 6.29.

It is shown that about 75% of the execution time is spent on the stereo FIR filtering. Users can also display the accumulated cycle counts by right-clicking on the profiling window and selecting **View Sample Count**. Users can even break down the details of percentage execution time within each function by double-clicking on the function name. Now, press the **SW7** switch for the BF533 (or **SW10** for the BF537) to deactivate FIR filtering. What happens? Note that the statistical profiling gives an accumulated result up to the current time. It is not meant to give an exact execution cycle count at every sample or block of samples. Users still need to use the CYCLE registers as illustrated in the preceding experiment to profile the required clock cycles.

Besides processing speed, the processor's memory also plays an important role in selection of the right processor. In most DSP processors, data and instruction memories are addressed in separate memory spaces. The Blackfin processor has on-chip data and instruction memory bands, which can be independently configured as SRAM or cache as shown in the BF533/BF537 memory map displayed in Figure 5.17. The data memory requirement can be estimated based on the algorithm. However, the actual requirement of program memory can only be known once the program is written and compiled.

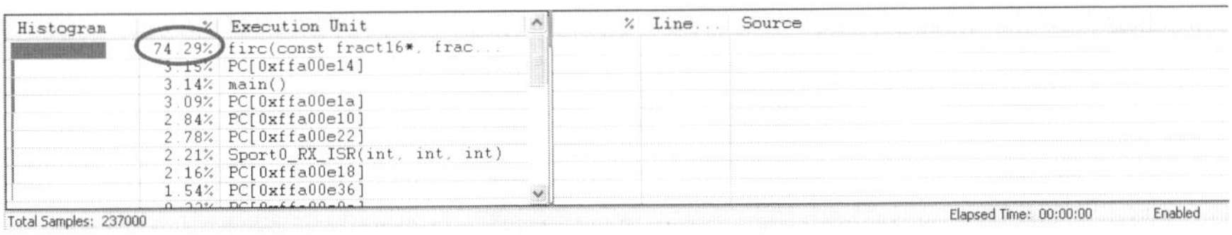

Figure 6.29 Statistical profiling of the filtering process

EXAMPLE 6.11

In a digital system, a 250-tap FIR filter is used at a sampling frequency of 8 kHz and with a resolution of 16 bits. Five minutes of the processed signal need to be saved in the external memory. The data memory required to store the coefficients and data samples is 250 and 249 words, respectively. If block processing mode and double buffering are used, we need an additional $4N$ memory locations, where N is the number of samples per block. In addition, we need to use an additional $8{,}000 \times 5 \times 60 = 2.4\text{M}$ words to store the five minutes of processed data. We need to examine the compiled memory map file to find out the memory size for the instructions.

HANDS-ON EXPERIMENT 6.13

This experiment examines the size of the main program before and after optimization. From the existing project loaded into VisualDSP++ in preceding experiment, click on **Project → Project Options**. In the new window, select **General** under **Link**. Click on **Generate Symbol Map** under **Additional Output** to produce an XML file that displays the address and size of different program and data memories. Check whether optimization is used by clicking on **General(1)** under **Compile**. In the first compilation, we disable optimization by unchecking the **Enable optimization [0]** box. Build the project and display the XML (.map) file in the debug directory. Search for firc to find the memory size used to implement the filter. Next, enable optimization by checking the **Enable optimization [0]** box. Make sure that the slider bar is pointing to the **Speed** position. Rebuild the project and observe the memory size used to implement firc. Do you find any change in the memory size after optimization? There is always a trade-off between speed and memory size optimization. Find out the smallest memory size with the optimization setting.

In power-sensitive portable multimedia applications, the power consumption of the processor and its power management capability are becoming the most critical selection factor. The Blackfin processor supports a multitiered approach to power management, and the processor is able to operate in five different operating modes: full-on, active, sleep, deep sleep, and hibernate. Power consumption is computed with the following formula:

$$P = \frac{CV^2 f_{clk}}{2}, \qquad (6.3.7)$$

where P is the power in watts, C is the capacitance in farads, V is the supply voltage in volts, and f_{clk} is the clock frequency in hertz. Therefore, power consumption can be reduced by lowering the capacitance, supply voltage, and clock frequency. The Blackfin processor allows the operating clock frequency for the core (CCLK) and system (SCLK) to be varied. At the same time, the Blackfin processor also allows core voltage to be changed in tandem with frequency change. This frequency-voltage scaling reduces the power consumption of the Blackfin processor.

The on-chip voltage regulator of the Blackfin processor generates internal voltage levels from 0.8 to 1.2 V from an external supply of 2.25 to 3.6 V. The voltage regulator allows the voltage level to be adjusted in 50-mV increments. Depending on the current model of the Blackfin processor, the core clock frequency can range from 350 to 700 MHz. The Blackfin processor offers low power consumption of 0.15 mW/MMAC at 0.8 V. Therefore, if a maximum MMAC of a Blackfin processor is 1,000, the power consumption is approximately 0.15 W. In Chapter 8, we show the steps in adjusting the voltage and core frequency of the Blackfin processor.

However, power consumption may not be a good comparison benchmark between processors. This statement is valid when processors are being deployed in portable devices, because batteries have limitation of energy [21]. A better benchmark is energy consumption. Energy consumption is defined as the time integral of power consumption. To estimate the energy required for a processor to complete a given task, the task execution time is multiplied by the estimated power consumption. The savings in energy consumption can be expressed by the following equation

$$\frac{E_n - E_r}{E_n} = \left(1 - \frac{f_{rc}}{f_{nc}} \times \left(\frac{V_r}{V_n}\right)^2 \times \frac{t_r}{t_n}\right) \times 100\ (\%), \qquad (6.3.8)$$

where E_r/E_n is the ratio of the reduced energy to nominal energy, f_{rc} is the reduced core clock frequency, f_{nc} is the nominal core clock frequency, V_r is the reduced internal supply voltage, V_n is the nominal internal supply voltage, t_r is the duration of task running at reduced core frequency, and t_n is the duration of task running at nominal core frequency.

EXAMPLE 6.12

The nominal core clock frequency is 600 MHz, and the reduced core clock frequency is set to 200 MHz. The nominal supply voltage is set to 1.2 V on power up and reduced to 0.8 V after the clock frequency is reduced to 200 MHz. Consequently, the duration taken to complete the task at reduced core frequency and voltage has increased from 1 to 3 se. Therefore, the savings in energy consumption is computed with Equation 6.3.8 as $(1 - (200/600) \times (0.8/1.2)^2 \times (3/1)) \times 100 = 55\%$ saving.

QUIZ 6.10

1. A processor's supply voltage changes from 3.6 to 1.2 V. What is the reduced power consumption in percentage?
2. The clock frequency of a Blackfin processor drops from 700 to 350 MHz, and the supply voltage is reduced from 3.6 to 0.8 V. What is the reduction in power consumption in percentage?
3. The execution times of processors A and B are 0.1 and 0.33 ms, respectively. However, the power consumption of processor A is twice that of processor B. Which processor consumes less energy?

6.4 INTRODUCTION TO THE IIR FILTER-BASED GRAPHIC EQUALIZER

The FIR filter-based eight-band graphic equalizer was introduced in Section 5.3. In this section, we replace FIR filters with IIR filters to implement the eight-band graphic equalizer. We only discuss the differences of using IIR filters as compared to FIR filters. We will go through the same process of designing the IIR filter in floating-point format, analyzing its fixed-point performance, and then porting the coefficients into the Blackfin processor.

HANDS-ON EXPERIMENT 6.14

This experiment designs eight filters based on the elliptic IIR filter to meet the filter specifications stated in Section 5.3. Open FDATool in the MATLAB window. Specify the passband frequencies according to Table 5.4. Users can specify their own stopband frequencies. Select IIR Elliptic and Minimum order from the **Design Method** and **Filter Order**, respectively. Set all the stopband attenuation to 40 dB and passband ripple to 1 dB. Click on **Design Filter** to complete the filter design for the first frequency band, and repeat the design process for the remaining seven bands. For reference, these eight IIR filters are saved in directory c:\adsp\chap6\exp6_14, using band-i.fda (where i represents the frequency band from 1 to 8) as file names for the eight frequency bands. Users can load these files into FDATool to examine the magnitude and phase responses for each band. Observe the characteristics of the IIR filters by filling in Table 6.10 and compare them with the characteristics of the FIR filters listed in Table 5.6. The implementation cost includes the number of multiplications and additions and the memory used to save the filter coefficients and signal buffers of the IIR filters.

Note that the IIR filter is seldom implemented with the high-order direct-form IIR structure shown in Figure 3.3 because of its sensitivity to coefficient quantization error. In practical applications, the high-order IIR filter is implemented as a cascade of 2nd-order IIR filter sections shown in Figure 4.17. This cascade structure reduces coefficient quantization and round-off errors. For example, a 6th-order IIR filter can be implemented as three 2nd-order IIR filters (or biquad) in a cascade structure as shown in Figure 6.30. Note that the coefficients are denoted as $\{b_{0k}\ b_{1k}\ b_{2k}\ a_{1k}\ a_{2k}\}$, where the subscript $k = 1, 2,$ or 3 specifies the section number. In Figure 4.17, a gain G is used at the input of the cascade IIR filter.

Table 6.10 IIR Filter Characteristics

IIR Filter Parameters and Responses	Characteristics
Average order of the IIR filter	
Magnitude response	
Phase response	
Group-delay response	
Impulse response	
Pole-zero plot	
Implementation cost	

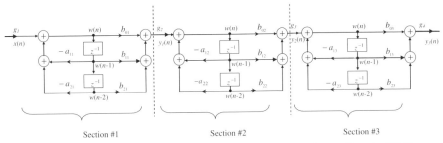

Figure 6.30 Cascade of three 2nd-order direct-form IIR sections to form a 6th-order IIR filter

This single gain ($G = g_1 \times g_2 \times g_3 \times g_4$) can also be distributed across all biquads as shown in Figure 6.30. At the input of each section, there is a gain g_k associated to each section. A final gain g_4 at the output of the final section is also available. These gains are generated from FDATool. The flexibility of the cascade IIR filter lies in the ability to arrange these 2nd-order sections to further minimize quantization errors. This step can be done by clicking on **Edit → Reorder and Scale Second Order Sections** ... of FDATool.

6.5 DESIGN OF IIR FILTER-BASED GRAPHIC EQUALIZER USING BLACKFIN SIMULATOR

This section introduces the process of completing fixed-point simulation of the IIR filter-based graphic equalizer with MATLAB and porting to the Blackfin VisualDSP++ simulator. As explained in Section 6.2.3.1, fixed-point simulation is very important in IIR filter design because the IIR filter is very sensitive to its pole locations. FDATool can convert the double-precision floating-point IIR filters designed in Hands-On Experiment 6.14 to 16-bit fixed-point IIR filters and analyze the characteristics of these fixed-point IIR filters.

HANDS-ON EXPERIMENT 6.15

This experiment converts the floating-point filter coefficients derived from Hands-On Experiment 6.14 into 16-bit fixed-point coefficients. Because all the IIR filter coefficients are larger than 1 but less than 2, we can assign 2 bits as the integer bits and the remaining 14 bits as fractional bits (refer to Table 6.2). This fixed-point (2.14) format can be specified in FDATool under the **Set quantization parameters** option as shown in Figure 6.31. Click on **Apply** to simulate the fixed-point IIR filter. Check the quantized filter responses by comparing with the reference filter in floating-point format and confirm that this fixed-point IIR filter is stable. Repeat these steps for the remaining seven IIR filters.

We can directly port the coefficients in (2.14) format into the Blackfin memory. However, we have to perform IIR filtering with different formats because the signal samples are represented in (1.15) format. The advantage of using (1.15) format for multiplication as described

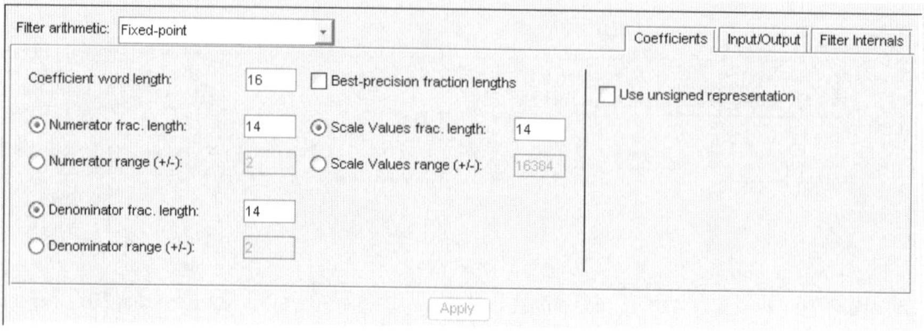

Figure 6.31 Filter arithmetic setting for using (2.14) format

in Section 6.1.1.2 cannot be achieved. Therefore, we have to convert the coefficients into (1.15) format before porting to the Blackfin memory. A simple method is to use saturated coefficients. This process saturates coefficients that exceed $(1 - 2^{-15})$ or -1 to their respective maximum values. This can be done by selecting a **Numerator range** and a **Denominator range** of (+/−) 1 in FDATool. However, this simple method greatly distorts the magnitude response of the IIR filter. Explore the characteristics of the IIR filter with saturated coefficients and compare it to the reference IIR filter using floating-point format.

A better way to fit the coefficients into (1.15) format is to scale all the numerator and denominator coefficients. In the IIR filter-based equalizer, we scale down all the coefficients by half so that these coefficients can fit into the number range of (1.15) format. However, scaling the IIR filter coefficients is not straightforward because the first feedback coefficient a_0 must always be 1. We can scale down all coefficients (including a_0) by half, perform the computation, and scale back the result by 2 to obtain the output signal $y(n)$ with the following equation:

$$2 \times \frac{1}{2} y(n) = 2 \times \left\{ \frac{1}{2} [b_0 x(n) + b_1 x(n-1) + b_2 x(n-2)] - \frac{1}{2} [a_1 y(n-1) + a_2 y(n-2)] \right\}. \quad (6.5.1)$$

In FDATool, click on **Target → Generate C Header**. . . . Select **Export Suggested** and click on **Generate** to complete the process. Open the generated header file for the 6th-order cascade IIR filter (band #1), and examine the numerator and denominator coefficients generated as shown below:

```
const int16_T NUM[MWSPT_NSEC][3] = {
    {
        162,          0,           0    ← Numerator gain for section #1
    },
    {
        16384,   -32688,      16384    ← Numerator for section #1
    },
    {
        16384,        0,           0   ← Numerator gain for section #2
    },
    {
        16384,   -32752,      16384    ← Numerator for section #2
    },
    {
        16384,        0,           0   ← Numerator gain for section #3
```

6.5 Design of IIR Filter-Based Graphic Equalizer Using Blackfin Simulator

```
    },
    {
        16384,   -32757,    16384    ← Numerator for section #3
    },
    {
        16384,       0,        0    ← Gain for output
    }
};
const int DL[MWSPT_NSEC] = { 1,3,1,3,1,3,1 };
const int16_T DEN[MWSPT_NSEC][3] = {
    {
        16384,       0,        0    ← Denominator gain for section #1
    },
    {
        16384,   -32524,    16142    ← Denominator for section #1
    },
    {
        16384,       0,        0    ← Denominator gain for section #2
    },
    {
        16384,   -32670,    16293    ← Denominator for section #2
    },
    {
        16384,       0,        0    ← Denominator gain for section #3
    },
    {
        16384,   -32741,    16366    ← Denominator for section #3
    },
    {
        16384,       0,        0    ← Output gain for section #3
    }
```

We can extract all the numerator and denominator coefficients from the header file and treat them as (1.15) format. For example, 16384 represents 1 in (2.14) format and represents 0.5 in (1.15) format. Therefore, if we treat the integer in (2.14) format with (1.15) format, we can directly perform IIR filtering in (1.15) format. However, because there is an inherent factor of two between the (2.14) and (1.15) formats, the section gains in the above C header file needed to be scaled up by 2 in order to turn them into (1.15) format. We are now ready to port filter coefficients and section gains into a fixed-point C program to run on the Blackfin simulator.

HANDS-ON EXPERIMENT 6.16

In this experiment, we move from fixed-point simulation using MATLAB to fixed-point implementation on the Blackfin processor. All the programs are written in C for ease of understanding. The first step is to port all the coefficients and gains in (1.15) format into the Blackfin memory. Two data files, equalizer_iir_coefs.dat and equalizer_iir_scales.dat in directory c:\adsp\chap6\exp6_16, are needed to be included in the main C program, main_EQ.c. Note that the coefficients in the coefficient data file equalizer_iir_coefs.dat are arranged as $b_{0k}, b_{1k}, b_{2k}, a_{1k}, a_{2k}$, where k represents the section number (see Fig. 6.30). The coefficient a_{0k}, is not required in the computation of the

IIR filtering as shown in Equation 6.5.1. Users can refer to the IIR filter function in `iirc.c` and find the IIR filter structure and the number of delay buffers used in this program. Because of the arithmetic sensitivity of the IIR filter, we use `fract32` for the delay buffer and perform built-in 32-bit multiplication and addition functions to reduce the quantization noise.

The DSP run time library also consists of the cascade IIR filter function `iir_fr16`. However, this function imposes the restriction that a_{0k} must be 1 and greater than both a_{1k} and a_{2k}. Because the coefficients derived from Hands-On Experiment 6.15 do not allow the restriction for arithmetic in (1.15) format, we have to write the IIR filter function `iirc.c`, which can implement the IIR filter expressed in Equation 6.5.1.

This experiment performs the IIR filtering routine eight times as shown in `main_EQ.c`. Because each frequency band requires different orders of IIR filter, we need to specify the stage number in the program. This information is included in the data file `equalizer_iir_stages.dat`. As in the FIR filter-based equalizer, we also include the same range of band gains for the IIR filter-based equalizer.

Build the project `equalizer.dpj` and open the following debug windows: `FreqResponse.vps` and `ImpResponse.vps`. Run the project and observe the frequency and impulse responses as shown in Figure 6.32(a) and (b), respectively. The impulse response is obtained by injecting an impulse of magnitude 32,767 into eight IIR filters and summing these eight impulse responses into a single impulse response over a duration of 256 samples. The frequency response is obtained by computing the FFT of these 256 samples with the built-in graphical feature in VisualDSP++. Observe the differences between the impulse

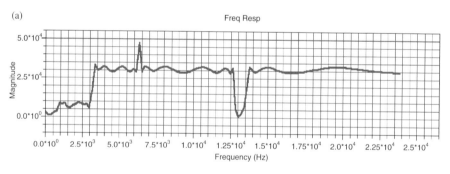

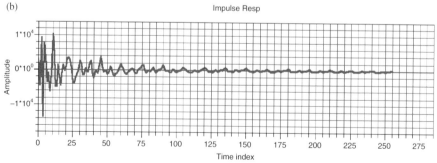

Figure 6.32 Impulse response (a) and frequency response (b) of combined IIR filter

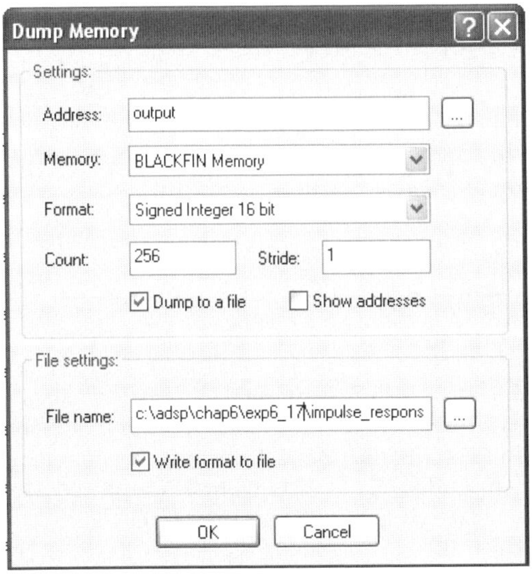

6.5 Design of IIR Filter-Based Graphic Equalizer Using Blackfin Simulator

Figure 6.33 **Dump Memory** window

responses of the IIR graphic equalizer and the FIR graphic equalizer shown in Figure 5.35. Note that the spikes and nulls in Figure 6.32(a) are caused by the unequal attenuation in the transition band between adjacent frequency bands.

To compare the impulse response of the combined IIR filter with that obtained in MATLAB with the floating-point format, we dump the specific Blackfin memory into a file and read this file into MATLAB workspace for comparison. To dump the memory, click on **Memory → 2 Dump** in VisualDSP++. A **Dump Memory** window opens, and we can fill in the parameters as shown in Figure 6.33. Specify the file name used to save the output data for comparison.

EXERCISE 6.1

1. Modify the C code to enhance the low-frequency bands (band #1 to band #3) to +12 dB, middle-frequency bands (band #4 to band #6) to 0 dB, and high-frequency bands (band #7 to band #8) to −12 dB.

2. Instead of using the elliptic filter design method, use the Butterworth filter design method. Obtain the frequency and impulse responses of the combined filter with the new filter design method. Observe any difference with the plots obtained in Hands-On Experiment 6.16.

3. Modify the C code to read data in the input file `sineNoise3sec_48k.dat` and attenuate the noise by −12 dB with the eight-band graphic equalizer. Hint: Refer to Hands-On Experiment 2.3.

6.6 DESIGN OF IIR FILTER-BASED GRAPHIC EQUALIZER WITH BF533/BF537 EZ-KIT

This section explores the real-time implementation of the IIR filter-based eight-band graphic equalizer with the Blackfin BF533/BF537 EZ-KIT. The real-time performance of this IIR equalizer is examined and compared to the FIR filter-based equalizer in Chapter 5.

HANDS-ON EXPERIMENT 6.17

This experiment modifies the C file used in Hands-On Experiment 6.16 to perform a real-time graphic equalizer on a stereo signal. The processed signals are sent to loudspeakers for playback. Activate VisualDSP++ for the BF533 (or BF537) EZ-KIT and open the project file exp6_17_533.dpj (or exp6_17_537.dpj) in directory c:\adsp\chap6\exp6_17_533 (or c:\adsp\chap6\exp6_17_537). Build the project, and the executable file exp6_17_533.dxe (or exp6_17_537.dxe) is automatically loaded into the memory of the EZ-KIT. As in Hands-On Experiment 5.8, we have programmed four push buttons and six LEDs to the same functions as shown in Table 5.9. Adjust the gain of the equalizer and note the performance.

A main difference in the real-time implementation of the FIR filter-based and IIR filter-based equalizers is that IIR equalizer must carry out the IIR filtering eight times.

EXERCISE 6.2

1. Find the cycle counts required to run (1) the interrupt service routine and (2) IIR filters and (3) compute the combined IIR filter coefficients. Note that the IIR filter function is written in C and has not been optimized with assembly code.
2. Turn on the statistical profiler and examine the percentage of computation load for running the IIR filter-based eight-band graphic equalizer.

6.7 IMPLEMENTATION OF IIR FILTER-BASED GRAPHIC EQUALIZER WITH LABVIEW EMBEDDED MODULE FOR BLACKFIN PROCESSORS

The IIR filter is commonly used in equalizer designs and can be implemented with many of the same principles discussed in Chapter 5. The following experiments explore an IIR filter-based graphic equalizer. The first experiment uses LabVIEW for Windows to simulate an eight-band equalizer, and the second experiment uses the LabVIEW Embedded Module for Blackfin Processors to create a four-band equalizer on the Blackfin processor. Push-button interrupts are introduced in the

LabVIEW Embedded Module for Blackfin Processors example for more efficient and functional user interaction.

HANDS-ON EXPERIMENT 6.18

This experiment explores the implementation and behavior of a multiband IIR filter-based equalizer with a LabVIEW for Windows simulation. Navigate to the compiled simulation IIR_EQ_Sim.exe located in directory c:\adsp\chap6\exp6_18 and launch the executable code. The user interface for this application, shown in Figure 6.34, is nearly identical to that of FIR_EQ_Sim.exe given in Chapter 5. Run the application and switch to the **Frequency Bands** tab. Note the shape of each frequency band.

Compare the frequency response of each band with the response seen in Chapter 5. What do you notice about the comparative sharpness of the IIR and FIR filter methods? What might account for this difference? Explain.

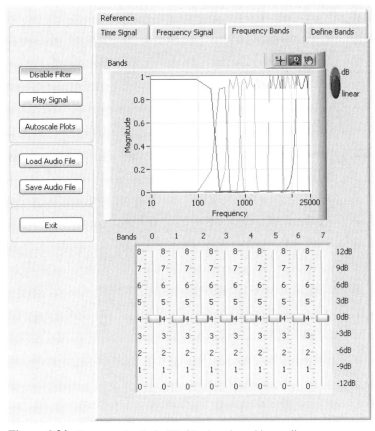

Figure 6.34 Frequency bands for IIR filter-based graphic equalizer

The IIR filter-based equalizer cannot be computed with the same method that was used in the FIR equalizer case, because impulse responses of IIR filters cannot be determined from their coefficients alone. Therefore, the IIR filter band equalizer is computed by applying several IIR filters to the input signal in parallel, one for each band. The selected gain is then applied to each band, and the results are summed to produce the overall response. Switch to the **Frequency Signal** tab of the simulation and choose the **Linear Display** option. The combined response of the eight filters is shown along with that of the input signal. Note that spikes occur in the plot corresponding to the transitions between each frequency band. Experiment with the slider values for each frequency band to see the relationship between the individual bands and the combined response. Note how greater amounts of overlap in the transition bands on the **Frequency Bands** tab correspond to larger spikes in the frequency response.

Each of the eight filters is defined by a set of filter parameters accessible on the **Define Bands** tab of the simulation. The **Filter Band Specifications** array contains specification parameters consisting of high and low cutoff frequencies, the desired order of the filter, the passband ripple, and stopband attenuation parameters (in dB). The first filter, at index 0, is a low-pass filter, and all others are bandpass filters. The parameters for each band are preloaded to match the filter specifications used in Section 5.3.

Load an audio file to hear the effects of the equalizer. The speech_tone_48k.wav file used in previous experiments is ideal for this purpose, because it contains noise centered at 1 kHz. Experiment with the various filter options to attenuate the noise while retaining as much of the voice signal as possible. What settings meet these criteria? Do the audio results differ from those heard in the eight-band FIR filter simulation?

HANDS-ON EXPERIMENT 6.19

In this experiment, we combine many of the concepts discussed previously, including audio input and output, the use of subVIs, and the **Inline C Node**, to create a four-band IIR filter-based equalizer within the LabVIEW Embedded Module for Blackfin Processors. This experiment introduces interrupt handling, which allows us to use the Blackfin EZ-KIT pushbuttons for updating the band gains during execution on the Blackfin processor.

Open the IIR Equalizer-BF5xx.lep LabVIEW embedded project in directory c:\adsp\chap6\exp6_19. Open the block diagram for IIR_4_Band_Equalizer-BF5xx.vi, as shown in Figure 6.35, which contains the top-level processing structure for the IIR equalizer.

The graphical code makes it easy to recognize the parallelism of band implementations. Each band is individually initialized with coefficients that are passed to the cascaded IIR filter block. Each filter block passes both the original input parameter and its result to the next stage. When all four stages have been completed, the output is passed back to the **Audio Write/Read VI**.

The filter used within the cascaded IIR filter block is nearly identical to that introduced in the previous VisualDSP++ experiment. The **Inline C Node** is used to combine the textual implementation of the filter with the graphical code. The output of each stage is accumulated with the previous stages rather than creating an intermediate array for each band and subsequently applying band gains. This structure allows for a pipelined approach where the output of the nth band's **IIR filter subVI** is the sum of the weighted IIR filter outputs of the

6.7 Implementation of IIR Filter-Based Graphic Equalizer

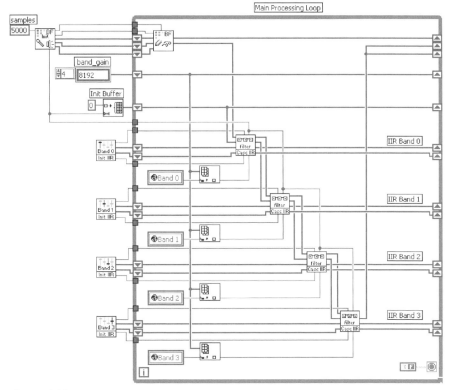

Figure 6.35 Block diagram of IIR filter-based equalizer (IIR_4_Band_Equalizer-BF5xx.vi)

first *n* bands. This method is advantageous because additional frequency bands can be added without modifying any of the underlying processing code. To add another frequency band, simply add another stage to the pipeline. Global variables and structures were removed to make the code more modular. When working with embedded applications, small improvements to repetitious code can often contribute to large performance enhancements.

Connect the Blackfin EZ-KIT audio input to the audio output jack of your computer and connect the loudspeakers to the Blackfin EZ-KIT audio output. Click on the **Run** button to execute the IIR equalizer project on the Blackfin processor. With an audio file playing on the computer, adjust the equalizer band gains by pressing buttons 1 through 4 on the Blackfin EZ-KIT. Each button corresponds to one of the four frequency bands and their levels. In the BF533 (BF537), **SW4** to **SW7** (**SW13/PB1** to **SW10/PB4**) correspond to band 0 to band 3, respectively. When a button corresponding to a band is pressed, the gain applied to that band will be incremented until the gain reaches the maximum value, at which point it will reset to the smallest value. You can keep track of the current level with the LED display, which shows a binary representation of the gain level of the last modified band.

Unlike the FIR filter-based equalizer in Chapter 5, the IIR filter-based equalizer uses interrupts to determine when a button has been pressed. In past experiments, the button values were checked during each iteration of the loop, which is a polling method. Handling

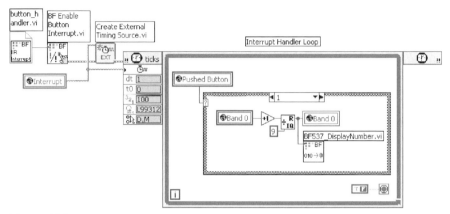

Figure 6.36 Interrupt handler loop in `IIR_4_Band_Equalizer-BF5xx.vi`

interrupts with the LabVIEW Embedded Module for Blackfin Processors is straightforward and makes your applications significantly more interactive and responsive. On the block diagram of `IIR_4_Band_Equalizer-BF5xx.vi`, scroll down to the second loop, as shown in Figure 6.36. This loop structure handles all of the processing necessary for the button-pressing functionality and only executes when a button-based interrupt is received by the Blackfin processor.

When button interrupts are enabled with `BF Enable Button Interrupt.vi`, an interrupt flag causes the timed loop to iterate one time. In this case, the number of the button pressed corresponds to a specific frequency band. The gain for that band is incremented unless it was already at the maximum level, in which case it will reset to the lowest value. The binary representation of the gain level, a value from 0 to 7, is then output to the LED display. The numeric result is shared with the main processing loop through a global variable, as seen in Figure 6.36.

Recall how the IIR filtering occurs in a pipelined fashion, with each frequency band calculated sequentially and accumulated with the rest of the filtered signal. How would you implement an eight-band equalizer with this architecture? What code would need to be added to the existing VI, and what code could be reused?

6.8 MORE EXERCISE PROBLEMS

1. Can multiplicative overflow occur in multiplication of two numbers in (2.14) format?
2. Equation 6.2.1 shows the quantization level in terms of full-scale (or peak to peak) voltage. Express the quantization level in terms of root-mean-square voltage V_{rms}.
3. Use Equations 6.2.1 and 6.2.3 to derive an expression for the SQNR for a signal with an input power P_x. Plot the SQNR in dB versus the input power in dB for a wordlength of 4, 8, and 16 bits.
4. Compute the variance of the quantization noise and the signal energy. Express the SQNR and show that each bit contributes to 6 dB of the system performance.

6.8 More Exercise Problems

5. The exponential bits in the floating-point number allow gain in multiples of decibels to be applied to the number. In the case of the IEEE-754 format, there are 8 exponential bits. Determine the dynamic range introduced by the 8 exponential bits, assuming that each bit contributes to 6 dB of dynamic range.

6. The IEEE single-precision floating-point format has a 23-bit mantissa and an 8-bit exponent. How many quantization levels are available in this format? What happens to the quantization level and dynamic range when the combined exponential value increases?

7. If a system uses an 80-dB ADC and DAC, what is a possible dynamic range for a processor, and how much degradation in SQNR is possible? State a way to increase the internal dynamic range of the processor.

8. Double-precision arithmetic such as that shown in Section 6.1.2 can be used to increase the dynamic range of internal processing. Examine the computation and memory overhead to perform a 10-tap FIR filter compared with the single-precision method.

9. A 120-dB, 24-bit ADC is used to digitize 24-bit samples in (1.23) format to a processor. If the processor generates 4 bits of quantization noise, suggest a working precision in the processor to ensure that quantization noise will never be seen by a 120-dB, 24-bit DAC. How can we prevent multiplicative and accumulative overflows?

10. Figure 6.24 shows the various sources of quantization error for a direct-form FIR filter. If the transposed-form FIR filter shown in Figure 6.37 is used, are the sources of quantization errors similar or different from the direct-form FIR filter?

11. A typical MP3 player comes with the following features: (1) voice recording in MP3 format, (2) equalization of MP3 sound track, (3) image storage and display, (4) bass enhancement of MP3 sound track, and (5) selection of sound track. State whether these features require real-time or offline processing.

12. In a speech processing system, a 20-ms window is used to analyze the spectrum of the speech signal sampled at 8 kHz. If 50% overlap of data samples are used, derive a double-buffering technique for this digital system.

13. The deadline constraint of a digital system is set at 2.7 ms. A Blackfin processor is operating at a core clock frequency of 540 MHz. If a 128-tap FIR filter is profiled as requiring 1,411,000 cycles, is there any violation of the real-time constraint? What happens when the core clock frequency is reduced to 270 MHz?

14. Write a simple MATLAB program to convert any number into any fixed-point format specified by the user. The input and output arguments for M-file should include the following:

    ```
    function [qnum,sat_flag, qerror] = quant(number,w,f)
    % qnum = quantized number
    % sat_flag = indicates whether saturation has occurred
    % qerror = number - qnum
    ```

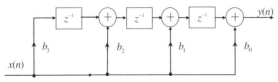

Figure 6.37 A 4-tap transposed FIR filter

```
% number = scalar or vector of input numbers
% w = total wordlength in bits
% f = number of fractional bits
```

Test the function with different numbers and wave files under different fixed-point formats.

15. In Example 6.7, different norms have been applied to the 2nd-order IIR filters. Use the MATLAB file given in Problem 14 to implement the fixed-point version of the IIR filter in (1.15) format. Verify whether the output of the filter (using different norms) will overflow when a sine wave with a normalized frequency of 0.5 is input to the IIR filter. Apply saturation or overflow mode when overflow occurs, and listen to the output signal.

16. Repeat Problem 15 with the Blackfin processor. A template project file is located in directory `c:\adsp\chap6\problem6_16\` to start this problem. The comments inside the code provide guidelines for implementing the IIR filter to run on different scaling norms. Verify whether overflow can still occur under different norms by using graphical display.

17. Investigate the error performance of midtread and midrise quantizers for the (1.15) format number range of −1.0 to +1.0 in steps of 0.1. Two MATLAB M-files, `midtread_q.m` and `midrise_q.m`, are provided to implement midtread and midrise quantizers, respectively. In addition, two other MATLAB M-files, `midtread_dq.m` and `midrise_dq.m`, are available to dequantize the respective quantized values.

18. Generate a sine wave of $f = 100\,\text{Hz}$ @ $f_s = 1{,}000\,\text{Hz}$ and save in (1.15) and (1.7) formats. Perform FFT on these two sine waves and compare the quantization noise floor. Verify whether there is a 48-dB difference in the quantization noise floor.

19. Generate a sine wave of $f = 100\,\text{Hz}$ @ $f_s = 1{,}000\,\text{Hz}$ in (1.15) format. A signal can be scaled by a small constant, α, and results in reduction of $20\log_{10}\alpha\,\text{dB}$ SNR. Determine the value of α that will result in the signal being buried under the quantization noise.

20. Repeat Hands-On Experiment 6.7 with the Blackfin BF533/BF537 simulator. Starting from using (1.15) format for representation of the sine wave, use the built-in shift-right function `fract16 shr` to simulate the quantization of the sine wave to 8, 4, and 2 bits.

21. Describe how biased and unbiased rounding is carried out on the Blackfin processor.

22. In Section 6.1.2, 32-bit multiplication is carried out. Extend the operation to 64-bit multiplication, using single-cycle 32-bit multiplications.

23. Design three-band IIR filters with floating-point arithmetic. The three-band filter split the input signal sampled at 48 kHz into three bands: [0–7.5 kHz], [8–15.5 kHz], and [16–24 kHz]. The output of each filter contains only the frequency components that fall into that filter's passband. This three-band filter can be extended to include adjustable gain at the filter output to control the level of amplification or attenuation. The passband ripple and stopband attenuation are set to 1 and 50 dB, respectively.

24. Convert the floating-point IIR filters in Problem 23 into fixed point. Take the necessary steps in porting the coefficients into the Blackfin memory using the (1.15) format. A chirp signal saved in data file `chirp_960.dat` (using 960 samples or 20 ms @ $f_s = 48\,\text{kHz}$) can be used to verify these IIR filters.

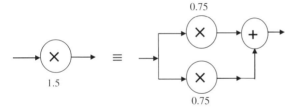

Figure 6.38 Splitting of single multiplier into two

25. IIR filter coefficients that are greater than 1 but less than 2 can be split into two equal coefficients whose values are less than 1, as shown in Figure 6.38. This division of selected coefficients allows all coefficients to be represented in (1.15) format.

    Modify the program `main_EQ.c` in directory `c:\adsp\chap6\exp6_16` to split those coefficients that exceed 1 with the splitting approach. State the problem in doing this splitting for the IIR filter-based eight-band graphic equalizer.

Chapter 7

Memory System and Data Transfer

We introduced several techniques to process digital signals in sample and block processing modes in Chapter 6. This chapter describes how to transfer data between the ADC and memory, within memory spaces, and between memory and peripherals with DMA. We present the unique caching mechanism of the Blackfin processor to speed up the transfer of program and data from external memory to internal memory.

7.1 OVERVIEW OF SIGNAL ACQUISITION AND TRANSFER TO MEMORY

This section presents the operations of the CODEC and its interface with the Blackfin processor. We use a simple talk-through program to illustrate a typical real-time signal processing chain from converting analog signal to digital samples to processing the data, and reconstructing the processed digital signal back to the analog form.

7.1.1 Understanding the CODEC

A CODEC consists of both ADC and DAC with associated analog antialiasing and reconstruction low-pass filters. For example, the Analog Devices AD1836A [41] shown in Figure 7.1 is a single-chip CODEC that provides two stereo ADCs and three stereo DACs. These ADCs and DACs can operate in 16-, 18-, 20-, or 24-bit resolution. An N-bit ADC produces 2^N digital output numbers, and an N-bit DAC has 2^N analog output levels. To achieve a fine resolution to encode a small signal, the value of N must be carefully chosen. For example, a 5-V peak-to-peak (or full scale) signal can be applied to a 16-bit CODEC to obtain a voltage resolution of $5/2^{16} = 76.29\,\mu V$. This

---

Embedded Signal Processing with the Micro Signal Architecture. By Woon-Seng Gan and Sen M. Kuo
Copyright © 2007 John Wiley & Sons, Inc.

7.1 Overview of Signal Acquisition and Transfer to Memory 275

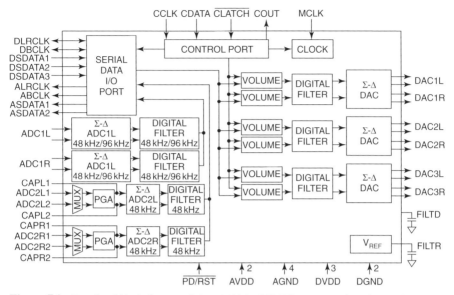

Figure 7.1 Functional block diagram of the AD1836A CODEC (courtesy of Analog Devices, Inc.)

resolution is 0.0015% (−96 dB) of full scale. The definitions of dynamic range, signal-to-quantization noise, and precision are presented in Section 6.2.2.

A more accurate signal-to-quantization-noise ratio (SQNR) in decibels can be expressed as

$$\text{SQNR} = 6.02B + 1.76 + 10\log_{10}[f_s/(2W)], \qquad (7.1.1)$$

where B is the number of bits, f_s is the sampling frequency, and W is the signal bandwidth. If f_s is oversampled by M times, SQNR = $6.02B + 1.76 + 10\log_{10}(M)$. Therefore, we can increase the SQNR by oversampling the analog signal. The oversampling technique increases SQNR by spreading out the quantization noise across a wider frequency band. Equation 7.1.1 shows that the quantization noise power is reduced by a factor of $10\log_{10}(M)$ dB. For example, if the sampling frequency is oversampled by $M = 4$, the quantization noise is reduced by 6 dB; this is equivalent to an increase of 1-bit precision.

Four ADC channels on the AD1836A can be configured as primary ADC stereo channels with differential inputs, a programmable secondary stereo pair with differential mode, differential mode with programmable input gain of up to 12 dB, and single-ended multiplex mode. The AD1836A can support time-division multiplex (TDM) mode, where receive or transmit data from different channels are allocated at different time slots in a TDM frame. In the TDM mode, the AD1836A operates at a 48-kHz sampling rate. The AD1836A also supports I²S mode, which can operate at 96 kHz with only the stereo primary channels active.

The AD1836A CODEC contains six DAC channels, which are arranged as three independent stereo channels. These channels are fully differential, and each channel has its own programmable attenuator of up to 60-dB attenuation. The resolution of the ADC and DAC in the AD1836A can be programmed as 16, 20, or 24 bits (default). A set of ADC and DAC control registers is given in Table 7.1 to determine the operation of the AD1836A.

HANDS-ON EXPERIMENT 7.1

This experiment uses the settings listed in Table 7.1 and the naming convention given in Figure 2.23. Set switch pins 5 and 6 of SW9 on the BF533 EZ-KIT to **OFF**. Modify the talk-through program in the project exp7_1.dpj (in directory c:\adsp\chap7\exp7_1) for the following configurations:

1. Connect the input to the ADC2 and the output to the DAC2. Set the ADC2L (left) gain to 0 dB and the ADC2R (right) gain to 12 dB. Verify these settings by listening at both left and right channels. The addresses of the CODEC registers are located in the talkthrough.h file, and the volume can be adjusted by changing the registers' values in main.c. From Figure 7.1, note that only ADC2 has a programmable gain amplifier (PGA) that allows users to select the gain of the ADC.

2. Reset ADC2R gain back to 0 dB. Attenuate DAC2L output to −10 dB and leave DAC2R unchanged at 0 dB (or value 1023). Again, verify the settings by listening. (Hint: Derive the volume value by taking the antilog of the desired dB value.)

3. Read the ADC2L peak-level data registers by inputting a sine wave of 2-V_{pp} to the ADC2L input of the EZ-KIT. The 2-V_{pp} signal can be derived from any signal generator. Extract bit 4 to bit 9 of the ADC2L peak-level data register and see whether it is equal to −3.0 dBFS.

7.1.2 Connecting AD1836A to BF533 Processor

The BF533 processor has two serial ports: SPORT0 and SPORT1. Each serial port provides synchronous serial data transfer and supports full-duplex communications (i.e., simultaneous data transfer in both directions). Figure 7.2 shows the connection between the AD1836A CODEC and the BF533 processor. SPORT0 receives serial data on its primary (DR0PRI) and secondary (DR0SEC) inputs and transmits serial data on its primary (DT0PRI) and secondary (DT0SEC) outputs simultaneously. SPORT1 can provide another two input and two output channels. Together, these two SPORTs support four input (stereo) channels and four output (stereo) channels. Note that transmit data are synchronous to the transmit clock (TCLKx) and receive data are synchronous to the receive clock (RCLKx), where x = 0 or 1 represents the selected SPORT.

Table 7.1 Control and Data Registers for the AD1836A CODEC

(a) DAC Control Register 1

Address	RD/WR	Reserved				Function			
			De-emphasis	Serial Mode	Data-Word Width	Power-Down	Interpolator Mode	Reserved	
15, 14, 13, 12	11	10	9, 8	7, 6, 5	4, 3	2	1	0	
0000	0	0	00 = None 01 = 44.1 kHz 10 = 32.0 kHz 11 = 48.0 kHz	000 = I²S 001 = RJ 010 = DSP 011 = LJ 100 = Packed Mode 256 101 = Packed Mode 128 110 = Reserved 111 = Reserved	00 = 24 Bits 01 = 20 Bits 10 = 16 Bits 11 = Reserved	0 = Normal 1 = PWRDWN	0 = 8× (48 kHz) 1 = 4× (96 kHz)	0	

(b) DAC Control Register 2

Address	RD/WR	Reserved		DAC Mute					
			DAC3R	DAC3L	DAC2R	DAC2L	DAC1R	DAC1L	
15, 14, 13, 12	11	10, 9, 8, 7, 6	5	4	3	2	1	0	
0001	0	00000	0 = On 1 = Mute	0 = On 1 = Mute	0 = On 1 = Mute	0 = On 1 = Mute	0 = On 1 = Mute	0 = On 1 = Mute	

(*continued*)

Table 7.1 (continued)

(c) DAC Volume Registers

Address	RD/WR	Reserved	Function
			Volume
15, 14, 13, 12	11	10	9:0
0010: DAC1L 0011: DAC1R 0100: DAC2L 0101: DAC2R 0110: DAC3L 0111: DAC3R	0	0	0 to 1,023 in 1,024 Linear Steps

(d) ADC Control Register 1

Address	RD/WR	Reserved	Function				
			Filter	Power-Down	Sample Rate	Left Gain	Right Gain
15, 14, 13, 12	11	10, 9	8	7	6	5, 4, 3	2, 1, 0
1100	0	00	0 = DC 1 = High Pass	0 = Normal 1 = PWRDWN	0 = 48 kHz 1 = 96 kHz	000 = 0 dB 001 = 3 dB 010 = 6 dB 011 = 9 dB 100 = 12 dB 101 = Reserved 110 = Reserved 111 = Reserved	000 = 0 dB 001 = 3 dB 010 = 6 dB 011 = 9 dB 100 = 12 dB 101 = Reserved 110 = Reserved 111 = Reserved

7.1 Overview of Signal Acquisition and Transfer to Memory

(e) ADC Control Register 2

Address	RD/WR	Reserved	Master/Slave AUX Mode	SOUT Mode	Word Width	ADC Mute			
						ADC2R	ADC2L	ADC1R	ADC1L
15, 14, 13, 12	11	10	9	8, 7, 6	5, 4	3	2	1	0
1101	0	0	0 = Slave 1 = Master	000 = I²S 001 = RJ 010 = DSP 011 = LJ 100 = Packed Mode 256 101 = Packed Mode 128 110 = Packed Mode AUX	00 = 24 Bits 01 = 20 Bits 10 = 16 Bits 11 = Reserved	0 = On 1 = Mute	0 = On 1 = Mute	0 = On 1 = Mute	0 = On 1 = Mute

(f) ADC Control Register 3

Address	RD/WR	Reserved	Clock Mode	Function					
				Left Differential I/P Select	Right Differential I/P Select	Left MUX/PGA Enable	Left MUX I/P Select	Right MUX/PGA Enable	Right MUX I/P Select
15, 14, 13 12	11	10, 9, 8	7, 6	5	4	3	2	1	0
1110	0	000	00 = 256 × f_s 01 = 512 × f_s 10 = 768 × f_s	0 = Differential PGA Mode 1 = PGA/MUX Mode (Single-Ended Input)	0 = Differential PGA Mode 1 = PGA/MUX Mode (Single-Ended Input)	0 = Direct 1 = MUX/PGA	0 = I/P 0 1 = I/P 1	0 = Direct 1 = MUX/PGA	0 = I/P 0 1 = I/P 1

(continued)

Table 7.1 (continued)

(g) ADC Peak-Level Data Registers

Address	RD/WR	Reserved	Peak Level Data (10 Bits)	
			6 Data Bits	4 Fixed Bits
15, 14, 13, 12	11	10	9:4	3:0
1000 = ADC1L 1001 = ADC1R 1010 = ADC2L 1011 = ADC2R	1	0	000000 = 0.0 dBFS 000001 = −1.0 dBFS 000010 = −2.0 dBFS 000011 = −3.0 dBFS 111100 = −60 dBFS Min	0000 The 4 LSBs are always zero.

Note: dBFS refers to the level in decibels with reference to the digital full-scale level.
Courtesy of Analog Devices, Inc.

7.1 Overview of Signal Acquisition and Transfer to Memory

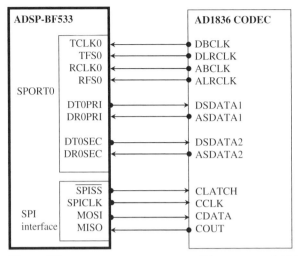

Figure 7.2 Serial connection between BF533 processor and AD1836A CODEC (only SPORT0 is shown)

Figure 7.2 shows that these serial clocks are input from the CODEC. In addition, frame synchronization signals for receive (RFSx) and transmit (TFSx) are provided by the external device to signal the start of serial data words. A serial peripheral interface (SPI) port in the BF533 processor is used to program the internal control registers of the CODEC with the configuration listed in Table 7.1. The SPI control port is a four-wire control port consisting of two data ports (MOSI and MISO), one device select pin (SPISS), and a clock pin (SPICLK). Like SPORT, SPI supports full-duplex operation. In the connection with the AD1836A shown in Figure 7.2, SPI allows reading of the ADC peak signal levels through the ADC-peak level data registers shown in Table 7.1(g). The DAC output level can be independently programmed with the DAC volume register shown in Table 7.1(c).

As explained in the previous section, the AD1836A CODEC can operate in the TDM mode. In this mode, two ADC left channels (L0 and L1) and two right channels (R0 and R1) occupy slots #1, #2, #5, and #6 of ASDATA1, respectively, as shown in Figure 7.3. Six DAC channels within the AD1836A occupy the six time slots of DSDATA1 as shown in Figure 7.3. A special TDM auxiliary mode allows two external stereo ADCs and one external stereo DAC to be interfaced to the AD1836A to form a total of eight input and eight output transfers. These external CODECs' time slots are marked with "AUX." Each time slot is 32 bits wide and is most significant bit (MSB) first. Because AD1836A has a maximum of 24-bit resolution, the least significant 8 bits in the time slot are filled with zeros.

By default, the frequency of the master clock, BCLK, is 32 bits/slot × 8 slot/frame × 48 frame/s = 12.288 Mbits/s (or MHz). Therefore, the period of the frame sync, FSTDM, is 20.8 μs (or f_s = 48 kHz), and eight input and eight output data streams can be received and transmitted during this period.

Besides the TDM mode, the data format of the AD1836A CODEC can also be configured as I^2S, right-justified (RJ), left-justified (LJ), or DSP mode according to

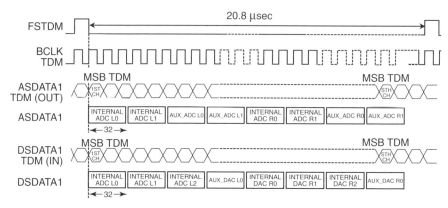

Figure 7.3 TDM timing diagram (courtesy of Analog Devices, Inc.)

ADC control register 2 and DAC control register 1, as shown in Table 7.1. By default, the data format for the AD1836A is the I^{2}S format. The I^{2}S is a three-wire serial bus standard protocol developed by Philips for transmission of two channels of pulse code modulation (PCM) digital data. The I^{2}S protocol is similar to the TDM mode, with only two time slots for left and right channels.

HANDS-ON EXPERIMENT 7.2

This experiment compares the settings used in configuring the AD1836A CODEC as TDM mode or I^{2}S mode. The project files exp7_2_tdm.dpj (in directory c:\adsp\chap7\exp7_2_tdm) and exp7_2_i2s.dpj (in c:\adsp\chap7\exp7_2_i2s) configure the CODEC to the TDM and I^{2}S modes, respectively. In this experiment, we use the BF533 EZ-KIT. In the TDM mode, pins 5 and 6 of SW9 on the EZ-KIT are switched to the **OFF** position, whereas pins 5 and 6 of SW9 are set to the **ON** position in the I^{2}S mode. Note that the BF533 EZ-KIT allows up to four input and six output channels in the TDM mode and supports only four input and four output channels in the I^{2}S mode. Examine the differences in these two main.c files.

7.1.3 Understanding the Serial Port

This section examines the serial port (SPORT) of the BF533 processor [23]. Figure 7.4 shows a block diagram of a single SPORT. Serial data from the ADC are connected to the primary receive (DRPRI) pin and secondary receive (DRSEC) pin and shifted bit by bit into the receive-primary (RX PRI) shift register and receive-secondary (RX SEC) shift register, respectively. The primary and secondary serial data bits are synchronized to the receive clock (RCLK), which can be generated from the internal clock generator in the processor or from an external clock source.

7.1 Overview of Signal Acquisition and Transfer to Memory

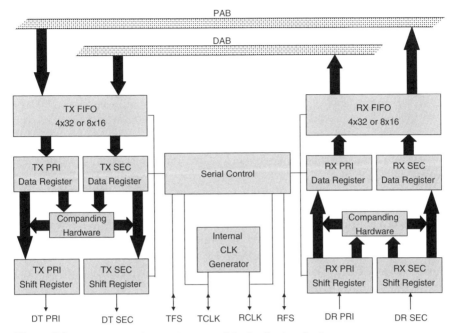

Figure 7.4 SPORT block diagram (courtesy of Analog Devices, Inc.)

The receive frame synchronization signal (RFS) indicates the start of the serial data. An optional companding hardware block supports the A-law or μ-law companding algorithm to reduce the number of bits before storing the received word in the RX PRI and RX SEC data registers. The primary and secondary data can be stored in the 8×16 bits (or 4×32 bits) receive first-in first-out (RX FIFO) in an interleaved manner. Finally, the data can be retrieved by the data address generator (DAG).

The transmit section of the SPORT transmits data from the processor to the DAC. Data from the Blackfin register are written to the 8×16 bits (or 4×32 bits) transmit first-in first-out (TX FIFO) in an alternating manner of primary-secondary channel. Again, data from the TX PRI and TX SEC can be optionally compressed by the companding hardware and transferred to the TX PRI and TX SEC shift registers. Finally, the bits in the shift registers are shifted out to the DAC via the data transmit-primary (DT PRI) and data transmit-secondary (DT SEC) pins. The transmit clock (TCLK) synchronizes the transmit data bit, and the transmit frame synchronization signal (TFS) indicates the start of transmission.

Both RX FIFO and TX FIFO are 16 bits wide and 8 words deep. These FIFOs are common to both primary and secondary data arranged in an interleaved manner, with primary first and then secondary. Therefore, there are four possible data arrangements for the FIFO:

1. When the data length is less than or equal to 16 bits and only the primary channel is enabled, a total of 8 primary words can be stored into the FIFO.

2. When the data length is greater than 16 bits and only the primary channel is enabled, a total of 4 primary words can be stored into the FIFO.

3. When the data length is less than or equal to 16 bits and both primary and secondary channels are enabled, a total of 4 primary and 4 secondary words can be stored into the FIFO.

4. When the data length is greater than 16 bits and both primary and secondary channels are enabled, a total of 2 primary and 2 secondary words can be stored into the FIFO.

The serial port must be configured before transmit or receive of data. There are two transmit configuration registers (SPORTx_TCR1 and SPORT_TCR2) per serial port for setting up the SPORT transmit as shown in Figure 7.5. Similarly, two registers (SPORTx_RCR1 and SPORTx_RCR2) per serial port are used for configuring the receive side as shown in Figure 7.6. These configuration registers can only be changed while the SPORT is disabled by setting TSPEN/RSPEN = 0 (in TCR1 and RCR1 registers). From Figures 7.5 and 7.6, users can select the desired settings for both transmit and receive operations. We use an example to set transmit and receive configuration registers.

EXAMPLE 7.1

Set transmit and receive channels of SPORT0 to operate on external clock and external frame sync (active low) (see Fig. 7.5 and 7.6). Also, transmit the most significant bit (MSB) of the 16-bit word first and turn on the stereo frame sync. In addition, enable the secondary side of the serial port. The settings are shown below:

```
write(SPORT0_TCR1, 0x5400)    /* setup for TCR1 register */
write(SPORT0_TCR2, 0x030F)    /* setup for TCR2 register */
write(SPORT0_RCR1, 0x5400)    /* setup for RCR1 register */
write(SPORT0_RCR2, 0x030F)    /* setup for RCR1 register */
```

Note that the serial wordlength = SLEN + 1. Therefore, the value of SLEN is set to 0xF for 16-bit word transmit/receive. SLEN is limited to a maximum of a 32-bit word and a minimum of a 3-bit word. Refer to the Blackfin processor hardware reference manual [23] for a more in-depth definition of the settings.

An important block of the serial port is the transmit/receive clock. The serial clock frequency can be generated from an internal source in the Blackfin processor or from an external source. The system clock (SCLK) is used for the internally generated clock. The transmit/receive clock can be derived by dividing the SCLK with the 16-bit serial clock divider register (SPORTx_TCLKDIV and SPORTx_RCLKDIV) as follows:

$$\text{SPORTx\_TCLK frequency} = \frac{\text{SCLK frequency}}{2(\text{SPORTx\_TCLKDIV} + 1)}$$

$$\text{SPORTx\_RCLK frequency} = \frac{\text{SCLK frequency}}{2(\text{SPORTx\_RCLKDIV} + 1)} \quad (7.1.2)$$

7.1 Overview of Signal Acquisition and Transfer to Memory

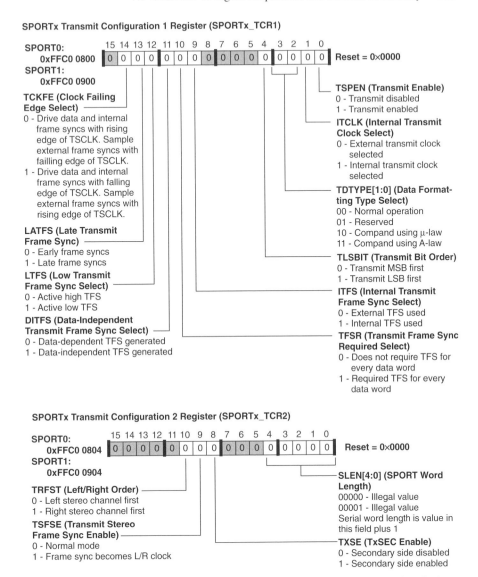

Figure 7.5 SPORT transmit configuration 1 and 2 registers (courtesy of Analog Devices, Inc.)

These equations show that the maximum and minimum serial clock frequencies are SCLK/2 and SCLK/(2^{17}), respectively.

In addition, the serial port generates internal frame syncs to initiate periodic transfer of data in and out of the serial port. The number of transmit serial cycles between frame sync pulses is SPORTx_TFSDIV + 1. Similarly, the number of receive serial cycles between frame sync pulses is SPORTx_RFSDIV + 1. Figure 7.7 illustrates the clock and frame sync timing relationship. To find the

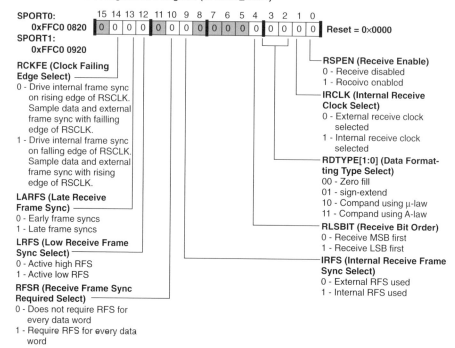

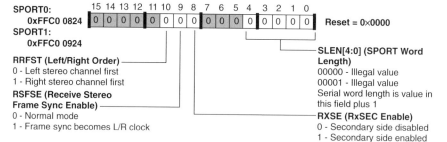

Figure 7.6 SPORT receive configuration 1 and 2 registers (courtesy of Analog Devices, Inc.)

desired frame sync frequencies for transmit and receive frames, we can use the following equations:

$$\text{SPORTx\_TFS frequency} = \frac{\text{TCLK frequency}}{\text{SPORTx\_TFSDIV} + 1}$$
$$\text{SPORTx\_RFS frequency} = \frac{\text{RCLK frequency}}{\text{SPORTx\_RFSDIV} + 1} \quad (7.1.3)$$

Note that the value of TFSDIV (or RFSDIV) must be greater than SLEN − 1.

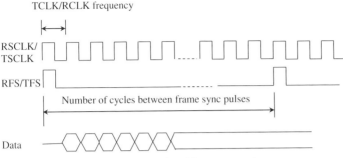

Figure 7.7 Timing diagram of clock and frame sync pulse

EXAMPLE 7.2

Compute the values to be written to the receive a serial clock divider register for a serial port frequency of 13.3 MHz. The maximum SCLK of 133 MHz is used in the BF533 processor. If a 48-kHz receive frame sync rate is desired, the values of the frame sync divider register are

$$\text{SPORTx\_RCLKDIV} = \frac{133\,\text{MHz}}{2 \times 13.3\,\text{MHz}} - 1 = 4$$

$$\text{SPORTx\_RFSDIV} = \frac{13.3\,\text{MHz}}{48\,\text{kHz}} - 1 = 276$$

7.2 DMA OPERATIONS AND PROGRAMMING

This section presents techniques for transferring data between SPORT and on-chip memory. Data can be transferred in either single-word or block transfers with DMA. In the single-word transfer, SPORT generates an interrupt every time it receives or transmits a data word. The drawback of this approach is the frequent interruption of the processing core, thus reducing the core's ability to process more complicated tasks. A more efficient data transfer is to configure and enable the SPORT DMA channel for receiving or transmitting an entire block or multiple blocks of data before interrupting the core processor. Once a block of data is transmitted or received, the interrupt service routine (ISR) can operate on the whole block of data (as shown in Section 6.3.3) instead of a single sample mode.

In addition to transferring data between SPORT and internal memory via DMA, the Blackfin BF533 processor can also perform DMA transfers between (1) memory and serial peripheral interface (SPI), (2) memory and the universal asynchronous receive-transmit (UART) port, (3) memory and the parallel peripheral interface (PPI), and (4) memory and memory. In the following sections, we use an example to configure the SPORT DMA channel and describe the DMA operations. In Section 7.2.3, examples and exercises are given to explain memory DMA such as moving program and data from external memory into the internal memory.

288 Chapter 7 Memory System and Data Transfer

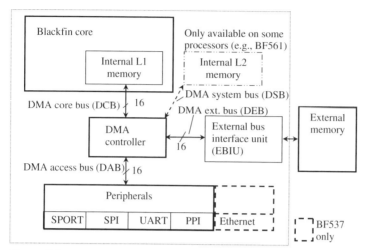

Figure 7.8 DMA controller and its connections

Figure 7.8 shows the connection of the DMA controller with the peripherals and external and internal memories. The three DMA buses include the following. (1) The DMA access bus (DAB) connects the peripheral to the DMA. (2) The DMA core bus (DCB) connects the DMA to the core. (3) The DMA external bus (DEB) connects external memory to the DMA. In some of the latest Blackfin processors such as BF561, an additional internal L2 memory is available and a DMA system bus (DSB) is used to connect the core to the L2 memory. The BF533 processor supports a total of six DMA-capable peripherals, including two SPORTs, one UART, one SPI, one PPI, and one memory. A total of 12 DMA channels are available in the BF533 processor as shown in Table 7.2. The BF537 processor has 4 more DMA channels, including 12 peripheral DMA channels to support seven DMA-capable peripherals of one Ethernet media access control, two SPORTs, two UARTs, one SPI, and one PPI. In addition, there are four memory DMA channels for transferring data between memories and between memory and off-chip peripherals. The latter DMA transfer is called the handshaking memory DMA, which enables external hardware to control the timing of individual data transfers or block transfers. This type of DMA transfer is particularly useful for asynchronous FIFO-style devices connected to the external memory bus.

Because the peripherals and memories are all connected to the DMA controller in the BF533 processor, the priority system allows the highest-priority channel to gain access to the DMA controller. As shown in Table 7.2, the BF533 processor gives the highest priority to the PPI and the lowest priority to memory DMA1 RX by default. The memory DMA channels are assigned to a lower priority than the peripheral DMA channels. In addition, the default peripheral mapping can be reassigned for all peripherals, except for the four memory DMA streams (8–11) that are rooted to the last four DMA channels. In the BF537 processor, the additional peripherals are Ethernet media access control and UART1, which have priority after PPI

Table 7.2 DMA Channels in the BF533 Processor and Their Default Priorities

DMA Channel	Default Peripheral Mapping
0 (Highest priority)	PPI
1	SPORT0 RX
2	SPORT0 TX
3	SPORT1 RX
4	SPORT1 TX
5	SPI
6	UART RX
7	UART TX
8	Memory DMA0 TX (destination)
9	Memory DMA0 RX (source)
10	Memory DMA1 TX (destination)
11 (Lowest priority)	Memory DMA1 RX (source)

and UART0. Similar to the BF533 processor, the memory DMA of the BF537 has the lower priority, and MDMA0 takes precedence over MDMA1.

7.2.1 DMA Transfer Configuration

The Blackfin processor provides two DMA transfer configurations: register mode and descriptor mode. The register-based DMA transfers allow the user to program the DMA control registers directly, whereas the descriptor-based DMA transfers require a set of parameters to be stored in the memory to initiate a DMA sequence. The latter transfer approach supports a sequence of multiple DMA transfers. Table 7.3 shows the DMA registers that are used to set up the DMA controller.

The register-based DMA provides two submodes: stop mode and autobuffer mode. The control registers are automatically updated with their initialized values (autobuffer mode) with zero overhead, or the DMA channel is automatically shut off after a single pass of DMA transfer (stop mode). We examine the autobuffer mode for handling the audio data streaming in and out of the processor in the next section.

In the descriptor-based DMA transfers, users are given flexibility in managing the DMA transfer. The DMA channel can be programmed to perform different DMA transfers in a sequential manner. There are three descriptor modes: descriptor array mode, descriptor list (small model) mode, and descriptor list (big model) mode. As shown in Figure 7.9, the descriptor array mode allows the descriptor to reside in consecutive memory locations. Therefore, there is no need to initialize the next descriptor pointer. However, in the descriptor list mode the descriptors are not required to locate in a consecutive manner. If the descriptors are all located within the 64K bytes in memory, a small-model descriptor list mode is used, where a single 16-bit field of the next descriptor pointer is used. When the descriptors are located

Table 7.3 DMA Registers

Generic DMA Registers [memory-mapped register name]	Description	Mode
Start address (lower and upper16 bits) [DMAx_START_ADDR]	Start address (source or destination)	Register and descriptor
DMA configuration [DMAx_CONFIG]	Control information of DMA	Register and descriptor
X_Count [DMAx_X_COUNT]	Number of transfers in inner loop	Register and descriptor
X_Modify [DMAx_X_MODIFY]	Number of bytes-address increments (signed and 2's complement) in inner loop	Register and descriptor
Y_Count [DMAx_Y_COUNT]	Number of transfers in outer loop	Register and descriptor
Y_Modify [DMAx_Y_MODIFY]	Number of bytes between end of inner loop and start of outer loop (signed and 2's complement)	Register and descriptor
Next descriptor pointer (lower and/or upper16 bits) [DMAx_NEXT_DESC_PTR]	Address of next descriptor	Only descriptor (descriptor list mode)

x denotes the DMA channel number (0, 1, ..., 7).

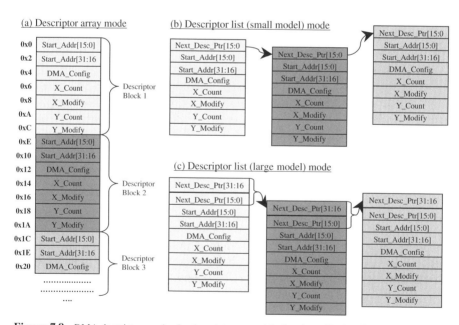

Figure 7.9 DMA descriptor modes for descriptor array (a), descriptor list (small model) (b), and descriptor list (large model) (c) (courtesy of Analog Devices, Inc.)

across the 64K-byte boundary, a large-model descriptor list mode is used to provide the full 32 bits for the next descriptor pointer as shown in Figure 7.9.

7.2.2 Setting Up the Autobuffer-Mode DMA

As stated above, the BF533 processor has 12 DMA channels. As shown in Figure 7.10, each DMA channel can be mapped to a different peripheral and memory with the peripheral map register (CTYPE = 0) and memory DMA map register (CTYPE = 1), respectively. In this section, we set up the DMA for transferring data in and out of the AD1836A CODEC with the Blackfin processor's serial port 0 (SPORT0). Therefore, we set up DMA channel 0x1 for serial port 0 receive (SPORT0 RX) and the DMA channel 0x2 for serial port 0 transmit (SPORT0 TX). This default setup also implies that SPORT0 RX has a higher priority than SPORT0 TX. The setup instructions are listed as follows:

```
DMA1_PERIPHERAL_MAP = 0x1000; // channel 1 for SPORT0 RX
DMA2_PERIPHERAL_MAP = 0x2000; // channel 2 for SPORT0 TX
```

After the DMA channels have been mapped to the peripherals, they can be configured by using their respective configuration registers. As shown in Figure 7.11, the 16-bit configuration register DMAx_CONFIG allows selections such as DMA transfer mode, data interrupt enable, transfer word size, DMA direction, DMA channel enable (DMA_EN), etc. During initialization, DMA_EN must be disabled and DMA parameters and modes are configured. The DMA channel can then be turned on and made ready for operation by writing DMA_EN = 1.

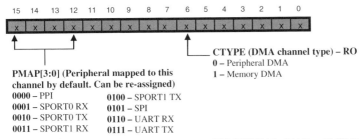

Figure 7.10 Peripheral map register (DMAx_PERIPHERAL_MAP and MDMA_yy_PERIPHERAL_MAP, where x = 0, 1, ... or 7, and yy = D0, S0, D1, or S1) (courtesy of Analog Devices, Inc.)

EXAMPLE 7.3

Set up the DMA configuration register (DMA1_CONFIG) to configure DMA channel 1 for SPORT0 RX and DMA channel 2 for SPORT0 TX with autobuffer mode. The transfer word size is set to 16 bits.

```
// set DMA channel 1 for autobuffer mode, enable data inter-
rupt, retain DMA buffer
// one-dimensional DMA using 16-bit transfers, DMA is a memory
write, and
// DMA channel 1 is not enabled.
DMA1_CONFIG = 0001 0000 1000 0110b
// set DMA channel 2 for autobuffer mode, disable data inter-
rupt, retain DMA buffer
// one-dimensional DMA using 16-bit transfers, DMA is a memory
read, and
// DMA channel 2 is not enabled.
DMA2_CONFIG = 0001 0000 0000 0100b
```

Note that DMA channel 2 does not interrupt the processor on completion of sending a block of data to the transmit register. This is due to the fact that receive and transmit channels are synchronized with the receive DMA channel 1.

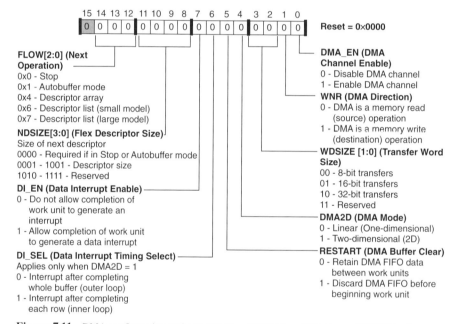

Figure 7.11 DMA configuration registers (DMAx_CONFIG or MDMA_yy_CONFIG, where x = 0, 1, ... or 7, and yy = D0, S0, D1, or S1) (courtesy of Analog Devices, Inc.)

7.2.2.1 Setting Up the Buffer Address, DMA Count, and DMA Modify Registers

Figure 7.12 shows a block diagram of receiving (or transmitting) data between the CODEC and the internal memory of the processor. SPORT0 (shown in Fig. 7.4) of the Blackfin processor is configured as the serial port for receiving and transmitting data. The DMA transfers data to the receive buffer via the DMA channel. Once the buffer is full, the DMA interrupts the processor core to process the received data.

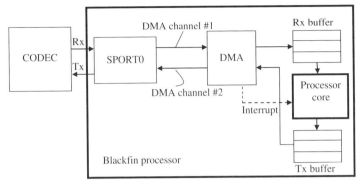

Figure 7.12 Block diagram of use of DMA for connecting CODEC and internal memory

In the opposite direction, the processed data are moved to the transmit buffer. Once the buffer is full, the data are transferred to SPORT0 via DMA with DMA channel 2. The transmit data is subsequently output to the CODEC. Because DMA operates on the system peripheral, the system clock (SCLK) is used to manage the data transfer. The SCLK is typically set up to 133 MHz and is lower than the core clock (CCLK), which can exceed 600 MHz. The peripheral DMA channels have a maximum transfer rate of one 16-bit word per two system clocks, per channel, in either direction. Therefore, if the SCLK = 133 MHz, DMA transfer rate = 16 × 133 M/2 = 1.064 Gbps.

After setting up the DMA configuration registers as shown in Example 7.3, we define the start addresses of the source and destination. In our case, the DMA channel 1 start address register, DMA1_START_ADDR, is set to a memory starting at the receive (Rx) buffer (sDataBufferRX), which is the destination address. The DMA channel 2 start address register, DMA2_START_ADDR, is set to a memory starting at the transmit (Tx) buffer (sDataBufferTX), which is the source address.

The X_COUNT register specifies the number of transfers that are required, and the X_MODIFY register specifies the number of byte increments after every data transfer. Note that the data transfer can be in 8, 16, or 32 bits. Therefore, X_COUNT is related to the number of words, and the word can be 8, 16, or 32 bits. However, X_MODIFY is always expressed in number of bytes. Blackfin processors allow one-dimensional (1D) and two-dimensional (2D) DMA modes. When the DMAx_CONFIG register shown in Figure 7.11 is set to operate in 1D mode, only the X_COUNT and X_MODIFY registers need to be set up. Otherwise, when 2D mode is set, Y_COUNT and Y_MODIFY registers must also be set up in addition to the X_COUNT and X_MODIFY registers. The 2D DMA can be considered as a nested loop, where X_COUNT and X_MODIFY specify the inner loop and Y_COUNT and Y_MODIFY specify the outer loop. The 2D DMA is particularly useful in implementing double buffers for block processing mode and in addressing 2D data like images. We show more examples on how to set up 2D DMA in Hands-On Experiments 7.4 and 7.5.

HANDS-ON EXPERIMENT 7.3

This experiment examines the DMA setup for connecting an external AD1836A CODEC with the BF533 processor on the BF533 EZ-KIT. We use a simple talk-through program in the project file exp7_3.dpj (located in directory c:\adsp\chap7\exp7_3) to acquire four input channels (L0, R0, L1, R1) from the CODEC. Similarly, four output channels are used to send the processed signals from the processor to the CODEC. The initialize.c file initializes the DMA as follows:

```
void Init_DMA(void)
{
    // set DMA1 to SPORT0 RX
    *pDMA1_PERIPHERAL_MAP = 0x1000;

    // configure DMA1
    // configure DMA1
    // 16-bit transfers, interrupt on completion, autobuffer mode
    *pDMA1_CONFIG = WNR | WDSIZE_16 | DI_EN | FLOW_1;

    // start address of data buffer
    *pDMA1_START_ADDR = sDataBufferRX;
    // DMA inner loop count
    *pDMA1_X_COUNT = 4;
    // inner loop address increment
    *pDMA1_X_MODIFY = 2;

    // set up DMA2 to transmit
    // map DMA2 to Sport0 TX
    *pDMA2_PERIPHERAL_MAP = 0x2000;

    // configure DMA2
    // 16-bit transfers, autobuffer mode
    *pDMA2_CONFIG = WDSIZE_16 | FLOW_1;
    // start address of data buffer
    *pDMA2_START_ADDR = sDataBufferTX;
    // DMA inner loop count
    *pDMA2_X_COUNT = 4;
    // inner loop address increment
    *pDMA2_X_MODIFY = 2;
}
```

SPORT0 RX and SPORT0 TX use 1D DMA. X_COUNT = 4 is used to define the four input and four output channels, whereas X_MODIFY = 2 is used to indicate a 2-byte increment for a 16-bit word transfer. An additional internal memory (sCh0LeftIn, sCh0RightIn, sCh1LeftIn, and sCh1RightIn) must also be used to move data from sDataBufferRX into these memory locations. Similarly, internal memory (sCh0LeftOut, sCh0RightOut, sCh1LeftOut, and sCh1RightOut) are used to transfer processed data to the sDataBufferTX memory. The processor's timing diagram is shown in Figure 7.13, which shows the operations within the ISR, including copy data into local memory, processing, and copy the processed data out of the processor. As explained in Section 6.3.2, this is the sample-by-sample processing mode. The real-time constraint for the sample processing mode is $T_{ISR} \leq T_s$, where T_{ISR} is the cycle time needed for completing the ISR plus the overhead of reading and writing data. This mode has higher overhead time compared to the block processing mode. In this experiment, we increase the processing load by increasing the loop counter in process_data.c as shown in Table 7.4. Do you

7.2 DMA Operations and Programming

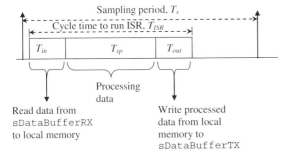

Figure 7.13 Sample-by-sample processing (input, process, and output 4 data channels within 1 sampling period)

Table 7.4 Benchmark Results of Sample-by-Sample Processing Mode

Loop Counter, N	ISR Processing Cycle Time (sample by sample)	% of Sampling Period
1		
10		
100		
1,000	Starts to fail	

observe any distortion of the output audio when increasing the loop counter? Profile the cycle time taken to run the ISR and use CCLK = 270 MHz to compute the time required to perform the loop N times. Compare this cycle time with the sampling period of $1/48{,}000 = 20.83\,\mu s$ and note the percentage of computation load. Complete the ISR cycle count in Table 7.4. Also, turn on the statistical profiler by clicking on **Tools → Statistical Profiling** to view the percentage of workload.

It is observed in Hands-On Experiment 7.3 that the overhead time associated with the sample-by-sample mode is longer compared to the block processing mode, which reduces overhead by performing setup and function call once every block. The following experiment uses block processing on the BF537 EZ-KIT.

HANDS-ON EXPERIMENT 7.4

As explained in Section 6.3.3, the block processing mode is more efficient for processing, but at the expense of higher memory usage. This experiment modifies the project file in Hands-On Experiment 7.3 with a double-buffer block processing mode. The program can be found in the project file, `exp7_4.dpj`, which is located in directory `c:\adsp\chap7\exp7_4`. Perform the following tasks:

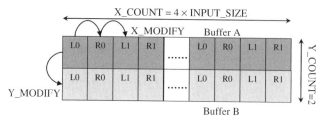

Figure 7.14 2D data buffer setup for moving data from DMA into the internal memory

1. Change the DMA from 1D to 2D.

   In init.c, change bit 4 of the DMA1_CONFIG registers to 1 (2D DMA). Set DMA1_X_COUNT = 4×INPUT_SIZE, where INPUT_SIZE = 100, to input 100 samples from L0in, R0in, L1in, and R1in (the CODEC on the BF533 EZ-KIT takes in 4 inputs). In other words, a window length of 100 samples is acquired for every block and put into the top row of the data buffer (Buffer A) as shown in Figure 7.14. Set DMA1_X_MODIFY = 2 to move from one 16-bit data to another. After the block of 100 data samples are acquired, an interrupt is generated to ask the processor to operate on these 100 samples. The DMA controller continues to bring in new samples and put them into the bottom row of data buffer (Buffer B). In this case, we need another set of registers (DMA1_Y_COUNT = 2 and DMA1_Y_MODIFY = 2) to specify the dual buffers and its selection between Buffer A and Buffer B. After Buffer B has been filled up, it generates another interrupt to ask the processor to operate on the new data set. Because the DMA is configured in autobuffer mode, the DMA1_X_COUNT and DMA1_Y_COUNT registers will be reloaded with the initialized values. A similar setup is used in the output data buffer to transfer the processed data with the DMA. Refer to the configuration for DMA channel 2 in init.c.

2. Transfer data from data buffer to local buffer for processing.

   After the input data have been transferred into the data buffer, sDataBufferRX can be reloaded into the local buffers sCh0LeftIn[i], sCh0RightIn[i], sCh1LeftIn[i], and sCh1RightIn[i], as shown in ISR.c. These four sets of memory contain 100 × 16-bit data samples of L0, L1, R0 and R1. Therefore, the programmer can process these data channels independently. A total of 3 × (4 × INPUT_SIZE) memory locations are required to store the incoming data samples in double-buffering mode. Another 3 × (4 × INPUT_SIZE) memory locations are required for saving processed data before passing out from the processor.

3. Processing in a block manner.

   The final change to the program is the use of block processing. Here, the processing needs to be repeated for the number of data samples in the block (INPUT_SIZE), as shown in the file process.c. The user can profile the time needed to process 100 samples and use Table 7.5 to document the results. A dummy inner loop is used to compare the same workload used in the sample-by-sample processing mode listed in Table 7.4. Change the inner-loop counter and observe when the block processing program starts to distort.

This experiment can also be carried out with the BF537 EZ-KIT in problems listed at the end of this chapter (see Problem 5).

Table 7.5 Benchmark Results of Block Processing Mode

Loop Counter (outer loop × inner loop)	ISR Processing Cycle Time (block processing time/ INPUT_SIZE)	% of Sampling Period
100 (100 × 1)		
1,000 (100 × 10)		
10,000 (100 × 100)	Starts to fail	

EXERCISE 7.1

Ping-pong (or double) buffering is the standard method of transferring a block of data samples from CODEC to the internal memory via DMA. As shown in Hands-On Experiment 7.4, this approach uses two buffers at the receive end and another two buffers at the transmit end. We can eliminate two buffers from the transmit end (sDataBufferTx) if we perform in-place DMA transfer. That is, only two buffers are required to save both the input data samples and the processed output data.

1. Modify the program in exp7_4.dpj to implement in-place buffering and computation in the BF533 EZ-KIT.
2. Perform the same for the BF537 EZ-KIT.

7.2.3 Memory DMA Transfer

Memory DMA transfers data between internal memory and external memory. A single memory DMA transfer requires a pair of DMA channels: one channel for source memory and the other for destination memory. In the BF533 processor, a total of four memory DMA channels (as shown in Table 7.2) are used for two simultaneous memory-to-memory DMA transfers. In addition, an 8-entry, 16-bit FIFO buffer is available for the source and destination channels to provide an efficient data transfer. A memory DMA is particularly useful in transferring data from the larger external memory to the smaller internal memory. This memory transfer avoids the need for stopping the processor from the current operation and fetching data from external memory.

Compared with the peripheral DMA, the memory DMA occupies the lowest priority as shown in Table 7.2. Also, we cannot reconfigure the priority of the memory DMA. In the BF533 processor, a round-robin access is provided in the memory DMA to allow one channel to gain access to the bus for a number of cycles before surrendering the bus access to the next channel. An added feature in the BF537 processor is the ability to program the DMA request enable (DRQ) bit field in the handshake memory DMA control register to make the priority become urgent.

In the following sections, we use examples and experiments to show how to set up the memory DMA and move data between memories with 1D and 2D memory DMA transfers.

7.2.4 Setting Up Memory DMA

As shown in Figure 7.10, the peripheral map registers in the BF533 processor for the memory DMA can be configured as follows:

```
MDMA_D0_PERIPHERAL_MAP = 0x0040;  // memory DMA desti-
                                  nation D0
MDMA_S0_PERIPHERAL_MAP = 0x0040;  // memory DMA source
                                  S0
MDMA_D1_PERIPHERAL_MAP = 0x0040;  // memory DMA desti-
                                  nation D1
MDMA_S1_PERIPHERAL_MAP = 0x0040;  // memory DMA source
                                  S1
```

In the memory DMA, destination stream 0 (D0) has the highest priority, and source stream 1 (S1) has the lowest priority.

Similar to the peripheral DMA, the memory DMA channels can be configured with the MDMA_yy_CONFIG (yy = D0, S0, D1, or S1) registers shown in Figure 7.11. However, the DMA configuration register for the source channel must be written before the DMA configuration register for the destination channel. The other parameter registers to be set up in the memory DMA channels are:

1. MDMA_yy_START_ADDR registers specify the start address of the data buffer for DMA access.

2. MDMA_yy_X_COUNT registers specify the number of elements (8, 16, or 32 bit) to be transferred. A value of 0x0 in this register corresponds to 65536 elements. X_COUNT is not necessarily the same as the number of bytes to be transferred.

3. MDMA_yy_X_MODIFY registers contain a signed, 2's complement byte-address increment. These registers allow different parts of elements to be extracted. In the 2D DMA configuration, the X_MODIFY register is not increased after the last element in each inner loop. The Y_MODIFY register is applied instead.

4. MDMA_yy_Y_COUNT registers are used only for 2D DMA. These registers specify the outer loop count.

5. MDMA_yy_Y_MODIFY registers are used only for 2D DMA. These registers specify a byte-address increment and are applied after each decrement of the Y_COUNT except for the last item in the 2D array.

7.2.5 Examples of Using Memory DMA

This section uses several experiments to illustrate how to move data around the internal memory without the intervention of the core processor.

HANDS-ON EXPERIMENT 7.5

This experiment uses the VisualDSP++ BF533/BF537 simulator. In the project file `exp7_5.dpj` (located in directory `c:\adsp\chap7\exp7_5`), a set of 15 (16 bit) sequential numbers is generated as shown in Figure 7.15. These numbers are saved in the source memory `s_MemDMA0_Src_Array`.

This experiment uses memory DMA to transfer data from the source memory to the destination memory at `s_MemDMA0_Dst_Array` with different transfer modes. Note that the memory DMA transfers below are configured as stop mode, 8-bit data transfers, and no interrupt enabled.

1. Using 1D-to-1D memory DMA transfer.

    Set `*pMDMA_S0_CONFIG = 0x0061` and `*pMDMA_D0_CONFIG = 0x0063` for source and destination memory channels, respectively. Because both memory DMA channels are configured as 1D, the Y_COUNT and Y_MODIFY registers are not used. Use X_COUNT = 0x10 and X_MODIFY = 0x2 for both source and destination channels. Modify the `main.c` with the above settings and build the program. Run and then halt the program. Click on **Memory → Blackfin Memory** and type in `s_MemDMA0_Src_Array` in the **Blackfin Memory** window to view the memory of the source and destination addresses. Observe the data in the source and destination memory.

    Change the destination modifier to pMDMA_D0_X_MODIFY = 0x1 (0x3 or 0x4). What happens to the destination memory? How can we modify the DMA setting for copying only the first 8 data from the source to the destination?

2. Using 1D-to-2D memory DMA transfer.

    The 1D-to-1D data transfer above can be modified to a 1D-to-2D data transfer by changing the `*pMDMA_D0_CONFIG = 0x0073`. Keep the source count and modify registers unchanged. Change the destination registers as follows:

    ```
    *pMDMA_D0_X_COUNT  =0x8;
    *pMDMA_D0_Y_COUNT  =0x2;
    *pMDMA_D0_X_MODIFY =0x2;
    *pMDMA_D0_Y_MODIFY =0x2;
    ```

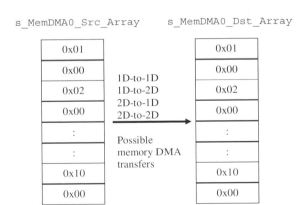

Figure 7.15 Memory DMA transfer from source memory to destination memory with different memory DMA transfers

Do you observe the same destination memory as in Part 1 for the 1D-to-1D DMA memory? Now change the destination registers as follows:

```
*pMDMA_D0_X_COUNT  =0x8;
*pMDMA_D0_Y_COUNT  =0x2;
*pMDMA_D0_X_MODIFY =0x2;
*pMDMA_D0_Y_MODIFY =0xFFF3;  // move back 13 bytes
```

What happens to the destination memory? It is important to make sure that the total number of bytes transferred on each side (source and destination) of the DMA channel are the same.

3. Using 2D-to-1D memory DMA transfer.

   If we need to access all odd numbers of data that follow an even number of data, we can use 2D memory DMA at the source memory to pick up the right set of data. Set up 2D-to-1D memory DMA as follows: *pMDMA_S0_CONFIG = 0x0071 and *pMDMA_D0_CONFIG = 0x0063. Change the source and destination registers as follows:

```
*pMDMA_S0_X_COUNT  = 0x8;
*pMDMA_S0_Y_COUNT  = 0x2;
*pMDMA_S0_X_MODIFY = 0x4;
*pMDMA_S0_Y_MODIFY = 0xFFE6;  // move back 26 bytes

*pMDMA_D0_X_COUNT  =0x10;
*pMDMA_D0_X_MODIFY =0x2;
```

   Do you observe the right sequence being displayed in the destination memory?

4. Using 2D-to-2D memory DMA transfer.

   We can move the data in the source memory to the destination memory with the 2D-to-2D memory DMA as shown in Figure 7.16.

   The source and destination registers can be initialized as follows:

```
*pMDMA_S0_X_COUNT  = 0x5;
*pMDMA_S0_Y_COUNT  = 0x3;
*pMDMA_S0_X_MODIFY = 0x6;
*pMDMA_S0_Y_MODIFY = 0xFFEA;  // -22 bytes

*pMDMA_D0_X_COUNT  =0x5;
*pMDMA_D0_Y_COUNT  =0x3;
*pMDMA_D0_X_MODIFY =0x1;
*pMDMA_D0_Y_MODIFY =0x6;
```

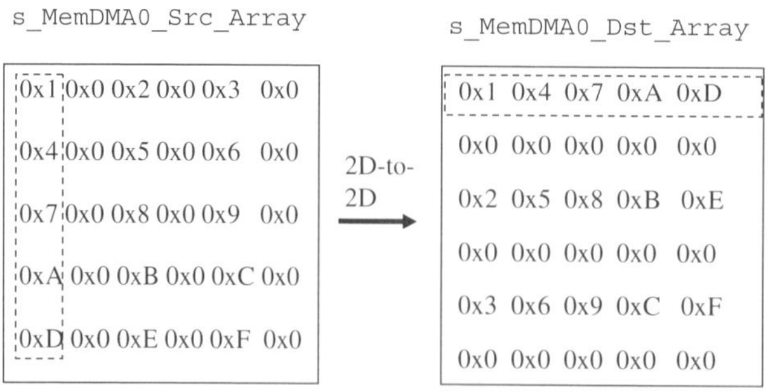

s_MemDMA0_Src_Array

0x1	0x0 0x2 0x0 0x3 0x0
0x4	0x0 0x5 0x0 0x6 0x0
0x7	0x0 0x8 0x0 0x9 0x0
0xA	0x0 0xB 0x0 0xC 0x0
0xD	0x0 0xE 0x0 0xF 0x0

2D-to-2D →

s_MemDMA0_Dst_Array

| 0x1 0x4 0x7 0xA 0xD |
| 0x0 0x0 0x0 0x0 0x0 |
| 0x2 0x5 0x8 0xB 0xE |
| 0x0 0x0 0x0 0x0 0x0 |
| 0x3 0x6 0x9 0xC 0xF |
| 0x0 0x0 0x0 0x0 0x0 |

Figure 7.16 Source and destination memory data arrangements

The 2D-to-2D memory DMA provides flexibility in defining data transfer from the source memory to the destination memory. Change the DMA data transfer above to 16 bits and 32 bits to observe any differences in the destination memory.

The above hands-on experiments show examples of moving data from the source memory to the destination memory. We further illustrate real-world applications in the following examples.

EXAMPLE 7.4

A video frame buffer of size 320 × 240 has a total of 76,800 pixels. Use a 2D DMA to retrieve a block of 16 × 8 bytes as shown in Figure 7.17.

The memory DMA registers for the source memory can be initialized as follows:

$$X\_COUNT = 16$$
$$X\_MODIFY = 1$$
$$Y\_COUNT = 8$$
$$Y\_MODIFY = 320-15 = 305$$

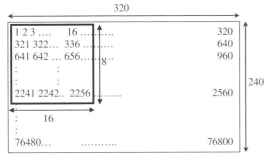

Figure 7.17 A video frame buffer of 320 × 240 pixels, where a block of 16 × 8 pixels are extracted into the data memory

EXAMPLE 7.5

A video data stream of bytes consists of red (R), green (G), and blue (B) of a color image of size $N \times M$ pixels as shown in Figure 7.18. It is desired to extract the R, G, and B of the corresponding pixels into destination memory as shown in the right-hand side of Figure 7.18.

The memory DMA registers for the source memory can be initialized as follows:

Source DMA	Destination DMA
X_COUNT = 3	X_COUNT = 3
X_MODIFY = $N \times M$	X_MODIFY = 1
Y_COUNT = $N \times M$	Y_COUNT = $N \times M$
Y_MODIFY = $-2(N \times M) + 1$	Y_MODIFY = 1

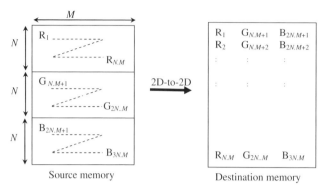

Figure 7.18 Transfer of R, G, and B data to the destination memory with 2D memory DMA

HANDS-ON EXPERIMENT 7.6

This experiment fills a large section of memory with some predefined values. A simple approach is to use the processor core for memory copy; thus the processor will not able to perform other tasks. A more efficient approach is to perform a 1D DMA in stop mode. The source memory DMA can be set up by fetching the predefined value (for example, 0x33) continuously, X_COUNT = number of transfer, and X_MODIFY = 0x0 to ensure that the source pointer keeps pointing to the predefined value. The destination memory DMA can be set up with X_COUNT = number of transfer and X_MODIFY = 0x1 for 1 byte/transfer. Modify main.c in the project file exp7_6.dpj located in directory c:\adsp\chap7\exp7_6 to initialize 128 bytes of destination memory with a value of 0x33. Create a new project and run it on the BF533/BF537 simulator to verify the results.

7.2.6 Advanced Features of DMA

We have introduced several DMA features including prioritizing DMA channels in previous sections to meet peripheral task requirements. For example, the parallel port DMA channel is given a higher priority (lower number DMA channel) over the serial port DMA channel. However, users have flexibility in configuring the priority of these DMA channels to optimize the data transfer flow in the system. Some of the latest Blackfin processors (such as the BF561) have multiple DMA controllers for transferring data to multiple processing cores.

The memory DMA channels always have lower priority than the peripheral DMA channels. Therefore, when both peripheral and memory DMA transfers exist, the memory DMA transfer can only take place during the unused time slots of the peripheral DMA. In addition, if more than one memory DMA transfer is enabled, only the highest-priority memory DMA channel is granted access. To allow the lower-priority memory DMA channel to gain access to the DMA bus, a round-robin period mechanism is available in the Blackfin processor to allow each memory DMA channel for a fixed number of transfers.

As applications are becoming more complex and involve multiple bidirectional data streams from audio and video devices, the DMA controller in the Blackfin processor provides advanced features such as a traffic controller to control the direction of data transfer. The reason for direction control is that each direction change of data transfer can impose several cycles of delay. Traffic can be independently controlled for each of the three DMA buses DAB, DCB, and DEB by using simple counters in the DMA traffic control counter period register, DMA_TC_PER. These counters state the cycle count for performing DMA data transfer in the same direction as the previous transfer. In other words, the DMA controller grants a DMA bus to peripherals and memory performing the same read/write direction for a period cycle count, until the counter time out occurs, or until the traffic changes direction. After the period counter decrements to zero, the preference is changed to the opposite flow direction.

Program and data can also be moved from the external memory to the internal memory with the cache mechanism. In the following section, we discuss cache memory concepts and how instruction and data caching can be carried out in the Blackfin processor. The comparison between cache and DMA access provides some guidelines in choosing between DMA and cache.

7.3 USING CACHE IN THE BLACKFIN PROCESSOR

We briefly introduced different types of memory in the Blackfin processors in Chapter 5. We described the internal program and data memories and split these memories between SRAM and cache. In this section, we examine the cache memory of the Blackfin processor in detail. Cache memory [46] can be thought of as copying a block of external memory to the internal memory that is closer to the core in order to speed up the memory access. Different methods are used for instruction and data caches, and these methods predict which blocks of memory need to be copied into the cache memory for optimum performance. The following sections explain cache memory concepts, cache terminology, the instruction cache, the data cache, and the memory management unit.

7.3.1 Cache Memory Concepts

The BF53x processor provides separate Level 1 (L1) caches for data and instruction. For time-critical applications, data and instructions are stored in the on-chip SRAM, which can be accessed in a single core clock (CCLK) cycle. However, if code and data are too large to fit into the internal memory, they are stored in larger but slower off-chip memory. Data transfer between internal and external memories can be carried out by using the DMA, as explained in the previous sections. Alternatively, some internal memories can be configured as cache to allow data transfer between

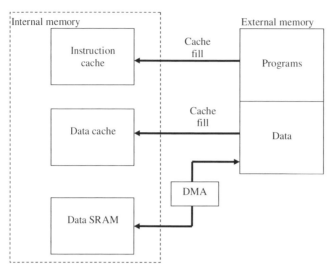

Figure 7.19 Memory configuration of the Blackfin processor

cache and external memories. Figure 7.19 shows the memory configuration of the Blackfin processor. As shown in the BF533's memory map (Fig. 5.17), the memory address from 0xFFA1 0000 to 0xFFA1 3FFF can be configured as either instruction SRAM or instruction cache. Similarly, the data memory address from 0xFF80 4000 to 0xFF80 7FFF and from 0xFF90 4000 to 0xFF90 7FFF can be configured as data SRAM or data cache. The rest of the internal memory of the BF533 processor can only be configured as SRAM. Note that the size of the cache is small (16 Kbytes). This small-size cache allows quick checking and finding of data/instruction.

The main reason for using cache is to reduce the movement of instructions and data into the processor's internal memory. Cache simplifies the programming model once it is properly set up. However, in many time-critical signal processing tasks, most programmers tend to avoid configuring memory as cache. The reason for this is to avoid movement of data and program code in and out of the internal memory that may degrade realtime performance, or even violate the real-time constraint. To solve this real-time issue for hard real-time code, the Blackfin processor provides a cache-locking mechanism that can lock time-critical code in the cache such that this code cannot be replaced. Other soft real-time code can be unlocked for possible cache fill. This caching mechanism is highlighted in subsequent sections.

If the entire program code and data can be fitted into the internal memory, there is no need to configure internal memory as cache. The cache is only configured if large program code or data needs to be extracted from the external memory. The cache memory system is able to manage the movement of program code and data from the external memory into the cache without any intervention from the programmer. This is unlike the memory DMA, where data transfer between external and internal memories is handled by the DMA controller. In addition, program overlay can be carried out by using the memory DMA to move code from external memory to internal memory when needed. Therefore, there is a need to understand the strengths and weaknesses of using cache and DMA and formulate a good strategy in selecting cache and DMA for data and code transfer. This issue is discussed in Section 7.4.

7.3.2 Terminology in Cache Memory

To discuss the cache mechanism in the Blackfin processor in detail, we need to understand some basic concepts and terms that are commonly used in describing cache memory. This introduction will be "just enough" to understand the operations of the cache mechanism and to set up the cache memory in the Blackfin processor.

Several memory spaces can be configured as cache or SRAM. The cache memory space can be divided into several fixed-size blocks, commonly known as cache lines. A cache line is the smallest unit of memory to be transferred from external memory to the cache memory on a cache miss. Figure 7.20 shows an example in which the cache memory is divided into six cache lines and maps to a bigger cacheable external memory that has 18 memory blocks (lines). Every cache line and memory line has the same number of bytes. In the Blackfin cache organization, each cache line has 32 bytes.

A "cache hit" occurs when the processor references a memory block that is already placed in the cache. The processor will access the data or instruction from the internal cache memory, instead of accessing from the external memory. In contrast, a "cache miss" occurs when the processor references a memory block that is not inside the cache. The consequence of a cache miss is that the cache controller needs additional access time to move the referenced memory block from the external memory into the cache memory. This process is called a cache line fill. Subsequently, when the same memory block is referenced by the processor again, the access will be from the faster cache memory.

Because the external memory is larger than the cache memory, there is a need to understand different ways of mapping the external memory to the cache memory. There are three schemes for mapping the external memory into the cache memory.

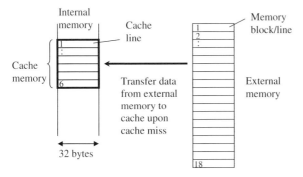

Figure 7.20 Fixed-size block/line for external memory and cache

1. **Direct-mapped cache:** In this mapping, every block in the external memory has only a fixed destination line in the cache. However, other blocks in the external memory can also share the same destination line in the cache. Therefore, this many-to-one mapping scheme is not suitable for control application, where the program constantly switches from one section to another, and results in many cache misses.
2. **Fully associative cache:** In this caching scheme, a block in the external memory can be mapped to any line in the cache memory. This mapping scheme is the opposite extreme of the direct-mapped cache, and has the least number of missing cache lines.
3. **Set-associative cache:** This caching scheme is used in the Blackfin processor. The cache memory is arranged as sets, and one set consists of several cache lines. For example, the Blackfin processor has four cache lines per set in the instruction cache, and this instruction cache is called a four-way set-associative cache. Therefore, any external instruction memory can be cached into any of the four cache lines within a set, as shown in Figure 7.21. The data cache in the Blackfin processor is a two-way set-associative cache.

The next commonly asked question is how the processor knows whether there is a cache hit or a cache miss. The answer can be derived by examining the tag field of the cache line. The 20-bit tag field (or address) is stored along with the cached data line to identify the specific address source in memory that represents the cache line. If the tag address matches the external memory, the processor checks the validity bit, which is a single bit that determines whether the cache line is valid or not. Only a valid cache line can be used directly. In subsequent sections, we will show in detail how to cache data and instructions into the internal memory.

Another question is what happens when all the cache lines are valid and a cache miss occurs. In this case, some of the cache lines need to be replaced, but how do

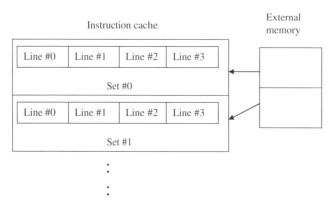

Figure 7.21 Caching of external memory into instruction cache using 4-way set-associative memory

we select and replace these cache lines? Some possible cache line replacement strategies are summarized as follows:

1. **Random replacement:** The destination cache line to be replaced is randomly selected from all participating cache lines. This strategy is the simplest but least efficient.
2. **FIFO replacement:** The oldest cache line is replaced first.
3. **Least recently used (LRU) replacement:** This method is based on the frequency of a cache line being accessed by the processor core. A recently used cache line is more likely to be used again. The replaced cache line is the one that has not been accessed for the longest time, or the least recently used.
4. **Modified LRU replacement:** Every cache line is assigned a low or high priority. If the incoming block from external memory is a low-priority block, only low-priority cache lines can be replaced. If the incoming block is a high-priority block, all low-priority cache lines are replaced first, followed by high-priority cache lines. This replacement strategy is only used in the instruction cache line.

In the following sections, we explain the detailed features of the instruction and data caches in the BF533 processor. The BF537 processor has the same cache memory and mechanism as the BF533 processor.

7.3.3 Instruction Cache

The BF533 processor has 80K bytes of on-chip instruction memory, in which 64K bytes can only be set as SRAM and the remaining 16K bytes (located at address 0xFFA1 0000 to 0xFFA1 3FFF) can be configured as either SRAM or cache. When the 16K bytes are enabled as cache, the cache is further arranged as four 4K-byte subbanks; each subbank consists of 32 sets, each set has four cache lines, and each cache line is 32 bytes, as shown in Figure 7.22. When the cache is disabled, only a single 64-bit instruction is transferred into the SRAM at a time. When the cache is enabled, instruction is fetched from the external memory via the 64-bit bus. Because the cache line is 32 bytes long, a burst of 4×64-bit (or 32 bytes) instruction data is transferred at a time.

When replacing the cache line from external memory during a cache miss, the cache line fill returns four 64-bit words, starting from the address of the missed instruction, and the next three words are fetched in sequential address order, as shown in Table 7.6. The advantage of using this fetching arrangement is to allow the processor to start executing the target instruction without waiting for the three unwanted instructions. This is made possible by the 4×64-bit line fill buffer.

As explained in the previous section, the tag field of the cache line consists of the tag address and the valid bit. In addition to this information, the tag field of the instruction cache line also consists of the LRU state field and the LRU priority

308 Chapter 7 Memory System and Data Transfer

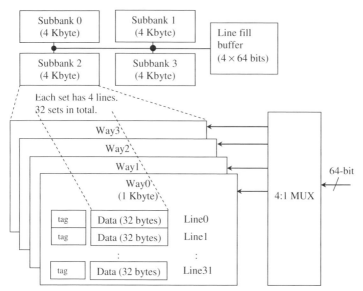

Figure 7.22 4-Way set-associative instruction cache of the BF533 processor

Table 7.6 Cache Line Word Fetching Order

Target Word	Fetching Order for Next Three Words
WD0	WD1, WD2, WD3
WD1	WD2, WD3, WD0
WD2	WD3, WD0, WD1
WD3	WD0, WD1, WD2

(LRUPRIO) bit. The LRU state field is used by the cache controller to indicate the frequency of the cache lines, and it replaces the cache line that is the least used. The LRU priority bit is used to assign a priority to each of the cache lines, and this scheme allows the high-priority cache line to be protected from replacement. The cache controller of the Blackfin processor can be configured to use either the modified LRU scheme or the LRU scheme for cache line replacement.

When addressing the instruction located in external memory, the 32-bit address (A0–A31) is partitioned into the following fields:

1. Twenty address bits, A31–A14 and A11–A10, are used to compare with the 20-bit tag address in the cache.
2. Two address bits, A13 and A12, are used to select one of the four 4K byte subbanks.
3. Five address bits, A9–A5, are used to select one of the 32 cache lines.
4. Five address bits, A4–A0, are used to select a byte within a given 32-byte cache line.

EXAMPLE 7.6

An access to an external memory at address 0x2010 2836 will result in the comparison of the tag address and cache selection as shown in Figure 7.23. By converting the hexadecimal address into its binary equivalent, the 20-bit address 0x20102 is compared to the tag addresses of the four cache lines in set 1 (out of the possible 32 sets) from subbank 2 (out of the possible 4 subbanks). If there is a match, a cache hit occurs and the 22nd byte of the selected cache line is accessed by the processor.

Example 7.6 showed how to select a cache line for comparison. However, if there is a cache miss, a cache line fill access is used to retrieve the cache line from the external memory. The cache line replacement unit is used to determine which cache line in the selected set can be replaced. The cache replacement scheme can be carried out in the following cases:

1. When only one invalid way is available in the set, the incoming external memory block replaces this cache line in the invalid way.
2. When more than one invalid way is available in the set, the incoming external memory block replaces the cache line in the following order of priority: Way0, Way1, Way2, and Way3.
3. If there is no invalid way in the cache, the LRU scheme explained above is used to replace the least recently used way. However, if the modified LRU scheme is used, ways with high priority cannot be replaced by the low-priority memory block. If all ways are high priority, the low-priority blocks cannot be cached, but high-priority blocks can be cached with the LRU scheme.

Finally, a particular way can be locked by using the instruction memory control register. The advantage of using this locking mechanism is to keep the time-critical code in one of the ways and allow the other three ways to response to a cache miss.

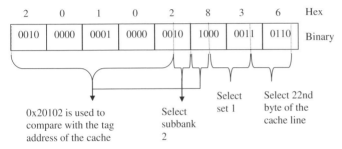

Figure 7.23 Mapping the external memory address space into the instruction cache memory space

HANDS-ON EXPERIMENT 7.7

This experiment demonstrates the concept of locking a particular way in the instruction cache memory. This locking concept is analogous to luring a mouse into the desired mouse trap. Suppose Way0 (the desired mouse trap) is to be locked and Way1, Way2, and Way3 are to be unlocked. This can be done by first unlocking Way0 (i.e., putting bait in Way0) and locking the remaining three ways. A dummy call to the time-critical function will force the function into Way0. This is similar to forcing the mouse into the Way0 trap. Subsequently, lock Way0 (to trap the mouse) and unlock the other ways. Any subsequent cache miss can only be channeled into Way1, Way2, or Way3 (i.e., other mice will be lured to other traps).

This concept is illustrated in main.c in the project exp7_7.dpj located in directory c:\adsp\chap7\exp7_7. Load this project file into the VisualDSP++ simulator and examine the following:

1. Lock_Control settings in main.c.
2. Use the **Blackfin Memory** window to locate the instructions in the external memory (0x0000 0000) and the internal memory (0xFFA1 0000).
3. Run the program and check the internal memory (0xFFA1 0000) again. Explain your observations.
4. Note that the ILOC bits in the IMEM_CONTROL register are used to control which way is to be locked. The program listing is shown in config_I_cache.c.

7.3.4 Data Cache

This section introduces the data cache configuration of the BF53x processor, which has 64K bytes of on-chip data memory. Half of the memory (32K bytes) is SRAM, and the other half can be configured as either SRAM or cache. The memory that can be configured as cache is separated into two independent memory banks of 16K bytes. The bank A address starts from 0xFF80 4000 to 0xFF80 7FFF, and bank B starts from 0xFF90 4000 to 0xFF90 7FFF. Therefore, these two banks can be separately configured as cache or SRAM, with the exception that bank B cannot be configured as cache if bank A is already configured as SRAM.

Unlike the instruction cache, the data cache is a two-way set-associative memory. Figure 7.24 shows the configuration of the data cache memory. The 16K-byte cache bank is configured as four 4K-byte subbanks. Unlike the instruction cache, each subbank in the data cache has 2 ways (2K bytes for each way) and consists of 64 cache lines per way. Similar to the instruction cache, each cache line has 32 bytes.

Like the instruction cache, the tag field of the data cache consists of the address tag, the valid bit, and the LRU field. However, the tag field of the data cache does not include LRUPRIO bits; instead, a dirty bit is included to indicate whether the cache line has been modified. Using the valid bit and dirty bit, the data cache line can occur in the following states: (a) invalid, (b) valid and clean, and (c) valid and dirty.

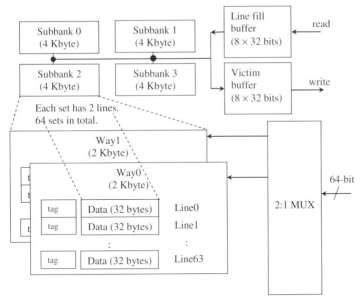

Figure 7.24 Two-way set-associative data cache of the BF533 processor

When addressing the data in external memory, the 32-bit address (A0–A31) is partitioned into the following fields:

1. Nineteen address bits, A31–A14 and A11, are used to compare with the 19-bit tag address in the cache.
2. Two address bits, A13–A12, are used to select one of the four (4K byte) subbanks.
3. Six address bits, A10–A5, are used to select one of the 64 cache lines.
4. Five address bits, A4–A0, are used to select a byte within a given 32-byte cache line.
5. If both data banks A and B are enabled as cache, bit 14 or bit 23 is used to determine which data bank.

Compared to the instruction cache, the data cache needs more complex programming tasks. These complexities are due to the read and write operations in the data cache. In addition to the line fill buffer (8 × 32 bits) for reading data from the external memory, the data cache has an extra victim buffer (8 × 32 bits) for writing data back to the external memory.

Two cache-write policies (write-through and write-back) are commonly used. In the write-through policy (also known as store through), data are written to both the cache line and the (external) source memory. Modification in the cache will also be written to the source memory as shown in Figure 7.25(a). However, when there is a data cache miss (the cache line is invalid), the write-through data cache only replaces the cache line and does not update the source memory.

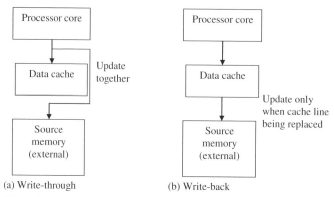

Figure 7.25 Write-through and write-back data cache

In contrast, in the write-back policy (also known as copy-back), data are only written to the cache line. The modified cache line is written to source memory only when it is being replaced, as shown in Figure 7.25(b). Therefore, as long as there is a cache hit, data are only modified in the cache line. When the cache entry is replaced during a cache miss, the source memory will also be updated. The dirty bit in the address tag of each cache line is used to indicate that the cache contains the only valid copy of the data, and these data must be copied back to the external memory before the new data are written into the cache line. The victim buffer shown in Figure 7.24 is used to hold the data that need to be written back while the processor writes the new data in the cache line.

A comparison between the two cache-write policies shows that the write-through mode allows source memory to be coherent (or consistent) with cache memory; thus multiple processors can access the updated data in the source memory. However, in the write-through mode, cache-write results in more communication overhead in transferring data between cache and source memory. The write-back mode is usually faster because of the fewer writes to the source memory unless the cache is being replaced. Nevertheless, the choice between write-through and write-back is not clear cut, and depends on applications. Therefore, the best approach in selecting a suitable mode is to test them separately for a given application.

HANDS-ON EXPERIMENT 7.8

This experiment uses the BF533 (or BF537) EZ-KIT to investigate the cycles needed to run an assembly program under different instruction and data cache modes. The project file exp7_8_533.dpj (or exp7_8_537.dpj) is located in directory c:\adsp\chap7\exp7_8_533 (or c:\adsp\chap7\exp7_8_537). Build the project and set a break point at the line "end:idle;" in main.asm. Click on **Register** → **Core** → **Data Register File** to open the **Data Register File** window. Change the format of display to Unsigned Integer 32-bit and observe the value of registers R0 (low word) and R1 (high word). These registers contain the cycle count to run myprogram.asm. By commenting out the selected

Table 7.7 Cycle Counts of Different Cache Modes

Mode	Cycle Count	Comment
Only data cache enable (default write through)		
Only data cache enable (write back)		
Only instruction cache enable		
Both data and instruction cache enable		
Data bank A and bank B as cache		

instruction in the header file, `cache_init.h`, various cache modes can be selected, and cycle counts are tabulated in Table 7.7. Comment on the performance.

7.3.5 Memory Management Unit

The Blackfin processor includes a memory management unit (MMU) that controls how to set up and access memory in the processor. However, the MMU is an optional feature that is disabled on reset. The MMU is important in embedded applications because it provides a means to protect memory at a page level and determines whether a memory page is cacheable.

The MMU in Blackfin processors consists of the cacheability and protection look-aside buffers (CPLBs). The BF533 processor has a total of 32 CPLBs, including 16 ICPLBs for instruction memory and 16 DCPLBs for data memory. The ICPLBs and DCPLBs are enabled by setting the appropriate bits in the L1 instruction memory control register (IMEM_CONTROL) and L1 data memory control (DMEM_CONTROL). A memory page size can be defined as 1K, 4K, 1M or 4M bytes. Therefore, users can program different pages in the Blackfin memory with different cacheability and protection properties. In a simple application, 32 CPLBs are sufficient to cover the entire addressable space of the application. This type of definition is referred to as the static memory management model. A quick calculation will show that we cannot specify all the 4G bytes of address of the Blackfin processor with just 16 pages of data and 16 pages of instruction, with a maximum page size of 4 Mbytes each. There will be instances when the processor accesses to a location without a valid CPLB. In this case, an exception error will occur and the exception routine must free up one CPLB and reinitialize the CPLB to that location. In a more complex application, a page descriptor table is often used to describe different memory management models. All the potentially required CPLBs are stored in the page descriptor table, and the relevant CPLBs are selected by the MMU.

Each CPLB has two associated registers: (a) start address for instruction (ICPLB_ADDRn) and data (DCPLB_ADDRn) pages and (b) instruction cache/protection properties (ICPLB_DATAn) and data cache/protection properties

(DCPLB_DATAn). The letter "n" in the registers indicates page numbers 0, 1, ..., 15 of the 16 pages. Typically, there are some rules for setting up the CPLBs:

1. Default CPLBs are set up for the system memory mapped registers and scratchpad memory. There is no need to use additional CPLB to set up these regions of memory.
2. CPLB must be configured for L1 data and L1 instruction memory as noncacheable.
3. Disable all memory other than the desired memory space.
4. Pages must be aligned on page boundaries that are integer multiples of their size.

Figure 7.26 and Figure 2.27 show the various bit fields and their functionalities in the ICPLB_DATA and DCPLB_DATA registers, respectively.

The VisualDSP++ IDDE allows users to set up a startup code when creating a new project. A startup code is a procedure that initializes and configures the processor on reset. One of these initialization routines is the configuration of the processor's cache and memory protection. The window shown in Figure 7.28 is obtained when a startup code option is enabled. In the **Cache and Memory Protection**

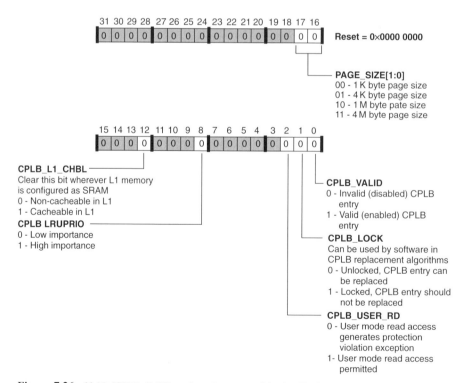

Figure 7.26 32-bit ICPLB_DATA register (courtesy of Analog Devices, Inc.).

7.4 Comparing and Choosing Between Cache and Memory DMA

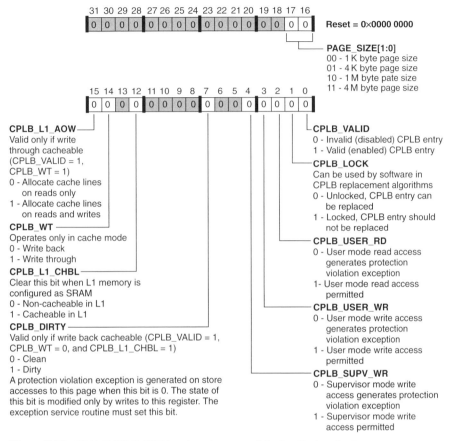

Figure 7.27 32-bit DCPLB_DATA register (courtesy of Analog Devices, Inc.)

option, users can specify the instruction cache memory and data cache memory options. There are three options under the instruction cache memory: (1) RAM with no memory protection, (2) RAM with memory protection, and (3) instruction cache. Under the data cache memory, there are four options: (1) RAM with no memory protection, (2) RAM with memory protection, (3) data cache (bank A), (4) data cache (banks A and B). The basiccrt.s file, which contains the user-specified memory settings and known machine states, is created in the new project file under the **Generated Files** folder.

7.4 COMPARING AND CHOOSING BETWEEN CACHE AND MEMORY DMA

As long as memory can be fitted into the internal memory of the processor, cache and memory DMA should not be used. However, in cases where data and/or program are larger than the available internal memory, part of the code or data must be

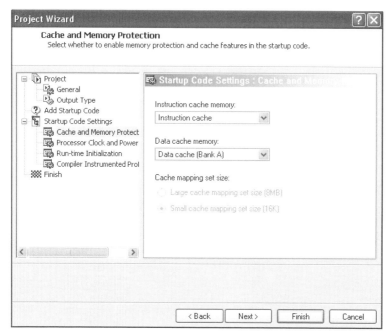

Figure 7.28 Startup code settings: **Cache and Memory Protection** window

allocated in external memory. Therefore, there is a need to decide whether DMA or cache is a better way of moving data or program into the internal memory of the processor. In this section, we analyze different cases for selecting cache or DMA as the memory transfer mode.

Case 1: When an application is required to move a large amount of data from external to internal memory, the memory DMA should be used because the cache memory is limited and DMA also prevents the intervention of the processor core.

Case 2: When data is static, data should be mapped into cacheable memory.

Case 3: If DMA is not part of the programming model, use cache memory.

Case 4: In the case of moving program into the internal memory, the preferred method is to use the instruction cache. However, when instruction cache (even with locking lines in critical code) fails to provide an acceptable performance, program code overlay via DMA is used.

Case 5: When a highly deterministic system is required, instruction and data DMA are the preferred option.

Case 6: When a system has no hard real-time constraints, it is preferred to use instruction cache and data DMA or data cache.

These cases are just for reference, and there is no perfect answer for selecting the right mode. A more detailed description of this topic can be found in [43]. For a more accurate evaluation of the DMA and cache performance, the statistical profiler in VisualDSP++ should be used on the actual application code.

7.5 SCRATCHPAD MEMORY OF BLACKFIN PROCESSOR

The scratchpad memory in the Blackfin processor is a dedicated internal memory. The BF53x processor has 4K bytes of scratchpad memory, which cannot be configured as cache or DMA access. Therefore, scratchpad memory is commonly used to house the user and supervisor stacks for fast context saving during interrupts. Alternatively, scratchpad memory can be used to store small program code.

7.6 SIGNAL GENERATOR USING BLACKFIN SIMULATOR

In this section, we implement the signal generator (as shown in Fig. 7.29) in the Blackfin VisualDSP++ simulator by using a look-up table stored in a data file. This application is important for generating digital signals internally to test a digital system without an external signal generator. A signal generator is programmed to take in the user selections and display the generated waveform with VisualDSP++ graphical display.

One period of 1-kHz sine wave samples (sampled at 48 kHz) are stored in a data file. These samples are extracted to form the desired waveform. For example, to form a 2-kHz sinewave, every other sample in the table is extracted. This technique can be applied to generate any sine wave at a multiple integer of the fundamental frequency.

To generate a sawtooth waveform, we add the fundamental frequency with several scaled harmonics as follows:

$$sawtooth(x) = \sin(x) + \frac{1}{2}\sin(2x) + \frac{1}{3}\sin(3x) + \frac{1}{4}\sin(4x) + \ldots, \quad (7.6.1)$$

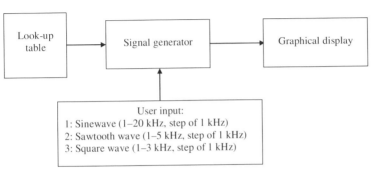

Figure 7.29 Signal generator for sine, sawtooth, and square waveforms at different frequencies

where $x = 2\pi f_0/f_s$ and f_0 is the fundamental frequency. Similarly, a square waveform can be generated as follows:

$$square(x) = \sin(x) + \frac{1}{3}\sin(3x) + \frac{1}{5}\sin(5x) + \frac{1}{7}\sin(7x) + \ldots, \quad (7.6.2)$$

In this section, we program the Blackfin processor to generate sine, sawtooth, and square waves. The Blackfin processor (BF533/BF537) provides a built-in function for performing circular addressing that can be effectively used in indexing the look-up table. We will use the circular buffer discussed in Section 5.1.3.2 in the signal generator program.

HANDS-ON EXPERIMENT 7.9

This experiment uses the project file exp7_9.dpj located in directory c:\adsp\chap7\exp7_9 to perform signal generation. There are three options: (1) a sine wave whose frequency can be varied from 1 to 20kHz, (2) a sawtooth wave whose frequency can be varied from 1 to 3kHz, and (3) a square wave whose frequency can be varied from 1 to 5kHz. The frequency increment step is 1kHz, and the sampling frequency is fixed at 48kHz.

The main program main.c uses a look-up table stored in the file sine1k_halved.dat. As shown in Figure 7.30, the look-up table consists of one period of 1-kHz sine wave with 48 samples. By extracting every other sample in a 1 kHz sine wave, we can generate a 2-kHz sine wave with a period of 24 samples. In theory, we can continue this process of generating a sine wave up to 24kHz, which is the Nyquist frequency. However, only zero-

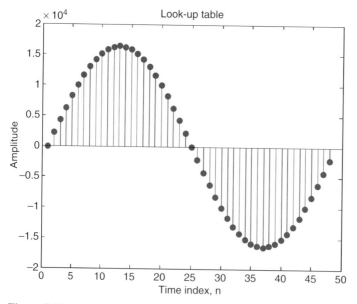

Figure 7.30 One period of a 1-kHz sine wave sampled at 48kHz

value samples are extracted from the 1-kHz sine wave table to generate a 24-kHz sine wave. Therefore, we limit our highest sine wave frequency to 20 kHz.

Activate the VisualDSP++ BF533 (or BF537) simulator and load and build the project. Make sure that the correct processor target is selected under **Project Options**.... Set a breakpoint at the "nop" operation at the end of the while loop before running the project. Select a particular sine wave frequency and view the graph with the waveout.vps file. Verify that the generated waveform is correct by viewing the FFT magnitude plot.

Examine the program section that generates the waveout array as shown below:

```
for (i=0; i<SAMPLESIZE; i++)
{
  waveout[i] = lookup_table[idx];
  idx = circindex(idx, freq, DATASIZE);
}
```

The build-in function circindex controls the step used to extract the samples from the sine wave table. The index idx wraps around once it exceeds 48 samples (DATASIZE) to form a circular buffer. The argument freq specifies the increment of the index.

We can generate other waveforms such as sawtooth and square waves by combining the sine wave with its harmonics as shown in Equations 7.6.1 and 7.6.2, respectively. However, because of the constraint of the Nyquist frequency, we can only combine a certain number of harmonics. Scaling of the harmonics can be done by multiplying the harmonic with the corresponding fractional value instead of performing division. The fractional value represented in (1.15) format is stored in a data file, scaler.dat. Verify the correctness of the generated waveform by examining the time-domain and frequency-domain plots as shown in Figure 7.31.

EXERCISE 7.2

Modify the main program (main.c) to implement the following tasks:

1. Cosine waveform
2. Dual tones of 1 kHz and 3 kHz
3. Triangular waveform

7.7 SIGNAL GENERATOR USING BF533/BF537 EZ-KIT

This section performs real-time implementation of the signal generator using the BF533/BF537 EZ-KIT to generate sine, sawtooth, and square waves. We can select the waveform and its frequency on the fly without reloading the program. In the EZ-KIT experiment, a linearly swept chirp signal is also generated.

HANDS-ON EXPERIMENT 7.10

Activate VisualDSP++ for the BF533 (or BF537) EZ-KIT and open exp7_10_533.dpj (or exp7_10_537.dpj) in directory c:\adsp\chap7\exp7_10_533 (or c:\adsp\chap7\

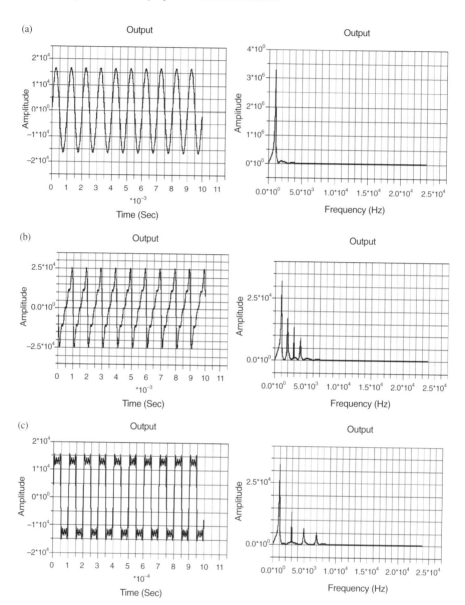

Figure 7.31 Generation of (a) 1-kHz sine wave (f_s = 48 kHz) and its frequency magnitude, (b) 1-kHz sawtooth wave (f_s = 48 kHz) and its frequency magnitude, (c) 1-kHz square wave (f_s = 48 kHz) and its frequency magnitude

7.8 Signal Generation with LabVIEW Embedded Module for Blackfin Processors

Table 7.8 Switch Control for Signal Generator

Mode	BF533 (control by SW7)		BF537 (control by SW13)	
	LED #8	LED #9	LED #5	LED #6
Sine wave	Off	On	Off	On
Sawtooth wave	On	Off	On	Off
Square wave	On	On	On	On
Chirp (linearly sweep from 0 to 10kHz)	Off	Off	Off	Off
Increase frequency*	SW4		SW10	
Decrease frequency*	SW5		SW11	

* Not applicable for chirp signal.

exp7_10_537). The waveform selection and frequency adjustment are carried out with the switches on the EZ-KIT as listed in Table 7.8.

An additional feature of this program running on the EZ-KIT is the generation of quadrature outputs (right channel is 90° phase shift with reference to the left channel) to the stereo output port of the EZ-KIT. Use a headphone to listen to the signals from the left and right channels. Alter the program such that the right channel is in phase with the left channel. Build the program and listen to the generated stereo signal again. Comment on any perceptual difference.

EXERCISE 7.3

Using the project files given in the preceding hands-on experiments, perform the following tasks:

1. Generate a pair of sine waves with the right channel 180° phase shifted with reference to the left channel. Connect the left and right channels to a pair of loudspeakers and face these two loudspeakers toward each other. Do you hear any reduction in volume? Explain the outcome.
2. Repeat Exercise 1 with no phase shift between the left and right channels.

7.8 SIGNAL GENERATION WITH LABVIEW EMBEDDED MODULE FOR BLACKFIN PROCESSORS

In previous chapters, we acquired input signals with the ADC, processed the sampled signals, and output them with the DAC. In this section, we create signals on the Blackfin EZ-KIT that can be used as test signals for other systems. Test signals are typically generated to contain specific frequency information or simulate real-world signals. These user-defined signals allow engineers to test systems for specific

	column 1 1209 Hz	column 2 1336 Hz	column 3 1477 Hz	column 4 1633 Hz
row 1 697 Hz	1	2	3	A
row 2 770 Hz	4	5	6	B
row 3 852 Hz	7	8	9	C
row 4 941 Hz	*	0	#	D

Figure 7.32 Touch-tone phone DTMF mapping

performance or real-world behavior while still in the lab. For instance, communication signals can be generated based on input data and adhering to a common standard agreed upon by both the transmitter and the receiver. Another usage of signal generation is the creation of control signals. For example, a pulse-width modulated (PWM) square wave can be used to control motor speed or heater voltage (or temperature) in an experiment.

In the following experiments, we focus on dual-tone multifrequency (DTMF) communication signals. DTMF is the standard protocol used for routing numbers pressed on a touch-tone telephone. A typical DTMF keypad is divided into rows and columns, where each row and column contains its own unique frequency as shown in Figure 7.32. When a key is pressed, two single-frequency signals are generated and added together to create the touch tone.

In this experiment, LabVIEW is used to simulate custom frequency DTMF generator (or encoder) and a DTMF receiver (or decoder). Each signal consists of two specific tonal frequencies that uniquely identify it. The LabVIEW Embedded Module for Blackfin Processors is then used to prototype and deploy a DTMF encoder on the Blackfin EZ-KIT. The interactive LabVIEW graphical interface is used for live debugging and interaction with the DTMF generator running on the Blackfin processor.

HANDS-ON EXPERIMENT 7.11

The purpose of this exercise is to gain an intuitive understanding of how a DTMF signal generator works and an understanding of its time- and frequency-domain characteristics. LabVIEW and the LabVIEW Embedded Module for Blackfin Processors are capable of generating many types of signals in addition to these sine waves and DTMF signals.

7.8 Signal Generation with LabVIEW Embedded Module for Blackfin Processors 323

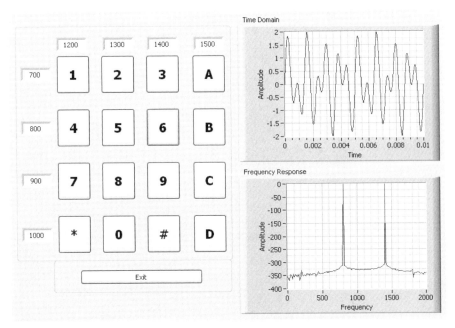

Figure 7.33 DTMF encoder (user interface for DTMF_Encoder_Sim.exe)

Open the program DTMF_Encoder_Sim.exe located in directory c:\adsp\chap7\exp7_11 to see the user interface shown in Figure 7.33. The array of 16 buttons is arranged in a 4-by-4 matrix as shown in Figure 7.32. Each row and column has a unique frequency assignment giving it a distinct dual-tone combination. Two tonal signals are added together to create a DTMF signal. The column and row tonal frequencies can be customized, although they may not be interpreted properly by most phone receivers. Note that we use different frequencies in Figure 7.33 instead of the standard DTMF frequencies defined in Figure 7.32.

In Figure 7.33, button "6" is pressed, causing sine waves of frequencies 800 and 1,400 Hz to be combined into a single signal. The resulting time-domain and frequency-domain responses can be seen in their respective graphs on the right side of Figure 7.33. Test other buttons to verify that the time-domain signal contains the correct corresponding frequencies. Attach loudspeakers or headphones to the audio output of the sound card to hear the resulting DTMF tones.

Now open the DTMF_Decoder_Sim.exe application located in directory c:\adsp\chap7\exp7_11. The interface shown in Figure 7.34 decodes DTMF signals according to the default frequencies specified in Figure 7.33.

The sound card output can be connected to the computer audio/microphone input to create a loop-back for the DTMF communications. This allows generating and decoding of signals when both simulations are running concurrently. The DTMF decoder interprets the detected frequencies as a row and a column in the format of Figure 7.33 and displays the corresponding button value. Both the time- and frequency-domain signals are shown on their respective graphs, and cursors show the values of the last acquired frequency spikes. **Device ID** specifies which sound card to use if more than one is detected in your machine. **Number**

324   Chapter 7   Memory System and Data Transfer

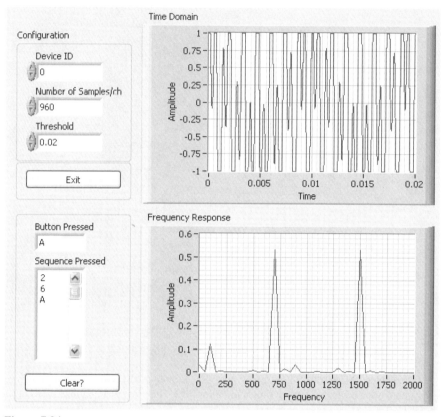

Figure 7.34   DTMF decoder (user interface for `DTMF_Decoder_Sim.exe`)

of Samples/ch can be varied to detect different pulse durations. Finally, **Threshold** is used for noise immunity and can be adjusted to reduce false readings.

Note that the acquired signals are not necessarily identical to those generated in each row and column combination. Experiment with different values for the DTMF encoder frequencies and listen to the tones with loudspeakers (or headphones). Are these results expected? Can the two frequencies for a button be placed too close together such that they are not distinguishable by the decoder? Explain.

HANDS-ON EXPERIMENT 7.12

This experiment implements the DTMF encoder on the Blackfin EZ-KIT using the LabVIEW Embedded Module for Blackfin Processors. Running the project in debug mode on the target gives us the ability to interact with the graphical front panel interface and dictate which tones are generated by the Blackfin processor. We can then use the DTMF

7.8 Signal Generation With Labview Embedded Module for Blackfin Processors 325

decoder application from Hands-On Experiment 7.11 to receive and decode the resulting touch-tone signals.

Open the DTMF-BF5xx.lep project located in directory c:\adsp\chap7\exp7_12. Note that there are four VIs in the project and the top-level VI is DTMF-BF5xx.vi. Double-click on this VI to open its front panel as shown in Figure 7.35.

Debug mode allows users to press one of the 16 Boolean buttons within the Boolean array input and generate the corresponding dual-tone signal. Open the block diagram for DTMF-BF5xx.vi to see how the DTMF encoder is implemented in the graphical LabVIEW code, as shown in Figure 7.36. Wires on the block diagram can also be probed to view the data being passed along them.

The data flow determines the execution order, which is generally from left to right. The process for outputting a key pressed on the keypad on the front panel follows several distinct steps. First, the program evaluates whether a key has been pressed by calculating a Boolean

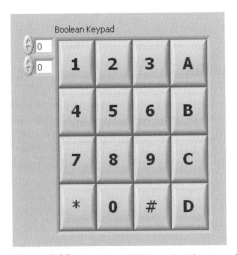

Figure 7.35 Blackfin DTMF encoder (front panel for DTMF-BF5xx.vi)

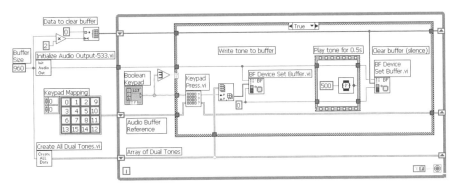

Figure 7.36 Blackfin DTMF encoder (block diagram for DTMF-BF5xx.vi)

logical OR of every element in the buttom array. If the result is true, the key press is detected and the resulting signals are combined and sent to the DAC output buffer. The buffer is allowed to output for the 500 ms, and then the output buffer is set to zero. In this instance, data flow is used to implement a 500-ms delay between the two output buffer write VIs by preventing the program from continuing until the timer within the sequence structure is complete. In this main VI, open each of the three subVIs one by one and explore their block diagrams to see how their functionality contributes to the overall DTMF encoder.

Execute the application on the Blackfin EZ-KIT in debug mode, and make sure that the audio output is wired to speakers (or headphones). Press buttons on the front panel to hear the tones created by the Blackfin processor. Next, wire the audio out from the Blackfin processor to the audio/microphone input on the computer. Open DTMF_Decoder_Sim.exe, which was used in the previous experiment to view and decode the signals generated by the Blackfin processor. Alternatively, implement the Blackfin DTMF decoder with the LabVIEW Embedded Module for Blackfin Processors and connect the audio out from the encoder to the audio in of the decoder in the Blackfin EZ-KIT.

7.9 MORE EXERCISE PROBLEMS

1. An FFT measurement of a single sine wave is commonly used to assess the performance of the ADC. A theoretical SQNR of a 16-bit ADC is 96 dB. However, an N-point FFT is used as an N-band spectrum analyzer; therefore, there is a processing gain of the FFT that pushes the noise floor lower than the ADC noise floor. Use Equation 7.1.1 to compute the overall noise floor. Determine the minimum level of distortion that can be picked up by a 1024-point FFT.

2. The SQNR of Equation 7.1.1 is based on the assumption that the input signal, x_{in}, is equal to the full-scale level of x_{max}. However, if the input signal is less than x_{max}, a new SQNR must be derived. In general, the quantization error power can be stated as $\sigma_e^2 \approx x_{max}^2/(3 \times 2^{2B})$. If the input signal power is x_{in}^2, derive the general expression of SQNR.

   (a) Plot the SQNR (in dB) versus input level (in dB) for 8-bit and 16-bit quantizers. The input power should range from −50 dB to 0 dB.

   (b) Explain the relationship between SQNR and signal power. Why is there a change of SQNR beyond −4.77 dB?

3. The BF537 EZ-KIT uses separate ADC (AD1871) and DAC (AD1854). The data specification of these ADC and DAC can be found in the ADI website. Examine their connections to the BF537 processor in the EZ-KIT and answer the following questions:

   (a) What are the resolutions and sampling frequencies of the ADC and DAC?

   (b) What are the possible gain settings of the ADC?

   (c) Draw the circuit connection from the ADC to the serial port of the BF537.

   (d) Draw the circuit connection from the serial port of the BF537 to the DAC.

   (e) Is the SPI port being used to program the ADC/DAC?

   (f) Examine and explain the timing diagrams of interface formats.

4. A pass-through project of the BF537 is provided in the file problem7_4.dpj located in directory c:\adsp\chap7\problem7_4. The ADC and DAC in the BF537 EZ-KIT

are configured for 24-bit receive and transmit. Subsequently, the received 24-bit data from the DMA receive buffer (iRxBuffer1) is truncated to 16-bit data before storing into the input memory. A simple block copy from the input memory to the output memory is performed in the file process_data.c. Finally, the 16-bit data from the output memory is extended to 24 bits before sending to the DMA transmit buffer (iTxBuffer1). Unlike the BF533 EZ-KIT, we cannot control the volume of the ADC and the DAC in the BF537 EZ-KIT as the configuration pins are hardwired to either Vcc or ground. Build and run the pass-through project in the BF537 EZ-KIT. Comment on the differences between the pass-through programs used in the BF533 and BF537 EZ-KITs.

5. Using the project files given in problem7_4.dpj located in directory c:\adsp\chap7\problem7_4, modify the program and repeat the same sample-by-sample processing and block processing experiments in Hands-On Experiments 7.3 and 7.4, respectively, with the BF537 EZ-KIT. Note that the ADC/DAC in the BF537 EZ-KIT is operating at 24-bit, and 32-bit DMA transfer is used.

6. In the digital domain, a 0 dBFS corresponds to 0x7FFF for the 16-bit wordlength. Therefore, any value with reference to full scale has a negative value of dBFS.

    (a) What is the dBFS for a value of 0.5 in (1.15) format?

    (b) What is the smallest dBFS for the (1.15) format number?

7. Explain why the instruction cache has more cache lines per set compared to the data cache.

8. This problem examines the different modes of using data and instruction caches in the BF533 processor. An FIR filter program is written with its code and data located in the external memory. The user can select one of the five cases shown in Table 7.9. Under the main.c file, select the cases one at a time and observe the cycle count after building and running c_cache_data.dpj in directory c:\adsp\chap7\problem7_8. In addition, choose different CPLB_DMYCACHE options in the cplbtab533.s file when data cache is used to examine the cycle counts of write-through and write-back cache. Discuss the performance of using different cache modes. Enable the **Cache Viewer** window (**View→ Debug Windows→ Cache Viewer**) in VisualDSP++ to visualize the activities of the processor's cache in simulation mode.

9. In Hands-On Experiment 7.10, the signal is generated continuously. Modify the program such that all signals are limited to 3 s.

10. Instead of storing a complete period (48 samples) of the sine wave, a quarter of the sine wave (i.e., only 12 samples) can be stored to reduce the memory usage. Modify the program in Hands-On Experiment 7.10 to implement a sine wave generator (based on

Table 7.9 Cycle Count of Different Cache Modes

Case	Cycle Count
Disable data and instruction cache	
Enable instruction cache only	
Enable instruction cache and data cache bank A	
Enable instruction cach e and data cache banks A and B	
Enable data cache bank A only	

partial data storage) with variable frequency from 1 to 20 kHz. What is the trade-off of using a quarter-period of data samples?

11. The touch-tone keypad was shown in Figure 7.32. The keypad numbers are encoded with combinations from the low-frequency group and the high-frequency group. A DTMF generation can be computed with the look-up table in Section 7.8 provided in LabVIEW software. Instead of using a look-up table to generate DTMF tone, we can design a pair of IIR filters that oscillate at the desired frequencies. See [11] for a detailed description of the implementation. A Blackfin program is written to perform this task on the BF533 (or BF537) EZ-KIT. These project files are located in directories c:\adsp\chap7\problem7_11_533 and c:\adsp\chap7\problem7_11_537 for the BF533 and BF537 EZ-KITs, respectively. Build and run the project. Read the comment section of the main.c file for the function of switch settings to test the DTMF generator. Determine the duration of the tones set in the program. Modify the program to construct a piano keyboard.

12. Besides using the IIR filters to generate the DTMF tone in Problem 11, we can also use a polynomial expansion for sine wave generation. A Blackfin program is also written to perform this task on the BF533 and BF537 EZ-KITs. These project files are located in directories c:\adsp\chap7\problem7_12_533 and c:\adsp\chap7\problem7_12_537 for the BF533 and BF537 EZ-KITs, respectively. Build and run the project. Examine how fixed-point polynomial expansion can be carried out with fixed-point arithmetic, and determine its computational load in the fixed-point Blackfin processor.

13. Three techniques of sine wave generation, look-up table (Section 7.6), IIR oscillator (Problem 11), and polynomial expansion (Problem 12), have been implemented in the Blackfin processor. State the advantages and disadvantages of using these techniques. In particular, examine and comment on the cycle count and memory requirement for each technique.

14. A 200-tap FIR low-pass filter is programmed to remove high-frequency noise; 256 samples are acquired every block, and the sampling frequency is 48 kHz. The FIR filter benchmark on the Blackfin processor is given as: [(number of samples/block)/2] * [number of taps +7] cycles.

    (a) What is the block rate? In other words, how many times is the buffer full per second?

    (b) How many MIPS does this FIR filtering algorithm require. When the Blackfin processor is operating at 300 MHz?

    (c) What is the percentage of the MIPS being used by the FIR filtering algorithm?

    (d) How many channels of simultaneous processing can be performed by the processor?

    The above questions only evaluate the CPU performance of the processor. The I/O that brings in the data from external devices must also be fast enough to handle the multiple channels. Analog signal is sampled and quantized with the ADC, and transfers to the internal memory via the serial port.

    (e) Compute the bit rate required for the serial port to handle the maximum channel density.

    (f) The serial port is operated at full duplex of 50 Mbps. Can the processor handle the maximum bit rate computed in (e)?

7.9 More Exercise Problems

- (g) There is another data movement from the serial port to the internal memory of the DSP processor via DMA. Can the DMA move the samples from serial port to the memory fast enough? The peripheral DMA transfer rate is around 1 Gbps, as stated in Section 7.2.2.1.
- (h) If double buffering is implemented in the block FIR filtering above, compute the internal data memory required. All the coefficients and data samples are stored as 16-bit words.
- (i) Is the internal data memory sufficient to meet the maximum channel density above? If not, what is your recommendation?

15. If the FIR filtering in Problem 14 is performed in sample processing mode, compute the following:
    - (a) Processing MIPS with sample processing mode. Comment on any difference from processing MIPS with block processing mode.
    - (b) Determine the maximum channel density with sample processing mode.
    - (c) Determine the memory required with sample processing mode and check whether it is able to fit into the internal data memory of the Blackfin processor.

Chapter 8

Code Optimization and Power Management

We have introduced arithmetic, memory architecture, programming, and real-time implementation issues for the Blackfin processors in previous chapters. This chapter focuses on code optimization and power management for developing efficient embedded systems with the Blackfin processor. We explore how to optimize a program with faster execution speed, efficient resource utilization, and power-saving features. At the end of the chapter, an FIR filter for sampling rate conversion is used for hands-on experiments.

8.1 CODE OPTIMIZATION

Because most embedded systems are real-time systems, code optimization in term of execution speed is an important performance index. Increasing execution speed with code optimization will result in decreasing power consumption; however, this may come at the cost of increasing memory usage. In other situations, reducing memory usage will result in lower power consumption due to fewer memory accesses, but may increase execution time. Speed, memory, and power optimizations determine the overall cost of the embedded systems. For example, speed optimization allows the choice of a slower but less expensive processor; memory optimization reduces external memory size; and power optimization means fewer cooling requirements and cheaper power supply.

In previous chapters, we have used a simple technique for optimizing C programs by turning on the **Enable optimization** option in the C compiler. In this chapter, we further optimize C programs by using special optimization settings and understanding the different optimization levels. We compare the execution speed and memory requirement of using normal C code versus using intrinsic functions. In addition, we explore different low-level assembly programming techniques that

---

Embedded Signal Processing with the Micro Signal Architecture. By Woon-Seng Gan and Sen M. Kuo
Copyright © 2007 John Wiley & Sons, Inc.

can further optimize the code. We use an FIR filter as an example to investigate the performance of these code optimization techniques.

C code optimization is very important in developing embedded systems. It is always easier to use the C compiler for optimization. For example, we can simply choose the optimization options in VisualDSP++ and rebuild the project for better performance. The performance gain from using C code optimization can be significant. However, in cases in which we have to save more processing MIPs and memory, assembly code optimization is the only way to achieve this level of performance. However, the development cost of writing the overall program with 100% assembly code far exceeds the performance gain. A better approach is to start with writing the code in C, perform a detail profiling to identify the time-critical sections of the code, and replace those code segments with assembly programming. A common 80/20% rule states that 80% of the processing time is spent on the 20% of the code. Therefore, if we can identify that 20% of the code and optimize it with assembly programming, significant performance gain can be achieved. A method to identify the hot spot of the workload is through the statistical profiler (for an EZ-KIT) or linear profiler (for a simulator) in the VisualDSP++ environment.

We use the FIR filtering to illustrate the concepts of optimizing both C and assembly programs. A pseudo code for implementing the FIR filter is shown below:

```
void firc (input, output, filter state)
  // FIR filter initialization
  // declare filter coefficients
  // declare pointer for reading delay buffer
  // declare pointer for updating delay buffer
  // declare number of input samples
  // perform multiply-add in a loop for N taps
  // update delay buffer
```

This pseudo code serves as a guideline to implement the FIR filter with different programming languages and styles. Both C and assembly programs are benchmarked in terms of execution cycles, processing time, and memory (data and code) usage. In implementing an FIR filter, the 20% of code that consumes 80% of the processing time comes from the multiply-accumulate operations and updating the delay buffer. Therefore, the optimization effort is concentrated on these bottlenecks. In the following sections, several versions of optimized C and assembly codes are developed and compared with the reference C and assembly programs without optimization.

8.2 C OPTIMIZATION TECHNIQUES

Hands-on experiments given in Chapters 5 and 6 show a large difference of execution speed between the C code with and without optimization. In particular, the IIR filter-based eight-band equalizer fails to operate properly when the compiler optimization is disabled. This is because the execution time exceeds the real-time constraint. The VisualDSP++ compiler's optimizer [35] is designed to generate efficient executable code for the program that has been written in a straightforward and simple manner. We discuss important programming guidelines and considerations in writing an efficient C program in the following sections.

8.2.1 C Compiler in VisualDSP++

Figure 8.1 shows the **Code Generation** window under the C compiler (**Compile: General(1)** option) of **Project Options** ... in VisualDSP++. Not that we can select the following parameters in the code optimization options:

1. **Enable optimization.** Turn on the optimization of C code. It is related to the sliding bar at the right column in Figure 8.1. The number indicates the optimization level, where **[100]** represents optimization for speed and **[0]** represents optimization for size. Other values of optimization lie between 100 and 0.
2. **Automatic inlining.** The compiler automatically inlines C/C++ functions that are not declared as inline in the source code. When this option is turned on, the compiler inlines the C/C++ function only when there is a reduction in execution time. More details on inlining are given in Section 8.2.4.
3. **Interprocedural optimization.** Most programs consist of more than one function, and it is advantageous for the compiler to perform optimization over the entire program. A two-stage compilation is performed when the interprocedural optimization is turned on. The first stage compiles the program and extracts information from all functions. In the second stage, the compiler is called from the linker stage to compile the program with the information obtained from the first stage.
4. **Generate debug information.** Allows users to debug programs and set breakpoints in C source code.

These optimizations can be applied to all functions in the project files. Sometimes, it may be useful that different optimization schemes are applied to different sections of the code. For example, a particular section of the code is optimized for memory usage, while another section is optimized for execution speed as shown below:

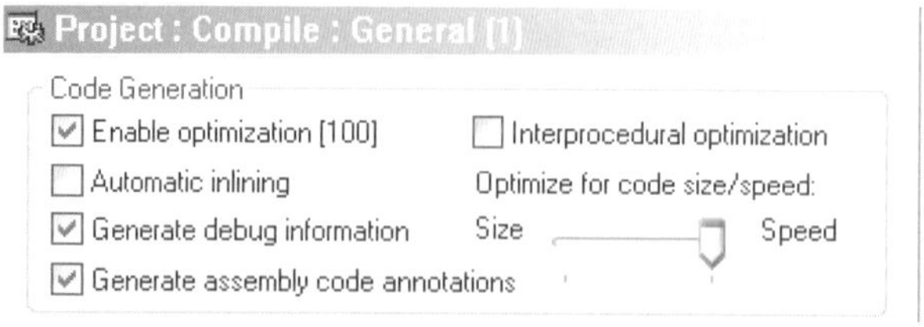

Figure 8.1 Code generation options in the **Project Option** menu

```
#pragma optimize_for_space
void op_for_size()  // code optimized for memory usage
#pragma optimize_for_speed
void op_for_speed()  // code optimized for speed
// subsequent function declarations optimized for speed
#pragma optimize off // optimization is turned off
```

The symbol `#pragma {instruction}` is placed before the code to be optimized. This symbol is called pragma, and it is an implementation-specific directive that modifies the compiler's behavior. Note that pragma does not produce any object code. The Blackfin C/C++ compiler supports many pragmas as user-specified instructions to the compiler for producing a more efficient code. These pragmas perform the following functions:

1. Align data.
2. Define functions that act as interrupt handler.
3. Change the optimization level in different sections of the program.
4. Change how to link an external function.
5. Loop optimization.
6. Header file configurations and properties.
7. Memory bank usage.

Some of these pragmas are discussed further in the following sections.

8.2.2 C Programming Considerations

This section examines several important C programming considerations [42] for the Blackfin processors.

8.2.2.1 Array versus Pointers

C uses an index or pointer to access a sequence of data from an array. There is no clear advantage in using one over the other, but the index is easier to understand. The pointer introduces additional variables that need more memory space. In contrast, the index must be transformed to a pointer by the compiler, which may degrade the performance. Therefore, the best strategy is to start with index style and move to pointer style if the former does not provide a good result.

HANDS-ON EXPERIMENT 8.1

This experiment examines the performance difference between the index-style and pointer-style FIR filter routines. The C program `process_data.c` implements the index-style FIR filter, which is located in directory `c:\adsp\chap8\exp8_1a_533` for the BF533 EZ-KIT (or c:\adsp\chap8\exp8_1a_537 for the BF537). A section of the FIR filtering program for indexing array is listed below. Note that there is a need to perform a right shift of 15 bits to save the 16 most significant bits (excluding the extra sign bit) of the fract32 variable acc to the fract16 variable out. The updating of the delay buffer is carried out by shifting the data in the wdelaybuf array iteratively.

```
for (i=0; i<IP_SIZE; i++)
   {
   wdelaybuf[0]=in[i];
   for (j=0; j<nc; j++)        // perform MAC
   {
      acc+=coef[j]*wdelaybuf[j];
   }
   out[i] = (fract16) (acc >> 15);
   for (k=(nc-1); k>=1; k-)  // update delay buffer
   {
      wdelaybuf[k]=wdelaybuf[k-1];
   }
   acc = 0;
   }
```

In the case of the pointer-style FIR filter, pointers point to the coefficient and delay buffers. However, pointers introduce additional variables that need extra memory locations. A section of the pointer-style FIR filter program is shown below. The project file is located in directory c:\adsp\exp8_1b_533 for the BF533 EZ-KIT (or c:\adsp\exp8_1b_537 for the BF537).

```
for (i=0; i<IP_SIZE; i++)
   {
   *wdelaybuf=in[i];
   for (j=0; j<nc-1; j++)      // perform MAC
   {
      acc += *coef++ * (*rdelaybuf++);
   }
   acc += *coef * (*rdelaybuf);
   out[i] = (fract16) (acc >> 15);
   tempdelaybuf = rdelaybuf;    // update delay buffer
   rdelaybuf-=1;
   for (k=(nc-1); k>=1; k--)
   {
      *tempdelaybuf-- = *rdelaybuf--;
   }
   acc = 0;
   coef=state.h;
   rdelaybuf=state.d;
   }
```

Build the project based on the index-style 32-tap FIR filter with 32 samples per block without optimization. The cycle count needed to perform 32 samples of FIR filtering is recorded by clicking on **View → Debug Windows → BTC Memory**. Select Cycle Counter from the **BTC Memory** window and select display format to hex32. Enable **Auto Refresh** and run the project. Click on **SW6** for the BF533 (or **SW11** for the BF537) to enable the FIR filtering. Now, select **Enable optimization[100]** and repeat the build and benchmark processes. Record the cycle count in Table 8.1. Repeat the experiment for the pointer-style FIR filter.

8.2 C Optimization Techniques

Table 8.1 Cycle Count Benchmark for 32-Tap FIR Filter (with 32 Data Samples/Block)

FIR Filter Program in C	Cycle Count	
	Optimization On	Optimization Off
Index style		
Pointer style		

Symbol	Demangled Name	Address	Size	Binding
_firc	firc	0xffa0119c	0xc6	GLOBAL
_Process_Data	Process_Data	0xffa01262	0x168	GLOBAL

(a)　Input section .\Debug\Process_data.doj(program)

Symbol	Demangled Name	Address	Size	Binding
__btc_nNumBtcMapEntries	__btc_nNumBtcMapEntries	0xff8002bc	0x0	GLOBAL
_lpf	lpf	0xff8002c0	0x40	GLOBAL
_ldelay	ldelay	0xff800300	0x40	GLOBAL
_rdelay	rdelay	0xff800340	0x40	GLOBAL

(b)　Input section .\Debug\main.doj(data1)

Symbol	Demangled Name	Address	Size	Binding
_iTxBuffer1	iTxBuffer1	0xff902234	0x200	GLOBAL
_iRxBuffer1	iRxBuffer1	0xff902434	0x200	GLOBAL

(c)　Input section .\Debug\main.doj(bsz)

Figure 8.2 Memory information for FIR filter program (a), data for FIR filter (b), and data for input and output block buffer (c).

HANDS-ON EXPERIMENT 8.2

The preceding experiment benchmarks the execution speed based on cycle count. This experiment uses VisualDSP++ to benchmark the memory usage of the program and data. Click on **Project → Project Options ... → Link → General** and enable **Generate symbol map**. An XML document containing the memory information will be generated after the build process. This file can be found in directory ..\Debug. Open the .map file and search for the FIR filter program memory as shown in Figure 8.2(a), the data memory as shown in Figure 8.2(b), and the buffer memory as shown in Figure 8.2(c).

Note that the size stated in the table is in byte units. Determine the length of the FIR filter and the number of data samples acquired per block. Benchmark the code size for the index-style and pointer-style FIR filters given in Hands-On Experiment 8.1. Experiment with different levels of optimization for execution speed and memory size and fill in the results in Table 8.2.

Table 8.2 Memory Benchmark for 32-Tap FIR Filter (Using 32 Data Samples/Block)

FIR Filter Program in C (firc)	Code Size in Bytes	
	Optimization On	Optimization Off
Index style		
Pointer style		

8.2.2.2 Using Loop Optimization Pragmas

In general, it is most effective to apply loop pragmas to the innermost loops. The loop pragmas are placed before the loop statement to supply additional information to the compiler to perform a more aggressive optimization. For example,

```
#pragma vector_for
for (i=0; i < 100; i++)
    a[i] = b[i]
    ...
```

The pragma directive notifies the optimizer that it is safe to execute two iterations of the loop in parallel. However, this pragma does not force the compiler to vectorize the loop. The final decision is still determined by the optimizer.

As seen in previous chapters, DSP algorithms usually involve extensive looping. Therefore, it is very important that the compiler can optimize the loop operations. The Blackfin C compiler can automatically perform loop unrolling and use multiple arithmetic units for loop optimization.

8.2.2.3 Using Data Alignment Pragmas

The data alignment pragmas are used for the compiler to arrange data within the processor's memory. For example, the following pragmas are used to align data to different byte boundaries:

```
#pragma align 1
char a;       // 1-byte alignment
#pragma align 2
short b;      // 2-byte alignment
#pragma align 4
int c;        // 4-byte (or 32-bit) alignment
#pragma align 8
int d;        // 8-byte (or 64-bit) alignment
```

In the Blackfin processor, a word is defined as 32-bit data. Data are often fetched with 32-bit loads and must be word (or 32-bit)—aligned to generate efficient code. In other words, the last two bits of the starting address of the data must be zero. Because data and coefficients in the FIR filter are 16 bits, we can use `#pragma align 2` to align these variables to a 2-byte boundary. Figure 8.3

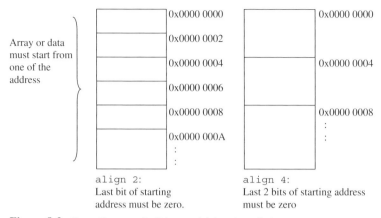

Figure 8.3 Data alignment for 2-byte and 4-byte boundaries

shows the possible starting addresses for the array aligned to 2-byte and 4-byte boundaries.

8.2.2.4 Using Different Memory Banks

The memory bank pragma allows the compiler to assume that groups of memory accesses in a loop based on different pointers reside in different memory banks. This pragma will improve memory access because the processor is capable of accessing multiple data from different banks in a single instruction. For example,

```
#pragma different_banks
for (i=0; i < 100; i++)
   a[i] = b[i]
   ...
```

The above pragma uses different memory banks to allow simultaneous accesses to arrays a and b.

8.2.2.5 No Aliasing

`#pragma no_alias` tells the compiler that the following loop has no load or store instruction that accesses the same memory. Using the no alias pragma leads to a better code because it allows any number of iterations (instead of two at a time) to be performed simultaneously.

HANDS-ON EXPERIMENT 8.3

This experiment investigates the performance gain from using some of the pragmas introduced in the preceding subsections. We use the index-style FIR filter example in Hands-On

Experiment 8.1 to test the optimization performance. Note that optimization must be turned on for the pragmas to be effective. Open the project file `exp8_3_533.dpj` for the BF533 EZ-KIT (or `exp8_3_537.dpj` for the BF537). This file is located in directory `c:\adsp\chap8\exp8_3_533` (or `c:\adsp\chap8\exp8_3_537`). Examine the `Process_data.c` file to see the additional pragma instructions. Benchmark the performance gain of the optimized C code with pragmas as comparing with the C code without optimization. Note that the instruction `__builtin_aligned( )` is similar to the data alignment pragma. Change the data alignment to four bytes and observe any difference in performance. Comment out the pragmas and rebuild the project with optimization. Observe any change in performance.

8.2.2.6 Volatile and Static Data Types

When declaring variables in the C source file, a volatile data type is declared to ensure that the optimizer is aware that their values may be changed externally at any time. The volatile data type is important for peripheral-related registers and interrupt-related data. For example, in the previous experiment, the transmit buffer `iTxBuffer` and the receive buffer `iRxBuffer` are declared as volatile because these variables are constantly updated by the DMA.

Static variables can only be accessed by the function in which they were declared as local variables. The static variable is not erased on exit from the function; instead, its value is preserved and is available when the function is called again. For example, the index variables, j and k, in the EX_INTERRUPT_HANDLER are declared as static because these values must be correctly maintained in the interrupt service routine.

8.2.2.7 Global versus Local Variables

Local variables are normally aliased to the core registers or internal memory. Local variables have an advantage over the global variables because the compiler does not need to maintain the local variables outside the scope. However, global variables reduce the size of the stack.

8.2.2.8 Arithmetic Data Types

As shown in Chapter 6, we use integer data types to represent fractional numbers for the fixed-point Blackfin processor. In Table 6.7, we observe that the integers can be represented as `char` (8 bit), `short` (16 bit), `int` (32 bit), and their unsigned formats. We can also use `fract16` and `fract32`, which are mainly reserved for fractional built-in or intrinsic functions. These data types are described in the next section.

For example, the following code performs a 256-tap FIR filter using the integer data type:

```
int filter(short *in, short *coeff)
{
    int i;
    int acc=0;
    for (i=0; i<128; i++) {
        acc += ((in[i]*coeff[i])>>15);
    }
    return acc;
}
```

Floating-point arithmetic is usually avoided for the Blackfin processor because it requires much a longer processing time. If floating-point arithmetic is required in certain sections of the code, it can be implemented with library routines. However, the optimizer is ineffective for optimizing floating-point code.

8.2.3 Using Intrinsics

The VisualDSP++ compiler supports intrinsic functions for efficient use of the Blackfin's hardware resources. These built-in C-callable functions generate assembly instructions that are designed to optimize the code produced by the compiler, or to effectively access system hardware. The intrinsic functions generally operate on single 16- or 32-bit values, and they support the following operations and functions:

1. Fractional value built-in functions
2. European Telecommunications Standard Institution (ETSI) support
3. Complex fractional data and operations
4. Viterbi history and decoding functions
5. Circular buffer functions
6. Endian-swapping functions
7. System built-in functions
8. Video operation functions
9. Misaligned data functions

Refer to the *VisualDSP++ 4.0 C/C++ Compiler and Library Manual for Blackfin Processors* [35] for details on using intrinsic function. In the following, we briefly introduce some of intrinsic functions.

The fractional value built-in function provides access to fractional arithmetic and the parallel 16-bit operations supported by the Blackfin instructions. For example, the intrinsic functions `add_fr1x32( )` and `mult_fr1x32( )` perform add and multiply of 32-bit fractional data, respectively. By using the intrinsic functions, the compiler can perform different optimization techniques to speed up the execution. In addition, the intrinsic functions support saturation arithmetic that prevents overflow. This is in contrast to the standard C program, which needs additional routines for checking overflows and setting saturation arithmetic.

The FIR filter routine using intrinsic functions can be written as follows:

```
#include <fract.h>
fract32 filter(fract16 *in, fract16 *coeff)
{
  int i;
  fract32 acc=0;
  for (i=0; i<128; i++) {
    acc = add_fr1x32(acc,mult_fr1x32(in[i],coeff[i]));
  }
  return acc;
}
```

The use of the intrinsic functions to implement the FIR filtering routine is illustrated in Hands-On Experiment 8.4.

HANDS-ON EXPERIMENT 8.4

This experiment evaluates the performance improvement using the intrinsic functions for implementing FIR filter on the Blackfin processor. The experiment files can be found in directory c:\adsp\chap8\exp8_4_533 for the BF533 EZ-KIT (or c:\adsp\chap8\exp8_4_537 for the BF537). Turn on the optimization option before building the project. Benchmark the cycle count and memory size of the C code using intrinsics and record the results in Table 8.3. Compare the performance with the index-style C code with optimization given in Hands-On Experiment 8.1.

In previous hands-on experiments, we used several intrinsic functions to perform fractional arithmetic such as addition, multiplication, absolute value, and shift. These intrinsic functions are categorized under the fractional value functions that support fractional arithmetic and parallel 16-bit operations. Special C data types are used with these built-in functions. Table 8.4 lists the fractional value C data types. The fract.h header file provides access to the definitions of the fractional value built-in functions.

Table 8.3 Cycle Count and Code Size for the C Code with Intrinsics

C Code Using Intrinsics (with optimization)
Cycle count
Code size (in byte) (firc)

Table 8.4 Fractional Value C Data Types

C Type	Number Representation
fract16	Single 16-bit signed fractional value in (1.15) format
fract32	Single 32-bit signed fractional value in (1.31) format
fract2x16	Double 16-bit signed fractional value in (1.15) format

Some of the commonly used fractional 16-bit and 32-bit functions are listed in Table 8.5.

The ETSI build-in functions are included in the `libetsi.h` header file. Similar to the fractional value built-in functions, the ETSI functions operate on both 16-bit and 32-bit fractional data. The ETSI functions include fractional multiply-accumulate, multiply-subtract, and divide.

The complex fractional built-in functions consist of `complex_fract16` and `complex_fract32` data types and support complex fractional multiply and accumulate, square, and distance. The header file `complex.h` must be included in the program to use these complex data types.

The VisualDSP++ compiler also provides intrinsic functions for automatic circular buffer generation, circular indexing, and circular pointer reference. These built-in functions are defined in the `ccblkfn.h` header file. Circular indexing was used in the previous hands-on experiment to update the pointer for a circular buffer. For example, the function `circindex(index, incr, nitems)` is used in the signal generator experiment in Section 7.6. The `index` and `incr` arguments are the circular buffer index and the step increment, respectively. The `nitems` argument is the length of the circular buffer. In addition, VisualDSP++ includes a function that performs a circular buffer increment of a pointer as follows:

```
void *circptr(void *ptr, long incr, void *base, unsigned long buflen);
```

Note that both `incr` and `buflen` are specified in bytes.

Table 8.5 Fractional 16-Bit and 32-Bit Built-In Functions

Built-in Functions	Operations
fract16 add_fr1x16	Add two 16-bit fractional numbers.
fract16 mult_fr1x16	Multiply two 16-bit fractional numbers. The result is truncated to 16 bits.
fract32 mult_fr1x32	Multiply two 16-bit fractional numbers and return the 32-bit result.
fract16 abs_fr1x16	Perform absolute value of the input number.
fract16 shl_fr1x16	Perform arithmetic left shift with sign extension for signed number.
fract16 shr_fr1x16	Perform arithmetic right shift.
fract32 add_fr1x32	Add two 32-bit fractional numbers.
fract32 mult_fr1x32x32	Multiply two 32-bit fractional numbers. The result is rounded to 32 bits.
fract16 sat_fr1x32	Saturate a 32-bit fractional number to 0x7FFF or 0x8000. Otherwise, return the lower 16 bits of the number.
fract16 round_fr1x32	Rounds a 32-bit number to 16-bit using biased rounding.
fract2x16 mult_fr2x16	Multiply two packed 16-bit fractional numbers, and truncate the result to 16 bits. Note that the `fract2x16` type represents two fractional 16-bit numbers packed into upper and lower halves and makes use of the dual multipliers.

* All fractional built-in functions are in saturation mode.

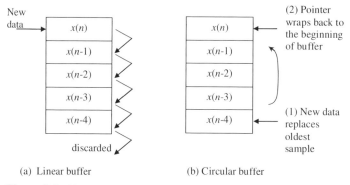

(a) Linear buffer   (b) Circular buffer

Figure 8.4 Linear buffer (a) and circular buffer (b)

HANDS-ON EXPERIMENT 8.5

In the previous experiments, the delay buffer was updated in a linear fashion as shown in Figure 8.4(a). All data in the buffer must be physically shifted by one position downward after the end of the multiply-accumulate operations of the FIR filtering. The oldest data are discarded and the new data sample is inserted at the top of buffer. Figure 8.4(b) shows the circular buffer, which does not require the physical moving of data. Instead, the circular buffer uses a pointer to replace the oldest data sample by the new data sample and updates the pointer in modulo fashion (wrap back) once the pointer reaches the bottom of the circular buffer.

This experiment implements the circular buffers for the FIR filtering. The experiment can be found in directory c:\adsp\chap8\exp8_5_533 for the BF533 EZ-KIT (or c:\adsp\chap8\exp8_5_537 for the BF537). The FIR filter program using the circular buffer is listed as follows:

```
for (i=0; i<IP_SIZE; i++)
  {
    *wdelaybuf = in[i];
    acc = mult_fr1x32(coef[0],*wdelaybuf);
    for (j=1; j<nc; j++)
    {
      wdelaybuf = circptr(wdelaybuf, 2, state.d, 2*TAPS);
      acc = add_fr1x32(acc,mult_fr1x32(coef[j],*wdelaybuf));
    }
    out[i] = (fract16)(acc >>15);
    acc = 0;
  }
```

This code uses the circular pointer to the delay buffer wdelaybuf. The pointer is increased by 2 bytes, and the length of the buffer is 2*TAPS bytes. Build the project with optimization and benchmark the performance against the FIR filter using a linear buffer in Hands-On Experiment 8.4. Fill in the benchmark results in Table 8.6.

8.2.4 Inlining

VisualDSP++ supports an "inline assembly" feature that allows programmers to insert small sections of assembly code within the C program. For example,

Table 8.6 Cycle Count and Code Size for the C Code Using Circular Buffer Intrinsic Functions

	C Code with Intrinsics with Optimization and Circular Buffer
Cycle count	
Code size (in byte) (firc)	

the C statement `asm("nop;")` inserts a `nop` assembly instruction into the C code. However, inline assembly must be avoided when built-in functions are available. In addition, a `volatil` keyword is used to prevent an `asm()` instruction from being moved, combined, or deleted, for example, `asm volatile("nop;")`.

VisualDSP++ uses the keyword `inline` to indicate that functions should have C code generated inline at the point of call. For example,

```
inline int mult(int a, int b) {
   return(a*b);
}
```

The use of this keyword has the advantages of reducing program-flow latencies, function entry and exit instructions, and parameter passing overheads. However, this keyword is only useful for small and frequently used functions, which gives the best gain in execution speed with little increment in code size.

Inline code is often used instead of a subroutine because a subroutine incurs overhead. Although a subroutine is more efficient in code size, inline code gains in execution speed. Therefore, software designers must make a careful decision on this memory-speed trade-off.

8.2.5 C/C++ Run Time Library

The C/C++ run time library is the collection of functions, macros, and class templates. This library simplifies the software development by providing many ready-to-use functions such as standard I/O routines, character and string handling, floating-point arithmetic emulation, mathematics, and others. Table 8.7 lists the standard C run time library header files.

8.2.6 DSP Run Time Library

The DSP run time library supports many general-purpose DSP algorithms such as filters, FFT, vector and matrix functions, math functions, statistical functions, window functions, and others. Table 8.8 lists the DSP library functions and their header files. These DSP functions were called from the main program in many

Table 8.7 Standard C Run Time Library Header Files

Functions	Header Files
Diagnostics	assert.h
Character handling	ctype.h
Macros and data structures for alternative device drivers	device.h
Enumerations and prototypes for alternative device drivers	device_int.h
Error handling	errno.h
Floating point	float.h
Boolean operation	iso646.h
Limits	limits.h
Localization	locale.h
Mathematics	math.h
Nonlocal jumps	setjmp.h
Signal handling	signal.h
Variable arguments	stdarg.h
Standard definitions	stdef.h
Input/output	stdio.h
Standard library	stdlib.h
String handling	string.h
Date and time	time.h

hands-on experiments in previous chapters. In addition, VisualDSP++ provides the source code of these DSP functions, which can be found in directory c:\Program Files\Analog Devices\VisualDSP 4.0\Blackfin\lib\src\libdsp. These source codes are available with .asm and .c extensions.

These functions support different data types including float, double, long double, and fract16. A detailed description of the DSP library functions can be found in the *VisualDSP++ 4.0 C/C++ Compiler and Library Manual for Blackfin Processors* [35]. Table 8.9 shows the cycle count and the code size of some commonly used DSP library functions.

EXAMPLE 8.1

Using the FIR filter benchmark results listed in Table 8.9, we can determine the memory required to perform a 256-tap FIR filter (with 16-bit arithmetic) using sample and block processing modes as follows. The FIR filter is sampled at 48 kHz.

1. Sample processing. Data memory includes 256 × 2 bytes for the coefficient buffer, 255 × 2 bytes for the delay buffer, and 2 × 2 bytes for input and output. From the benchmark results, 354 bytes are required for the code. Thus the total memory needed is 1,380 bytes.
2. Block processing (using a block size of 32 samples). Data memory includes 512 bytes for coefficients and 510 bytes for the delay buffer. There are 4 × 32 × 2 bytes required for the input and output buffers in the ping-pong buffering scheme. Again, 354 bytes are required for the code. Thus the total memory needed is 1,632 bytes.

8.2 C Optimization Techniques

Table 8.8 DSP Library Functions and Header Files

Functions	Header Files
Complex arithmetic function	`complex.h`
Cycle count function	`cycles.h`
Filters and transformations	`filter.h`
Math functions	`math.h`
Matrix functions	`matrix.h`
Statistical functions	`stats.h`
Vector functions	`vector.h`
Window functions	`window.h`

Table 8.9 Benchmark of Some DSP Library Functions

Functions	Cycle Count	Code Size (in bytes)
`fir_fr16` FIR filter in (1.15) format (N_b: number of input sample/block) (N_c: number of coefficients)	~ $64 + N_b/2 \times (3 + N_c)$	354
`iir_fr16` IIR filter in (1.15) format (B: number of biquad)	~ $58 + 3 \times N_i \times B + 1.5 \times N_i$	384
`div16` 16-bit integer divide	~35	32
`sin_fr16` Sine function in (1.15) format	~25	78
`sqrt_fr16` Square root function in (1.15) format (*In*: input value in (1.15) integer)	~ $33 + 8 \times \log_4(8192/In)$ for positive input value	116
`mean_fr16` Mean function in (1.15) format (N_b: number of samples)	~ $20 + N_b$	58

Extracted from the comment section of ADI source codes.

EXAMPLE 8.2

Based on Table 8.9, determine the cycle count and processing MIPS in implementing a 256-tap FIR filter (with 16-bit arithmetic) using sample and block processing modes. The FIR filter is sampled at 48 kHz.

1. Sample processing. The cycle count needed to perform FIR filtering for generating one output sample ~194 cycles. Thus MIPS needed for performing sample FIR filtering ~9.3 MIPS.

2. Block processing (using a block size of 32 samples). The cycle count needed to perform FIR filtering on 32 samples ~4,208 cycles. Thus, to perform 1 sample requires ~4,208/32 = 131.5 cycles. The required MIPS for performing the block FIR filter ~6.3 MIPS. Therefore, we can perform more tasks on the same processor with the block processing mode.

HANDS-ON EXPERIMENT 8.6

The DSP library functions are hand-optimized in either assembly or C. This experiment examines the performance of the FIR filter function `fir_fr16` given in the run time DSP library. This function has been used in many previous hands-on experiments. In this experiment, we build the project and benchmark its cycle count and code size. The experiment file is located in directory `c:\adsp\chap8\exp8_6_533` for the BF533 EZ-KIT (or `c:\adsp\chap8\exp8_6_537` for the BF537). Fill in the results in Table 8.10 and compare with the previous benchmark results using intrinsic functions. Comment on the results.

Table 8.10 Cycle Count and Code Size for C Code Using the Run Time DSP Library

C Code Using Run Time DSP Library Function `fir_fr16`
Cycle count
Code size (in byte) (`fir_fr16`)

8.2.7 Profile-Guided Optimization

Profile-guided optimization (PGO) allows the compiler to use data collected during program execution for the optimization analysis. Representative data sets are passed to the application program to profile which sections of the code are executed most frequently in order for the compiler to perform selective optimization. The process of performing PGO is stated as follows:

1. Compile the application program to collect profile information.
2. Run the application in the simulator using a representative data set.
3. The simulator accumulates the profile data and identifies the "hot spot" for optimization.
4. Recompile the application program by using the collected profile data.

8.2 C Optimization Techniques   **347**

In particular, PGO informs the compiler about the application that affects branch prediction, improves loop transformation, and reduces code size. Therefore, the compiler can perform more advanced and aggressive optimizations by using profiler statistics generated from running the application with representative data.

HANDS-ON EXPERIMENT 8.7

This experiment illustrates the process of using PGO in the VisualDSP++ simulator. This simple program multiplies the positive and negative input samples with different gain values, and the compiler decides the most common case in an `if...else` construct. Use the following steps to perform PGO experiment:

1. Activate the BF533 (or BF537) VisualDSP++ simulator and load the project file `exp8_7.dpj` in directory `c:\adsp\chap8\exp8_7`. Make sure that the target processor (in **Project Options** menu) matches the selected simulator type.

2. The C program uses three input data files. The first data file, `dataset_1.dat`, has 50% positive samples; the second file, `dataset_2.dat`, has 75% positive samples; and the third file, `dataset_3.dat`, has 100% positive samples. Users can read these data files by double-clicking on the filename. Because these files contain a bias toward the positive samples, more execution time is needed for the `if` branch. Therefore, this information will be used by the PGO to tune the compilation to optimize the `if` branch.

3. The data sets must be fed into the PGO by clicking on **Tools → PGO → Manage Data Sets**. Figure 8.5 shows the **Manage Data Sets** dialog box, which allows users to control the optimization level. Set **Optimization level** to **Fastest Code**. Click on **New...** to open the **Edit Data Set** dialog box. Replace the **Data set name** with a more descriptive one such as `50%p50%n` for a file with 50% positive numbers. Type in the file name `dataset_1.pgo` in **Output filename**.

4. Attach an input stream to the data set by clicking on the **New...** button in the **Edit Data Set** dialog box. An **Edit PGO Stream** dialog box is opened as shown in Figure 8.6. Key in the input file `dataset_1.dat` in the **Filename** field and other parameters as shown in Figure 8.6. Click on **OK** to return to the **Edit Data Set** dialog box. Click on **OK** to save the data set and close the dialog box.

5. Create the remaining two data sets by following Steps 3 and 4. We can also use the **Copy** button in the **Manage Data Set** dialog box to speed up the creation of these two data sets. Highlight the input file under **Input Streams...** and click on **Edit** to enter a new name in **Input Source File**.

6. Once the data sets have been configured, we can optimize the program by clicking on **Tools → PGO → Execute Data Sets**. Note that the project is built and run with different data sets. The simulator will monitor the number of cycles and the execution

348 Chapter 8 Code Optimization and Power Management

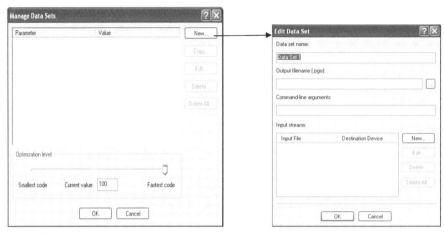

Figure 8.5 **Manage Data Sets** and **Edit Data Set** dialog boxes

Figure 8.6 **Edit PGO Stream** dialog box

path and store them in the respective .pgo files. Subsequently, the project is recompiled by using the information in the .pgo files to optimize the executable code. Finally, the executable code is run again with the data sets to measure the speed gain after optimization.

7. When the execution in Step 6 is completed, an XML report of the PGO result is generated and displayed as shown in Figure 8.7.

Data Set: 50%p50%n	
Command line:	
Input stream:	File: C:\ADSP\CHAP8\exp8_7\DataSet1.dat Device: 0xFFD00000-0xFFD00FFF
PGO output:	C:\ADSP\CHAP8\exp8_7\dataset1.pgo
Before optimization:	7111 cycles
After optimization:	7111 cycles
Cycle reduction:	0.00%

Data Set: 75%p25%n	
Command line:	
Input stream:	File: C:\ADSP\CHAP8\exp8_7\DataSet2.dat Device: 0xFFD00000-0xFFD00FFF
PGO output:	C:\ADSP\CHAP8\exp8_7\dataset2.pgo
Before optimization:	7943 cycles
After optimization:	6279 cycles
Cycle reduction:	20.95%

Data Set: 100%p0%n	
Command line:	
Input stream:	File: C:\ADSP\CHAP8\exp8_7\DataSet3.dat Device: 0xFFD00000-0xFFD00FFF
PGO output:	C:\ADSP\CHAP8\exp8_7\dataset3.pgo
Before optimization:	8775 cycles
After optimization:	5447 cycles
Cycle reduction:	37.93%

Figure 8.7 PGO results on the program in Hands-On Experiment 8.7

8.3 USING ASSEMBLY CODE FOR EFFICIENT PROGRAMMING

We have used different C programming techniques to enhance the performance of the C code. These techniques are under the first level of optimization and have the advantage of being easy to program and maintain. However, a C program does not give the programmer complete control over using registers, addressing modes, parallel instructions, zero-overhead loops, and managing multiple resources of the processor. Therefore, a C program may produce a larger code size and a slower execution speed.

The DSP run time library functions mentioned in Section 8.2.6 are mainly written in optimized assembly to take advantage of the unique features of the Blackfin processors. These source codes serve as important references in learning the art of optimizing the low-level programming in Blackfin processors. Because an assembly code requires more time to write and maintain, a good practice is to write the time-critical or resource-critical sections of code in assembly and leave most of the housekeeping and initialization tasks in C. This hybrid C-and-assembly code produces the best balance between development time and code efficiency. As discussed above, 20% of the code accounts for 80% of the execution time. This 20% of code can be written in assembly to optimize the most critical portions of the code.

In the following sections, we highlight some important optimization techniques for assembly programming [28]. We benchmark the performance of these techniques with the FIR filter. We begin by writing a linear assembly code without using any special feature of the Blackfin processor, and gradually add low-level optimization techniques, one at a time, to show the corresponding improvement achieved by different changes. Based on the pseudo code listed in Section 8.1, we can write the assembly code in the following three parts:

1. Initialization. The data sample and the corresponding filter coefficient are fetched to registers R0.L and R2.L, respectively. A delay line buffer is defined in memory. The initialization code is listed as follows:

```
E_FIR_START: R0.L=W[P0++P3];  // fetch x(n) from input
                                 array
      R2.L=W[I2];             // fetch filter coefficient
      T1=P4;                  // index to READ/WRITE pointer
      W[I0]=R0.L;             // store input sample into delay
                                 buffer
                              // restore delay line address in I1
```

2. Multiply-accumulate operation. Multiply-accumulate is the key operation in FIR filtering. The multiplication is carried out between the 16-bit data sample in R0.L and the filter coefficient in R2.L, and the 32-bit result is accumulated in the 40-bit register A0. The multiply-accumulate operation is repeated for the number of taps in the FIR filter. A software loop is used in the linear assembly code as follows:

```
              R6=P2;            // setup inner loop counter
E_MAC_ST:     R2.L=W[I2++];     // input sample
              R0.L=W[I1++];     // filter coefficient
              A0+=R0.L*R2.L;    // multiply add
              R6 += -1;         // decrement loop counter
              CC=AZ;            // is CC a zero?
E_MAC_END:    if !CC JUMP E_MAC_ST;
              R3.L=A0(S2RND);   // save in a register
```

Note that the register R6 is used as a counter for the software loop. This register is decremented at every iteration, and the CC flag is compared with zero. If this flag is not zero, the program returns to the start of the filtering loop (E_MAC_ST), and repeats the multiply-accumulate operation of the next coefficient and the corresponding data sample. This process continues until the loop counter equals the number of filter taps. The final filtering result in A0 is rounded and stored to register R3.L.

8.3 Using Assembly Code for Efficient Programming

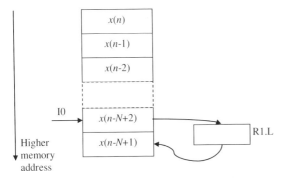

Figure 8.8 Updating of the linear delay line buffer

3. Update the delay line buffer. The final part of FIR filtering is to update the delay line implemented as the linear buffer. A software loop that performs the buffer update as follows:

```
R6=P2;                          // loop counter to update buffer
UPDATE_BUF_ST: W[I0-]=R1.L;     // update input samples
   I0-=2;
   R1.L=W[I0++];
   R6 += -1;                    // decrement loop counter
   CC=AZ;                       // is CC a zero?
UPDATE_BUF_END: if !CC JUMP UPDATE_BUF_ST;
```

Again, the counter R6 is initialized to the total number of taps. The oldest data sample is replaced by the next data sample in the delay line via register R1.L, and this linear replacement (or shifting) of the data samples is carried out until the end of the buffer. This operation is shown in Figure 8.8.

HANDS-ON EXPERIMENT 8.8

This experiment examines the performance of the assembly code for implementing the FIR filter. Open the project file exp8_8_533.dpj for the BF533 EZ-KIT (or exp8_8_537.dpj for the BF537). The project file is located in directory c:\adsp\chap8\exp8_8_533 (or c:\adsp\chap8\exp8_8_537). Build the project and benchmark the cycle count and code size and write the results in Table 8.11.

The assembly program uses a conditional branch for looping. The branch is executed based on the state of the CC bit. A conditional branch takes nine cycles to complete. However, the Blackfin processor provides a branch prediction that can accelerate the execution of a conditional branch. If the prediction is "branch taken", the branch will be executed and five cycles are required; otherwise, nine cycles are required. If the prediction is "branch not taken", only one cycle is required; otherwise, nine cycles are required. To implement branch prediction, the option bp is applied to the conditional branch instruction. For example,

```
if !CC JUMP E_FIR_START (bp);
```

Table 8.11 Cycle Count and Code Size for the Linear Assembly Code

	Cycle Count	Code Size in Bytes
Linear assembly (without branch prediction)		
Linear assembly (with branch prediction)		

Insert the branch prediction and note the changes in the cycle count and code size in the second row of Table 8.11.

8.3.1 Using Hardware Loops

As mentioned in Chapter 5, the Blackfin processor supports hardware loops. Unlike the software loop that performs counter decrement and checks for loop completion with a conditional branch, the hardware loop has zero overhead. The setup for the hardware loop is explained in Example 5.15. Here, we apply the hardware loop to the MAC operation and the delay buffer update as follows:

```
    // perform MAC
    LSETUP(E_MAC_ST,E_MAC_END)LC1=P2;
        // setup inner loop counter
        // P2 = number of filter taps
E_MAC_ST:   R2.L=W[I2++];       // input sample
            R0.L=W[I1++];       // filter coefficient
E_MAC_END:  A0+=R0.L*R2.L ;     // MAC

    // update delay line
    LSETUP(UPDATE_BUF_ST,UPDATE_BUF_END)LC1=P2;
            // loop counter to update buffer
UPDATE_BUF_ST: W[I0--]=R1.L;    // update input samples
            I0-=2;
UPDATE_BUF_END: R1.L=W[I0++];
```

HANDS-ON EXPERIMENT 8.9

This experiment examines the performance of the assembly code that uses the hardware loop instructions. Open the project file exp8_9_533.dpj for the BF533 EZ-KIT (or exp8_9_537.dpj for the BF537). The project file is located in directory c:\adsp\chap8\exp8_9_533 (or c:\adsp\chap8\exp8_9_537). Study the assembly program fir_asm_hardware_loop.asm and identify the hardware loops for MAC and delay line update. Build the project and benchmark the cycle count and code size to fill in Table 8.12.

8.3 Using Assembly Code for Efficient Programming

Table 8.12 Cycle Count and Code Size for Assembly Code Using Hardware Loop

	Cycle Count	Code Size in Bytes
Hardware loop		

Table 8.13 Cycle Count and Code Size for Assembly Code with Hardware Loop and Dual MACs

	Cycle Count	Code Size in Bytes
Hardware loop and dual MACs		

8.3.2 Using Dual MACs

After optimizing the loop, we turn our attention to optimizing the MAC operations in the assembly code. The Blackfin processor consists of two MAC units as described in Section 5.1.3. We can use these dual MAC units to speed up the MAC operations. One possible approach is to halve the loop count and perform two MAC operations in the loop as:

```
        LSETUP(E_MAC_ST,E_MAC_END)LC1=P2>>1;
                            // setup inner loop counter
E_MAC_ST:   R2=[I2++];      // input samples
            R0=[I1++];      // filter coefficients
            A1+=R0.H*R2.H;  // MAC1
E_MAC_END:  A0+=R0.L*R2.L;  // MAC2 -** UNROLL 2x **-
```

The loop counter `LC1` is reduced by half or, equivalently, the loop is now unrolled by a factor of two.

HANDS-ON EXPERIMENT 8.10

This experiment examines the cycle count and code size of the assembly program using the hardware loop and dual MACs. The program can be found in directory `c:\adsp\chap8\exp8_10_533` for the BF533 EZ-KIT (or `c:\adsp\chap8\exp8_10_537` for the BF537). Fill in the performance in Table 8.13.

8.3.3 Using Parallel Instructions

The Blackfin instructions can be executed in parallel. Up to three instructions can be executed in a single cycle with some limitations. A multi-issue instruction is

354   Chapter 8   Code Optimization and Power Management

| 32-bit ALU/MAC instruction | 16-bit instruction | 16-bit instruction |

Figure 8.9   Parallel issued combinations

64-bit long and consists of one 32-bit instruction and two 16-bit instructions as shown in Figure 8.9.

There are three possible ways to form parallel instructions with following syntaxes:

1. 32-bit ALU/MAC instruction || 16-bit instruction || 16-bit instruction.
2. 32-bit ALU/MAC instruction || 16-bit instruction.
3. MNOP || 16-bit instruction || 16-bit instruction.

The vertical bar (||) indicates that the instructions are issued in parallel. In Method 2, a 16-bit NOP is automatically inserted into the unused 16-bit slot. In Method 3, a 32-bit NOP instruction may be required in the parallel instruction, and the assembler can automatically insert it if needed [37]. In addition, there are some restrictions to the 16-bit instructions in a multi-issue instruction. Only one of the 16-bit instructions can be a store instruction; however, it is possible to have two load instructions. In addition, if these two 16-bit instructions are memory access instructions, one memory access must be an I-register and the other can be either an I- or a P-register.

We can further optimize the previous FIR filtering program by using parallel instruction for loading input samples and filter coefficients as follows:

```
E_FIR_START: MNOP  || R0.L=W[P0++P3]  || R2.L=W[I2];
       // parallel fetch x(n) from input array
       // and fetch filter coefficient
```

and

```
E_MAC_ST: MNOP  || R2=[I2++]  || R0=[I1++];
       // parallel fetch input and filter coefficient
```

The above multi-issue instructions are executed in a single cycle.

HANDS-ON EXPERIMENT 8.11

This experiment examines the cycle count and code size of the assembly program using multi-issue instructions, coupled with the hardware loop and dual MAC operations. The program can be found in directory `c:\adsp\chap8\exp8_11_533` for the BF533 EZ-KIT (or `c:\adsp\chap8\exp8_11_537` for the BF537). Fill in the performance in Table 8.14.

Table 8.14 Cycle Count and Code Size for Assembly Code with Multi-issue Instructions

	Cycle Count	Code Size in Bytes
Hardware loop, dual MACs, and multi-issue instruction		

8.3.4 Special Addressing Modes: Separate Data Sections

As discussed in Section 5.1.4, the Blackfin processor has separated data memory banks A and B. Programmers can allocate different variables in different memory banks and thus access data from these banks in a single cycle. In the case of the FIR filter program, we can assign the delay buffer in data bank A (L1_data_a), and the filter coefficients are placed in data bank B (L1_data_b). This arrangement can be programmed in C as follows:

```
section ("L1_data_a") fract16 ldelay[TAPS]={0};

section ("L1_data_b") fract16 lpf[TAPS] = {
   #include "coef32.dat"
};
```

HANDS-ON EXPERIMENT 8.12

This experiment examines the performance improvement when variables are allocated in different data banks. This optimization technique is applied to Hands-On Experiment 8.11. The program can be found in directory c:\adsp\chap8\exp8_12_533 for the BF533 EZ-KIT (or c:\adsp\chap8\exp8_12_537 for the BF537). Fill in the performance results in Table 8.15.

Examine the memory map of this project and note the starting address of the delay line buffer ldelay and the filter coefficient buffer lpf. Refer to the memory map of the BF53x processor and verify that different data banks are used for delay line and filter coefficient buffers. This separated memory allocation allows simultaneous accesses to support a single-cycle dual-MAC operation.

Table 8.15 Cycle Count and Code Size for Assembly Code Using Different Memory Banks

	Cycle Count	Code Size in Bytes
Hardware loop, dual MACs, multi-issue instruction, and different data memory banks		

8.3.5 Using Software Pipelining

Software pipelining is a technique to optimize loop code to further maximize utilization of the processor's functional units. This technique can be used if one section of the algorithm is not dependent on the others. For example, when performing the multiply-accumulate operation in the FIR filtering loop, the current iteration is independent of the previous one, and these two iterations can be overlapped.

We will use pseudo-assembly code to illustrate the performance of dual-MAC operations with and without software pipeline. The first pseudo code without using any software pipeline is listed as follows:

```
// dual MAC operations without software pipeline

Loop_start
  MNOP || Load input || Load coef
  Dual MAC
  Store result
Loop end
```

Software pipeline can be used to schedule loop and functional units more efficiently. Because the Blackfin processor has multiple hardware resources that can be operated in parallel, the programmer can enhance the parallelism by bringing the load instructions outside the loop. The software pipeline for the FIR filtering is shown in the following pseudo code:

```
// dual MAC operations with software pipeline

Prolog:      MNOP || Load input || Load coef

Loop kernel: DualMAC || Load input || Load coef

Epilog:      DualMAC || Load input || Load coef
  Store result
```

The software pipeline greatly improves the execution speed because the 32-bit MNOP (no operation) instruction that wastes the processor cycle is no longer inside the loop. As shown above, parallel-load instructions are carried out before entering the loop kernel. This overhead period is called the prolog. The loop kernel now contains only the dual-MAC operations in parallel with the loading of the next data sample and coefficient. An epilog is an overhead period after the loop kernel that is needed to complete the final tap of the FIR filtering. As a trade-off, the software pipeline technique has increased code size.

HANDS-ON EXPERIMENT 8.13

This experiment examines the cycle count and code size of the assembly program with the software pipeline technique. The program can be found in directory c:\adsp\chap8\exp8_13_533 for the BF533 EZ-KIT (or c:\adsp\chap8\exp8_13_537 for the BF537). Fill in the performance results in Table 8.16.

Table 8.16 Cycle Count and Code Size for Assembly Code with Software Pipeline

	Cycle Count	Code Size in Bytes
Hardware loop, dual MACs, multi-issue instruction, and software pipeline		

Table 8.17 Summary of Execution Cycle Counts and Code Sizes for Different Optimization Techniques

Optimization Technique	Cycle Count (firc)	Code Size (in bytes)
(a) Linear assembly code	$[28 \times N_c + 22] \times N_b + 66$	140
(b) Linear assembly code with branch prediction	$[20 \times N_c + 18] \times N_b + 66$	140
(c) Using hardware looping	$[6 \times N_c + 11] \times N_b + 66$	128
(d) Using dual MACs and (c)	$[5 \times N_c + 14] \times N_b + 66$	148
(e) Multiple instructions and (d)	$[4.5 \times N_c + 9] \times N_b + 67$	148
(f) Separate data sections and (e)	$[4 \times N_c + 10] \times N_b + 67$	148
(g) Software pipeline and (f)	$[3.5 \times N_c + 10] \times N_b + 69$	160

EXERCISE 8.1

Modify the assembly program fir_asm_simple_pipeline.asm in Hands-On Experiment 8.13 to further improve its cycle count and code size by using multi-issue instructions inside the loop kernel.

EXERCISE 8.2

As discussed in Section 8.2.6, an optimized FIR filter function fir_fr16.asm is provided in the run time DSP library. Compare the cycle count and code size of this function (in Table 8.10) with that of the optimized assembly code in Hands-On Experiment 8.13. Examine the fir_fr16.asm code and identify the optimization techniques used in this function.

8.3.6 Summary of Execution Cycle Count and Code Size for FIR Filter Implementation

We have studied different optimization techniques in assembly programming. A faster execution speed can be achieved by using parallel instructions and multiple hardware resources and applying software pipeline. The cycle counts and code size for these techniques are summarized in Table 8.17.

EXERCISE 8.3

Using the cycle count benchmark of Technique (g) in Table 8.17, tabulate the cycle count and memory size (for both data and code) for implementing an N_c-tap FIR filter using N_b samples per block as shown in Table 8.18(a) and (b), respectively.

8.4 POWER CONSUMPTION AND MANAGEMENT IN THE BLACKFIN PROCESSOR

Power consumption and management are important design considerations in many embedded systems, especially in portable devices where a battery is the energy source. This section discusses the power consumption behavior and power-saving features of the Blackfin processors.

8.4.1 Computing System Power in the Blackfin Processor

This Section studies the total power consumption of the Blackfin processors in detail. The power performance evaluated with mW/MMAC (in Section 6.3.4) does

Table 8.18(a) Cycle Count for Implementing Optimized Assembly Code with Optimization Technique (g)

$N_b \backslash N_c$	4	8	16	32	64	128	256	512	1024
1									
2									
4									
8									
16									
32									
64									
128									
256									
512									

8.4 Power Consumption and Management in the Blackfin Processor

Table 8.18(b) Code and Data Sizes (in bytes) for Implementing Optimized Assembly Code with Optimization Technique (g)

$N_b \backslash N_c$	4	8	16	32	64	128	256	512	1024
1									
2									
4									
8									
16									
32									
64									
128									
256									
512									

not reflect the actual operating conditions, in which cores, peripherals, and memory are turned on and off according to the state of the application code.

There are three power domains that consume power: the internal voltage supply, the external voltage supply, and the real-time clock supply. In the BF53x processors, the core operates in the voltage range of 0.8 to 1.32 V with a nominal voltage of 1.2 V. The I/O circuitry supports a range of 2.25 to 3.6 V with a nominal voltage of 2.5 V or 3.3 V. The real-time clock (RTC) can either be powered from the I/O supply or another external supply with the same nominal voltage.

Power dissipation is defined as the product of supply voltage and current drawn. The total average power P_{tot_av} dissipated by the Blackfin processor is defined as:

$$P_{tot\_av} = P_{int} + P_{ext} + P_{RTC}, \tag{8.4.1}$$

where $P_{int} = V_{int} \times I_{int}$ is the internal core power, $P_{ext} = V_{ext} \times I_{ext}$ is the external core power, and P_{RTC} is the RTC power. The terms V_{int} and V_{ext} are internal and external voltages, respectively. Similarly, I_{int} and I_{ext} are internal and external currents, respectively.

The internal power of the Blackfin processor consists of static and dynamic power components. The static component is derived from the leakage current when the core processor is disabled. The leakage current is proportional to the supply voltage and temperature. Therefore, we can save power by lowering the supply voltage when the core is disabled. On the other hand, the dynamic component is a function of both supply voltage and switching frequency. Dynamic power dissipation increases as the voltage and/or frequency increases.

The external power consumption is contributed by the number of enabled peripherals in the system. The equation for external power consumption is expressed as

$$P_{ext} = (V_{ext})^2 \times C \times f/2 \times (O \times TR) \times U, \qquad (8.4.2)$$

where C is the load capacitance, f is the frequency at which the peripheral runs, O is the number of output pins, TR is the number of pins toggling each clock cycle, and U is the ratio of time that the peripheral is turned on.

EXAMPLE 8.3

The serial port is operating at 4 MHz and consists of two pins. The capacitance per pin is 0.3 pF, and the external supply voltage is 3.65 V. The utilization and toggle ratios are 1. This implies that the output pin changes state every clock cycle. With Equation 8.4.2, the external power dissipation is computed as 1.6 mW.

The final source of power consumption in the Blackfin processor comes from the RTC power domain. Typically, the RTC can be powered between 2.25 and 3.6 V, and the current drawn is in the range of 30 to 50 μA. Therefore, the worst-case power consumption for the RTC is around 180 μW.

To compute the actual average power consumption, a statistical analysis of the application program must also be performed to determine what percentage of time (% time) the processor spends in each of the defined states (P_{state}) in an application (i). The total average power can then be derived as

$$P_{total} = \sum_{i=1}^{N} \% \text{ time}(i) \times P_{state}(i). \qquad (8.4.3)$$

This total average power only shows how much the Blackfin processor is loading the power source over time. However, the power supply must also support the worst-case requirement of a given application.

8.4.2 Power Management in the Blackfin Processor

The Blackfin processor provides several power-saving features that are commonly used in embedded systems. These features include:

1. Powering down the peripherals and internal functional units when they are not in use.
2. Reducing the power supply voltage.
3. Running the processor and peripherals at lower clock frequency.

The Blackfin processor has a dynamic power management controller and a programmable phase-locked loop (PLL) to control the operating modes, dynamic voltage, and frequency scaling of the core and the system. In the following sections, we investigate the power-saving blocks and the power modes of the Blackfin processors.

8.4.2.1 Processor Clock

In the Blackfin processor, a core clock (CCLK) and a system clock (SCLK) are generated by the clock generation unit. The core clock is responsible for instruction execution and data processing, and the system clock is used for the peripherals, DMA, and external memory buses. As shown in Figure 8.10, the processor uses a PLL with a programmable clock divider to set the clock frequencies for the core and system. The PLL is a circuit that is used to generate multiple frequencies from one or more reference frequencies. A fixed-frequency reference CLKIN is connected to a phase detect (or comparator) block, and the other phase detect input is driven from a divide-by-N counter (divider), which is in turn driven by the voltage-controlled oscillator (VCO). The input clock, CLKIN, which provides the necessary clock frequency and duty cycle, is derived from a crystal oscillator in the range of 10 to 40 MHz.

The PLL circuit operates on negative feedback, which forces the output of the internal loop filter to generate a frequency at $VCO = N \times CLKIN$. The CCLK and the SCLK are obtained by dividing the VCO frequency by their respective scalar values, CSEL and SSEL. The divider ratio for the core clock CSEL (2 bits) can be set to 1, 2, 4, or 8. However, the divider ratio for the system clock SSEL (4 bits) can be set to a wider range of 1 to 15. These bits are located in the PLL divide register,

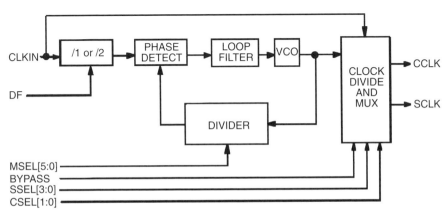

Figure 8.10 Functional blocks of the Blackfin processor's clock generation unit (courtesy of Analog Devices, Inc.)

PLL_DIV, and they can be changed on the fly to adjust the core and system clock frequencies dynamically.

EXAMPLE 8.4

A 27-MHz oscillator is built into the BF533 EZ-KIT to drive the CLKIN of the processor. If DF = 0 (i.e., no further divide of the CLKIN by 2) and MSEL = 14, the VCO frequency = MSEL × CLKIN = 27 × 10 = 270 MHz. The CSEL and SSEL in the PLL_DIV register are divided from the VCO frequency to obtain the CCLK and SCLK, respectively. Note that SCLK should be less than or equal to 133 MHz, and it must be smaller than CCLK. Therefore, SSEL must be set to a value of more than 2.

The dynamic power management controller allows users to program the processor in five operating modes and different internal voltages and to disable the clock to a disabled peripheral. In addition, the dynamic power management controller works with the clock generation unit to control the CCLK and SCLK. In the following subsections, we discuss the five operating modes and how to program the voltage and frequency of the Blackfin processor.

8.4.2.2 Power Modes

As stated in Section 6.3.4, the Blackfin processor works in five operating modes: (1) full on, (2) active, (3) sleep, (4) deep sleep, and (5) hibernate. Each of these modes has different power saving levels and clock activities as shown in Table 8.19.

In the full-on mode, the processor and all peripherals run at full speed determined by the CCLK and SCLK, respectively. The PLL is enabled, and the CCLK and SCLK are derived from the CLKIN and their respective divider ratios. DMA access to the L1 memory is allowed in this mode. In the active mode, the PLL is enabled but bypassed. Therefore, the CCLK and SCLK run at the same frequency as CLKIN, which is around 30 MHz. DMA access to the L1 memory is also avail-

Table 8.19 Operating Characteristics of Different Power Modes

Operating Mode	Power Saving	Core Clock	System Clock	PLL
Full on	None	Enabled	Enabled	Enabled
Active	Medium	Enabled	Enabled	Enabled/disabled Bypass PLL
Sleep	High	Disabled	Enabled	Enabled
Deep sleep	Maximum	Disabled	Disabled	Disabled
Hibernate	Maximum	Off	Off	Off

Table 8.20 Allowed Power Mode Transition

New Mode Current Mode	Full On	Active	Sleep	Deep Sleep
Full on	—	Allowed	Allowed	Allowed
Active	Allowed	—	Allowed	Allowed
Sleep	Allowed	Allowed	—	—
Deep sleep	—	Allowed	—	—

able in this mode. The active mode is important because it allows the user to program the PLL for different core and system frequencies.

In the sleep mode, significant reduction in power consumption can be achieved because the core clock is disabled. The system clock is still enabled, and DMA access can only take place between peripherals and external memories. In the deep-sleep mode, maximum power saving is achieved by disabling the PLL, CCLK, and SCLK. In addition, the processor core and all peripherals, except the real-time clock, are disabled. Therefore, DMA is not supported in this mode. The deep-sleep mode can be awaked by an RTC interrupt or hardware reset. Power dissipation still occurs through the processor leakage current. To further cut down the power dissipation, the hibernate mode in the Blackfin processor turns off the internal core voltage supply, thus eliminating any leakage current. However, the internal state of the processor must be written to external SDRAM before switching to the hibernate mode. The SDRAM is set to self-refreshing mode before entering the hibernate mode. Table 8.20 shows the possible mode transition between different power modes. The PLL status register (PLL_STAT) indicates the status of the operating mode.

Besides these different power modes, the Blackfin processor allows peripherals to be individually enabled or disabled. By default, all peripherals are enabled, but they can be selectively disabled by controlling the respective bits of the peripheral clock enable register PLL_IOCK.

8.4.2.3 Dynamic Voltage and Frequency Scaling

Power dissipation depends on clock frequency and voltage as stated in Equation 6.3.7. Changing the clock frequency (CCLK) affects the processing speed. A faster clock frequency needs a shorter period to complete a given task, resulting in longer idle time. We can set the device to sleep mode during this idle time, but power consumption still occurs because of the leakage current. A better approach is to reduce the idle time by lowering the clock frequency and complete the task in the required time period. When the frequency is reduced, the supply voltage must also be reduced. It is always recommended to use the lowest possible voltage that supports the required clock frequency. The lower voltage leads to significant power reduction, as power dissipation is proportional to the squared voltage.

Table 8.21 Voltage Selection Via VLEV

VLEV	0110	0111	1000	1001	1010	1011	1100	1101	1110	1111
Voltage	0.85 V	0.90 V	0.95 V	1.00 V	1.05 V	1.10 V	1.15 V	1.20 V	1.25 V	1.30 V

To implement the frequency-voltage scaling in a real-time system, we need to examine the deadline and execution time to perform real-time processing. As explained in Section 6.3, an event-triggered system (such as real-time audio processing) needs to interrupt the processor at regular intervals. The deadline time is basically determined by the time needed to acquire N data samples for block processing. For example, if the block size is set at 32 samples and the sampling frequency is 48 kHz, the deadline time is 666.67 μs. In the actual implementation, the deadline can be expressed as deadline cycle, which is deadline time multiplied by the core clock frequency as shown in Equation 6.3.4. For a core frequency of 270 MHz, the deadline cycle is 180,000 cycles. Note that the deadline time is always constant for the same sampling frequency and block size, but the deadline cycle varies with different core frequency. To guarantee real-time processing, the maximum execution time must be bounded to the worst-case execution time (WCET), which is the deadline time. The number of execution cycles needed for a given task is always constant, whereas the execution time depends on the core clock frequency. Based on the constant property of the deadline time $t_{deadline}$ and execution cycle N_{exe_cycle}, a suitable operating frequency F_{core} can be determined as

$$F_{core} = N_{exe\_cycle} / t_{deadline}. \qquad (8.4.4)$$

Note that the number of execution cycles, N_{exe_cycle}, includes the cycles needed for task initialization, overhead, and the actual task in the interrupt service routine. As shown in Figure 8.10, the CCLK (or F_{core}) and SCLK can be adjusted by changing the value in the PLL divide register. The adjustment of CCLK is performed with some fixed integer divisor. Therefore, F_{core} is quantized to some permissible operating frequencies.

The Blackfin processor provides an on-chip switching regulator controller that is used to generate internal voltage levels from the external voltage supply. There are 10 internal voltage levels that can be programmed into the 4-bit internal voltage level (VLEV) field of the voltage regulator control register (VR_CTL) as shown in Table 8.21. Once the core frequency has been selected, the minimum core voltage can be obtained from the Blackfin datasheet [26, 27]. In this way, the minimum core frequency and voltage are used for a given processing task.

HANDS-ON EXPERIMENT 8.14

This experiment implements the frequency-voltage scaling technique [5] for running a processing load. An FIR filter with variable filter length is implemented with a C program. The processing load can be varied by using different filter lengths of 32, 128, 512, and 1,024. Figure 8.11 shows the block diagram of this experiment. The setting of the DMA controller

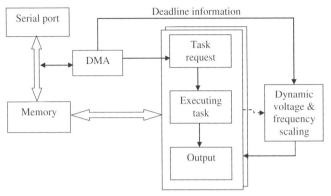

Figure 8.11 Block diagram of dynamic frequency-voltage scaling

Table 8.22 Benchmark Results of Execution Cycle, Core Frequency, and Voltage

Load (filter length)	Execution Cycle	Frequency	Voltage	Power Reduction
32				
128				
512				
1024				

provides the deadline time information to a dynamic frequency-voltage scaling routine. The cycle count is benchmarked in real time and furnished to the frequency-voltage scaling routine. Because we cannot benchmark the execution time on the same processing interval, we benchmark the current execution time and use it as the estimated execution time of the next processing frame. In this way, we are able to track the execution time of the processing load and adjust the frequency and voltage accordingly.

The project file is located in directory c:\adsp\chap8\exp8_14_533 for the BF533 EZ-KIT (or c:\adsp\chap8\exp8_14_537 for the BF537). The maximum core clock frequency is set to 459 MHz, and the core voltage is 1 V at startup.

Build and run the project. Enable the filter by pressing **SW6** on the BF533 EZ-KIT (or **SW12** on the BF537). Vary the load of this program by pressing **SW6** (or **SW12**). The execution cycles are displayed in the BTC memory window. Users can check the core frequency in the PLL_DIV register and the core voltage in the VR_CTL register. Click on **Register → Core → PLL Register File** to display these registers. Complete the tasks listed in Table 8.22.

8.5 SAMPLE RATE CONVERSION WITH BLACKFIN SIMULATOR

This section presents an application of the FIR filter for sample rate conversion. For example, we change an incoming digital signal sampled at 48 kHz to 8 kHz. This

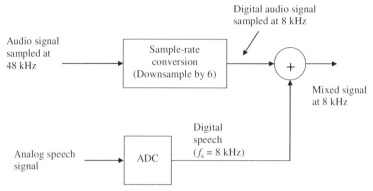

Figure 8.12 Sample rate conversion and mixing of signals

signal can be mixed with another digital speech signal sampled at 8 kHz as shown in Figure 8.12. The sample rate conversion block is downsampled (or decimated) by a factor of $D = 6$ from 48 kHz to 8 kHz. The downsampling process can be implemented by simply discarding every $D-1$ samples. However, decreasing the sampling rate by the factor D reduces the bandwidth by the same factor. To prevent aliasing, a low-pass filter with a cutoff frequency of $f_s/2D$ (4 kHz) must be applied to the original high-rate signal before the downsampling process. The FIR filter is commonly used to realize the low-pass filter for sampling rate conversion applications.

The sampling rate can also be increased by an integer factor, U. This process is called upsampling (or interpolation). For example, the sampling frequency of the digital speech can be upsampled from 8 kHz to 48 kHz before mixing with the audio signal sampled at 48 kHz. Upsampling can be realized by inserting additional zero samples between successive samples of the original lower-rate signal. For example, we insert $U-1$ samples when upsampling by a factor of $U = 6$ from 8 kHz to 48 kHz. Again, a low-pass filter is required to smooth out the interpolated samples.

This section investigates the decimation and interpolation processes with the Blackfin simulator. The run time DSP library provides several decimation and interpolation FIR filter functions. However, users need to select and design a suitable FIR filter before using these functions.

HANDS-ON EXPERIMENT 8.15

This experiment performs decimation and interpolation with the project file `exp8_15.dpj` located in directory `c:\adsp\chap8\exp8_15`. A 1-kHz sine wave sampled at 48 kHz is downsampled by a factor of 6 to 8 kHz, followed by upsampling to 16 kHz (by a factor of 2). This 16-kHz-sampled signal is further upsampled to 32 kHz by a factor of 2, and 48 kHz by a factor of 3, as shown in Figure 8.13. The objective of this experiment is to check whether the original signal sampling at 48 kHz can be recovered with decimation and a series of

8.5 Sample Rate Conversion with Blackfin Simulator

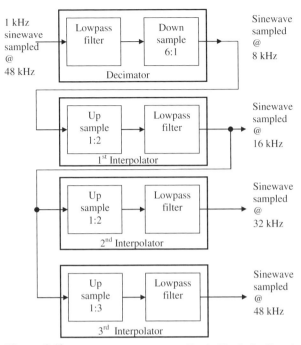

Figure 8.13 Downsampling and upsampling in Hands-On Experiment 8.15

interpolation processes. In addition, users can examine the signal sampled at different frequencies.

Because the number of zeros introduced between successive samples of the input signal during interpolation, only one out of every U samples input to the interpolation filter is nonzero. To efficiently implement the low-pass filter, we divide the FIR filter of length L to form U FIR filters, each with a length of L/U. These smaller FIR filters operate at lower sampling rate f_s instead of Uf_s after the upsampler. However, the interpolation filter coefficients must be rearranged for the smaller filters. These interpolation filters are provided in data files `interp_8kto16k_64taps.dat`, `interp_16kto32k_64taps.dat`, and `interp_16kto48k_96taps.dat`. Because the decimation FIR filter operates at the original sampling frequency as the input signal, there is no need to rearrange the coefficients of the filter.

The program `main.c` implements the decimation and interpolation processes shown in Figure 8.13. The programming steps are listed below:

```
fir_decima_fr16(inp_arr, out8k, INPUTSIZE, &filter_state_dec);
fir_interp_fr16(out8k,out16k,INPUTSIZE/6,&filter_state_interp_
8to16);
fir_interp_fr16(out16k,out32k,2*INPUTSIZE/6,&filter_state_interp_
16to32);
fir_interp_fr16(out16k,out48k,2*INPUTSIZE/6,&filter_state_interp_
16to48);
```

Build and run the project. Display the input and output signals at different stages with the plot feature in the debug window as shown in Figure 8.14.

368 Chapter 8 Code Optimization and Power Management

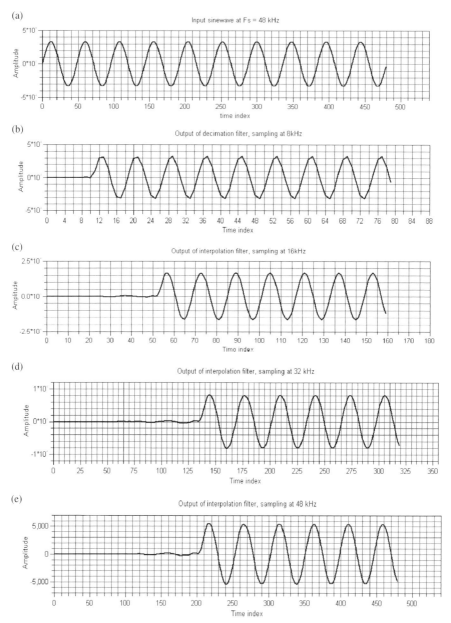

Figure 8.14 (a) Input sine wave ($f_s = 48\,\text{kHz}$), (b) decimator output ($f_s = 8\,\text{kHz}$), (c) 1st interpolator output ($f_s = 16\,\text{kHz}$), (d) output of cascade of 1st and 2nd interpolators ($f_s = 32\,\text{kHz}$), and (e) output of cascade of 1st and 3rd interpolators ($f_s = 48\,\text{kHz}$)

Figure 8.14 reveals some important observations:

1. The amplitude of the output signal from the interpolator has reduced compared with the original signal. The magnitude reduction is in the order of $1/U$. Users can compensate this amplitude reduction by amplifying the interpolator output by a gain of U.
2. The output from the decimator or interpolator has been delayed compared with the original signal. This is due to the group delay introduced by the decimation and interpolation filters.

Users can use the FFT with the correct sampling rate to verify the magnitude spectrum of the signals at different stages. Instead of using 1-kHz sine wave `sine1k.dat` as the input, generate a 5-kHz sine wave in MATLAB and save it in a data file, `sine5k.dat`. Repeat the same experiments with the 5-kHz sine wave and comment on the results.

EXERCISE 8.4

Modify the main program `main.c` to implement the following tasks:

1. Add a 16-kHz sine wave sampled at 48 kHz to a 1-kHz sine wave sampled at 8 kHz.
2. Mix the musical wave file `liutm_48k_mono.wav` sampled at 48 kHz with the speech wave file `timit.wav` sampled at 8 kHz. Hint: Convert the wave files into data files and make sure that the two data sequences have the same length before adding.

8.6 SAMPLE-RATE CONVERSION WITH BF533/BF537 EZ-KIT

This section implements sample rate conversion in real time with the Blackfin EZ-KIT. Because the CODEC in the EZ-KIT can only operate at either 48 kHz or 96 kHz, we sample the signal at 48 kHz and decimate it to different sampling frequencies. The decimated signal can be displayed in the display window and played via the computer sound card. In another experiment, we sample the signal at 48 kHz, down sample to 8 kHz and mix it with an internally generated signal sampled at 8 kHz, and upsample the mixed signal to 48 kHz before sending it to the CODEC. The input and output connections to the BF533 and BF537 EZ-KITs follow the default settings described in previous chapters.

HANDS-ON EXPERIMENT 8.16

Activate VisualDSP++ for the BF533 (or BF537) EZ-KIT target and open `exp8_16_533.dpj` (or `exp8_16_537.dpj`) in directory `c:\adsp\chap8\exp8_16_533` (or `c:\adsp\chap8\exp8_16_537`). This project samples the incoming signal at 48 kHz, downsamples it to 8 kHz, and upsamples it to 32 kHz. Because the sampling frequency of the EZ-KIT is 48 kHz, we can only save the downsampled signals to a data file or display them on the graphical display. In the `main.c` and `interpolation.h` files, determine the lengths and durations of the decimated signals sampled at 8 and 32 kHz. Build and run the program, and display the input signal at 48 kHz (`48k.vps`), downsampled signal at 8 kHz (`8k.vps`), and downsampled signal at 32 kHz (`32k.vps`). Instead of listening to these signals directly from the EZ-KIT, the downsampled signals can be exported to the sound card (steps explained in Exercise 3.18) with proper sampling frequency settings for playback. Benchmark the cycle count and memory size for implementing the interpolator and decimator.

HANDS-ON EXPERIMENT 8.17

In the second experiment, the project file `exp8_17_533.dpj` (or `exp8_17_537.dpj`) located in directory `c:\adsp\chap8\exp8_17_533` (or `c:\adsp\chap8\exp8_17_537`) is used to mix an incoming signal sampled at 48 kHz with an internally generated 1-kHz sine wave sampled at 8 kHz. The sine wave generation is based on the look-up table approach, which is explained in Section 7.6. Build and run the project. The switch settings for the BF533 and BF537 EZ-KITs are listed in Table 8.23.

Listen to the signals with the switch settings listed in Table 8.23. Do you hear any difference between different modes?

Table 8.23 Switch Settings for Different Modes

Mode	Switch on BF533 EZ-KIT	Switch on BF537 EZ-KIT
(1) Pass-through	**SW4**	**SW10**
(2) Downsample to 8 kHz → Mixing at 8 kHz → Upsample to 48 kHz	**SW5**	**SW11**
(3) Downsample to 8 kHz → Upsample to 48 kHz (without mixing)	**SW6**	**SW12**

EXERCISE 8.5

Based on the project files given in the preceding hands-on experiment and using either the BF533 or the BF537 EZ-KIT, perform the following tasks:

1. Mix an incoming 1-kHz sine wave sampled at 48 kHz with an internally generated 1-kHz sine wave sampled at 8 kHz. Both signals must be aligned without any phase shift. Listen to the mixed signal at 48 kHz.
2. Similar to Task 1, mix the two sine waves but with one sine wave at 180° out of phase with the other sine wave. Verify the result.

8.7 SAMPLE RATE CONVERSION WITH LABVIEW EMBEDDED MODULE FOR BLACKFIN PROCESSORS

In practical engineering applications, it is often necessary to change the sampling rate of signals within a system. For example, the Blackfin EZ-KIT operates with an audio CODEC that digitizes signals at 48 kHz, which is a common rate for capturing audio signals. It is often desirable to process signals with a lower sampling rate, such as 8,000 Hz, which minimizes the amount of computation necessary by reducing the number of samples. Sample rate conversion is commonly used in industrial applications for analyzing and manipulating communication signals, mixing signals from multiple sources, and noise cancellation.

The following experiments use LabVIEW to illustrate the concepts of sample rate conversion. We control the sampling rates of signals and observe what design considerations are affected by resampling. The interactive simulation allows you to see and hear how decimation and interpolation filters can be used to manipulate and process audio signals. This understanding will further reinforce these concepts when using the Blackfin EZ-KIT target.

HANDS-ON EXPERIMENT 8.18

This experiment reinforces the concepts of sample rate conversion. This conversion allows us to combine and process signals sampled at different rates within the same system. This experiment begins with an input signal sampled at a frequency of 48 kHz and then analyzes the results of applying decimation and interpolation filters to decrease and increase the sampling rate, respectively. Remember from Section 8.5 that the downsampling process requires low-pass filtering of the input signal to prevent aliasing by ensuring that the signal satisfies the Nyquist rate.

Open the compiled LabVIEW simulation Resample_Sim.exe located in directory c:\adsp\chap8\exp8_18 to explore sample rate conversion. The graphical user interface is similar to those in earlier chapters and is shown in Figure 8.15. The interface shares many controls, indicators, and functions used in previous hands-on LabVIEW experiments.

The tabs on **Tab control** are unique, allowing the user to choose which signals to view, including input, downsampled, and upsampled signals in both time and frequency domains. The **Upsampled Output** tab enables the **Upsample To** rate selector, which allows each of the upsampled rates to be seen and heard. The frequency slider allows the frequency of the default test sine wave to be manipulated. The sine wave input signal is sampled at 48 kHz to simulate that of the Blackfin EZ-KIT, downsampled by a factor of 6 to 8 kHz, and then upsampled to 16, 32, or 48 kHz. The 32-kHz and 48-kHz sampling rates are achieved by first upsampling by a factor of 2 to 16 kHz and then upsampling the resulting signal by a factor of 2 or 3, respectively.

When you open the simulation, a default sine wave signal with 1-kHz frequency is used as the input. Explore the different tabs to gain an understanding of how the signal changes when decimation and interpolation filters are applied. Also, manipulate the signal frequency while observing its time- and frequency-domain data. Pay attention to the effects of sample rate conversion on audio bandwidth. What observations can you make about the time-domain plots of the input signal after its sampling rate has been altered? Can you hear these effects

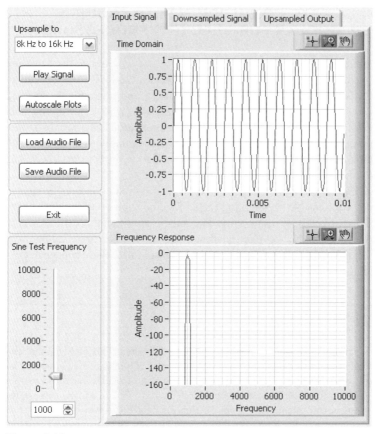

Figure 8.15 User interface for Resample_Sim.exe

when playing back the altered signals? Experiment with different input sine wave frequencies by changing the value of the **Sine Test Frequency** slider control.

Now load an audio file to experiment with the effects of decimation and interpolation. You can use the speech_tone_48k.wav file from Hands-On Experiment 2.5. Play the signal at different sampling rates to determine how sample rate conversion affects the audio quality. What observations can you make? How might consumer electronics, digital audio media, or the music recording industry utilize these concepts?

HANDS-ON EXPERIMENT 8.19

This experiment examines a LabVIEW Embedded Module for Blackfin Processors project that uses decimation and interpolation filters to mix signals of different sampling rates. LabVIEW programming structure and code are reused from previous exercises to maintain a similar program structure and more efficiently develop a sample rate conversion application that can be deployed in the Blackfin EZ-KIT.

8.7 Sample Rate Conversion with LabVIEW Embedded Module

Open the `Resample_Ex-BF5xx.lep` project located in directory `c:\adsp\chap8\exp8_19`. Open the top-level VI named `Resample_Ex - BF5xx.vi` and view its block diagram, shown in Figure 8.16. This VI provides the ability to acquire live audio at 48 kHz, downsample it by a factor of 6 to 8 kHz, mix the 8-kHz signal with a 1-kHz sine wave sampled at 8 kHz, and then upsample back to 48 kHz for audio output. Which portions of the block diagram look familiar?

Note that the structure for writing and reading audio is common to most previous LabVIEW Embedded Module for Blackfin Processors audio exercises. In particular, `Init Audio BF5xx.vi` and `BF Audio Write-Read (BF5xx).vi` set the audio buffer sizes and generate and acquire the audio signals. After acquiring a buffer of data, the sample rate is converted, mixed, and output in the next loop iteration. As in previous experiments, the graphical program is logically organized with subVIs that allow program modules to be reused, thus reducing development time.

The three custom subVIs, found in the lower left of the block diagram, initialize the filters and sine wave generation before the **While Loop** begins. `Init Decimation Filter.vi` and `Init Interpolation Filter.vi` both store filter coefficients and create the necessary data buffers to perform the decimation and interpolation FIR filtering inside the processing loop. A 1-kHz sine wave sampled at 8 kHz is generated from an eight-point look-up table by `Gen 1k Sine.vi` and is mixed with the downsampled live audio when the correct case is executed. These subVI functions are part of the project and can be seen in the **Embedded Project Manager** window below the main VI.

The **Case Structure** in the processing loop determines which of the three processing modes is used. The three modes include audio talk-through, resample, and resample with mix. The button selections for the EZ-KIT are shown in Table 8.24.

The **Case Structure** in the center of the VI contains the three cases seen in Figure 8.17. One of the three cases is executed in each iteration of the loop, depending on the button

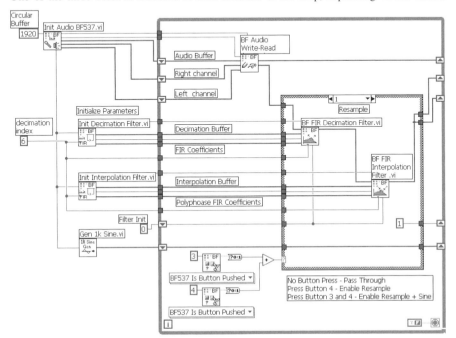

Figure 8.16 Block diagram of `Resample_Ex - BF5xx.vi`

Table 8.24 Switch Selections for Different Modes

Mode	Switch on BF533 EZ-KIT	Switch on BF537 EZ-KIT
1 Pass-through	None	None
2 Downsample to 8 kHz → Mixing at 8 kHz → Upsample to 48 kHz	**SW4** and **SW5** pressed together	**SW10/PB4** and **SW11/PB3** pressed together
3 Downsample to 8 kHz → Upsample to 48 kHz (without mixing)	**SW4**	**SW10/PB4**

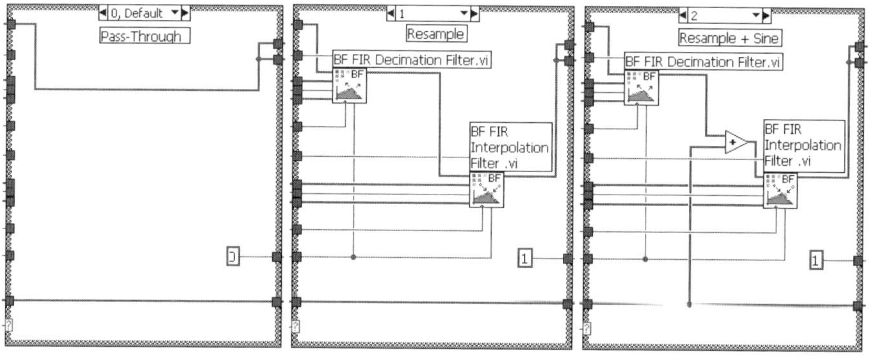

Figure 8.17 Cases in the `Resample_Ex - BF5xx.vi` graphical code.

combination being pressed. Several wires are required to configure the decimation and interpolation filters, but the general flow of the input signal is easily followed.

Connect the audio input of the Blackfin EZ-KIT to the computer sound card or CD player. Then compile and run the LabVIEW Embedded project on the Blackfin processor. Press different switch combinations to hear each of the three audio modes. What differences are heard between audio pass-through and the resampled audio signals? Explain. Can you hear the mixed 1-kHz tone?

8.8 MORE EXERCISE PROBLEMS

1. In Hands-On Experiments 8.1 and 8.2, the implementation is carried out with a 32-tap FIR filter with 32 samples per block. Benchmark the cycle count and code/data size when implementing different filter lengths given in Table 8.25. These filter coefficients are also included in the project. Comment on the performance with different filter taps.

2. Problem 1 uses block processing of 32 samples per block. The block size can be varied by changing the value of `IP_SIZE` to 1 (sample mode), 4, 8, 16, 32, 64, and 128. The

8.8 More Exercise Problems

Table 8.25 Benchmark Table (IP_SIZE = 32 Samples/Block)

Number of Taps	Index-Style C (optimize)		Pointer-Style C (optimize)	
	Cycle count	Data and code size (byte)	Cycle count	Data and code size (byte)
16				
32				
64				
128				
256				

Table 8.26 Benchmark Table (Filter Taps = 32)

Number of Samples per Block (IP_SIZE)	Index-Style C (optimize)		Pointer-Style C (optimize)	
	Cycle count	Data and code size (byte)	Cycle count	Data and code size (byte)
1				
4				
8				
16				
32				
64				
128				

FIR filter length is fixed at 32 taps. Fill in Table 8.26 and comment on the performance versus the different block sizes.

3. Instead of using global optimization for the project in Hands-On Experiment 8.1, implement local optimization (in Section 8.2.1) on the filtering and delay buffer update sections of the code. Benchmark its performance and compare to the results in Table 8.1. Also, experiment with the loop optimization pragma in Section 8.2.2.2 and observe its performance.

4. Extend Hands-On Experiment 8.6 for different samples per block and different filter lengths. Comment on the following:

   (a) Cycle count per sample for different block sizes

   (b) Trade-off between processing gain and block size

   (c) Filter length and block size

5. Implement a symmetric FIR filter based on Equation 2.3.10 in C with intrinsic functions on the Blackfin processor. Build the project by enabling the optimization in the VisualDSP++ compiler. Benchmark on the cycle count, data, and code size of the symmetric FIR filter and compare the results with the direct-form FIR filter. Does the symmetric FIR filter always result in a better performance as compared to the direct-form FIR filter? If not, why?

Table 8.27 Benchmark Results Using 32-Tap FIR Filter

$N_b =$	1	8	16	32	64
Cycle count/input sample					
Processing time/input sample					

Table 8.28 Total External Power Consumption

Peripheral	Frequency	No. of Pins	Capacitance/ Pin	Toggle Ratio	Ratio of Time Peripheral Is On	V_{ext}	External Power
PPI	27 MHz	9	30 pF	1	1	3.65 V	
SPORT0	4 MHz	2	30 pF	1	1	3.65 V	
UART	0.115 MHz	2	30 pF	1	0.25	3.65 V	
SDRAM	133 MHz	36	30 pF	0.25	0.50	3.65 V	
Total							

6. Repeat Problem 5 by implementing the symmetric FIR filter with assembly programming.

7. The cycle count for performing FIR filter in linear assembly code without branch prediction is $[28 \times N_c + 22] \times N_b + 66$ cycles, where N_c is the number of coefficients in the FIR filter and N_b is the number of samples per block. The 66 cycles are overhead for setting up the FIR filter. Compute the cycle count/input sample and processing time/input sample (Table 8.27) in implementing a 32-tap FIR filter with different data samples per block. The Blackfin processor is operating at 270 MHz.

8. Open the optimized IIR filter function `iir_fr16.asm` in the run time DSP library. Identify the techniques used in this optimized assembly code. How do these optimized techniques differ from the FIR filter function `fir_fr16.asm`?

9. Open the optimized radix-2 FFT function `r2fftnasm.asm` in the run time DSP library. Examine and list the techniques used in this optimized assembly code.

10. Compute the total external power of the Blackfin processor based on the following peripherals and their activities listed in Table 8.28.

11. Develop a real-time speech and audio recorder with playback using the BF533 or BF537 EZ-KIT. A microphone can be connected to the input port of the EZ-KIT. The output port of the EZ-KIT is connected to loudspeakers or a headset. There are four operating modes in this recorder:

Mode I: Normal pass-through mode without recording

Mode II: Normal recording

Mode III: Recording with low-pass filtering (cutoff frequency is 4 kHz)

Mode IV: Playback

The switches and LEDs on the EZ-KIT are user control and indicator on the above modes. In addition, the recorded data must be stored in the external SDRAM. Determine the maximum recorded period at a sampling rate of 48 kHz. Downsample the signal to 8 kHz and determine the maximum recorded period. Finally, develop a stand-alone system.

12. Perform cycle count and memory benchmark on the different modes in the recording system developed in Problem 11.

Part C
Real-World Applications

Chapter 9

Practical DSP Applications: Audio Coding and Audio Effects

Audio coding exploits unique features of audio signals to compress audio data for storage or transmission. Today, digital audio coding techniques are widely used in consumer electronics such as portable audio players. This chapter introduces basic concepts of audio coder and decoder (codec) based on moving pictures experts group (MPEG) layer 3 (MP3) and implements a more recent, license-and-royalty-free Ogg Vorbis decoder using the BF533/BF537 processors. In addition, two audio effects and their implementations are presented and used for experiments.

9.1 OVERVIEW OF AUDIO COMPRESSION

The overall structure of audio codec is illustrated in Figure 9.1. An audio codec is an algorithm or a device that compresses the original digital audio signal to a lower bit rate and decompresses the coded data (bit stream) to produce a perceptually similar version of the original audio signal. The goals of audio coding are to minimize the bit rate required for representing the signal, to maximize the perceived audio quality, and to reduce system cost in terms of memory and computational requirements of codec. In general, bit rate, quality, and cost are conflicting issues, and trade-offs must be resolved based on a given application.

The human ear can perceive frequencies of sound between 20Hz and 20kHz. According to the sampling theorem defined in Equation 2.2.5, we have to sample a sound (analog signal) at least twice the highest frequency component in the signal to avoid aliasing. Therefore, the sampling frequency of 44.1kHz is commonly used in digital consumer products such as CD players; and 48kHz is used in many pro-

---

Embedded Signal Processing with the Micro Signal Architecture. By Woon-Seng Gan and Sen M. Kuo
Copyright © 2007 John Wiley & Sons, Inc.

Figure 9.1 An overall audio codec structure

fessional audio systems. In data converters, the signal amplitude is quantized into a binary value represented by a finite number of bits. The noise introduced by this quantization process is called quantization noise, which sets the maximum signal-to-noise ratio (SNR) that can be achieved by representing analog signal with digital values. This signal-to-quantization noise ratio (SQNR) stated in Equation 7.1.1 can be approximated as

$$\text{SQNR} = (6.02N + 1.76)\, \text{dB}, \tag{9.1.1}$$

where N is the number of bits (wordlength) used to represent digital samples.

The bit rate of digital signals can be computed as

$$\text{Bit rate} = C \times N \times f_s \text{ bit/s}, \tag{9.1.2}$$

where C is the number of channels and f_s is the sampling rate. For example, music in CDs is recorded at 44.1 kHz with 16 bits per sample, which requires a bit rate of 1,411.2 kilobits per seconds (kbit/s) for stereo (2 channels) audio. This bit rate poses a challenge on channel bandwidth or storage capacity for emerging digital audio applications such as Internet delivery. The increasing demand for better-quality digital audio, such as multichannel audio coding (5–7 channels) or higher sampling rate (96 kHz), requires more sophisticated encoding and decoding techniques to minimize the transmission cost and provide cost-efficient storage. For example, the Ogg Vorbis coder is typically used to encode audio at bit rates of 16–128 kbit/s per channel.

EXAMPLE 9.1

MPEG-1 has three operation modes for audio, layer 1, layer 2, and layer 3, with increasing complexity and sound quality. MPEG-2 is the second phase of MPEG, which provides extension of MPEG-1 to support multichannel audio coding and lower sampling frequencies. MP3 is a popular digital audio encoding format standardized in 1991 for Internet audio delivery and portable audio devices. A new standard called MPEG-2 AAC (advanced audio coding) was developed in 1997 to improve coding efficiency. More recent developments of audio coding are based on AAC with enhanced quality.

MP3 supports variable bit rates and sampling frequencies for different sound files. Bit rates available for MP3 are 32, 40, 48, 56, 64, 80, 96, 112, 128, 160, 192, 224, 256, and 320 kbit/s, and the available sampling frequencies are 32, 44.1, and 48 kHz. MP3 provides a range of bit rates and supports the switching of bit rates between audio frames. MP3 at 320 kbit/s produces perceptually comparable quality with CD at 1,411.2 kbit/s. MP3 files encoded with a lower bit rate will have a lower quality. A compression ratio of approximately 11:1 is most commonly used for MP3 to achieve a sampling rate of 44.1 kHz and a bit rate of 128 kbit/s. Similar to MP3, the Ogg Vorbis coder is also a variable bit-rate codec, with the bit rate varied from sample to sample.

EXERCISE 9.1

1. Compute the number of bytes required to store an hour of music in the CD format for stereo audio.
2. Redo Exercise 1 for MP3 music with the bit rates listed in Example 9.1.
3. Redo Exercise 1 for the most widely used multichannel audio coding with the 5.1-channel configuration.

In general, the efficient speech coding techniques such as linear predictive coding using a vocal track model are not applicable to audio signals. In addition, we have to deal with stereo or multichannel signal presentations, higher sampling rate, higher resolution, wider dynamic range, and higher listener expectation. There are many different audio compressing standards, and most use the principles of psychoacoustics [67]. That is, the hearing system adapts to dynamic variations in the sounds, and these adaptations and masking effects form the basic principles of psychoacoustics. The basic structure of the MP3 encoder illustrated in Figure 9.2 includes three important functions:

1. Filterbank splits the full-band signals into several subbands uniformly or according to the critical band model of the human auditory system, and a modified discrete cosine transform (MDCT) converts time-domain signals to frequency-domain coefficients. Note that MP3 uses a hybrid of filterbank and MDCT, and MPEG-2 AAC and Ogg Vorbis use MDCT with a longer length for splitting full-band signals.
2. The psychoacoustic model consists of critical bands, calculates the masking threshold according to the human auditory masking effect from the spectral coefficients, and uses the masking threshold to quantize the MDCT coefficients. Psychoacoustic analysis is done in the frequency domain with the computational efficient FFT introduced in Chapter 3.
3. Quantization allocates bits to indexes for quantizing the MDCT coefficients at each subband based on the masking threshold provided by the psychoacoustic model in the encoder and converts these indexes back to the spectral coefficients in the decoder. The coded bit stream redundancy can be further removed with entropy coding. For example, a fixed Huffman codebook is used in MP3. In contrast, the Ogg Vorbis coder does not define a set of precomputed codebooks. Instead, the quantized data are entropy

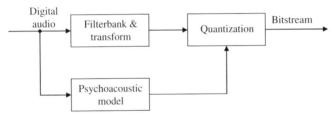

Figure 9.2 Basic structure of audio encoder

coded with a codebook-based vector quantization algorithm, and codebooks are transmitted as a part of the audio stream.

In addition, the output bit stream includes side information of bit allocation information needed for the decoder, which is packed by a multiplexer.

The threshold of hearing is also called the absolute threshold in quiet, which represents the lowest audible level of any sound at a given frequency. Sound lying below the threshold curve cannot be perceived by our hearing, and thus can be removed before quantization for reducing the bit rate. In addition, quantization noise introduced by codec below this level will not be perceived by humans. This is the fundamental principle of psychoacoustic compression, which permits more noise in sounds close to strong masking sounds. Allowing additional noise in the sampling process is equivalent to quantization with fewer bits, and this achieves data compression as shown in Equation 9.1.2.

The threshold of hearing can be approximated with the following equation [67]:

$$T_h(f) = 3.64(f/1000)^{-0.8} - 6.5e^{-0.6(f/1000-3.3)^2} + 10^{-3}(f/1000)^4, \quad (9.1.3)$$

where f is frequency in hertz.

EXAMPLE 9.2

This example demonstrates the threshold of hearing with `example9_2.m`. The plot of human threshold in quiet is shown in Figure 9.3. It shows that human hearing is relatively sensitive in the frequency range of 2 to 5 kHz, but we can tolerate more noise outside this range. For example, this threshold increases rapidly at frequencies higher than 10 kHz and changes slowly in the frequency range from 500 Hz to 2 kHz. Note that the threshold of hearing changes with age and an older person has a higher threshold.

Every narrowband signal has a masking curve, and various spreading functions can be used to estimate the masking capability of each tone. Typically, the masking curves extend farther toward higher frequencies. The slope of the frequency-masking curve depends on the sound pressure level (SPL) of the tone and the frequency at which it is present [55]. A basic spreading function is a triangular function with a steeper slope for lower frequencies and a negative, gradually falling slope for the higher frequencies. For example, the Schroeder spreading function has been used in [56], which can be expressed as:

$$10\log_{10} F(dz) = 15.81 + 7.5(dz + 0.474) - 17.5\left(1 + (dz + 0.474)^2\right)^{1/2}, \quad (9.1.4)$$

where dz is the Bark scale difference between the maskee (neighboring frequencies) and the masker (loud tone) frequency. The SPL of the tone is added to the spreading function to obtain the masking curve [55]. The bark scale gives the conversion between frequency in hertz and the critical bands of the human auditory system as follows [56]:

9.1 Overview of Audio Compression

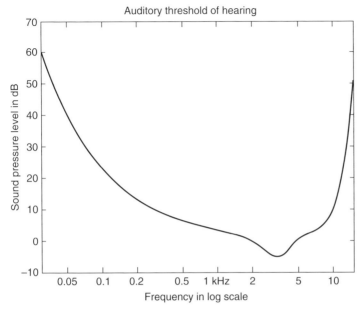

Figure 9.3 A threshold of hearing curve in log scale

$$z(f) = 13\tan^{-1}(0.00076f) + 3.5\tan^{-1}\left[(f/7500)^2\right] \quad (9.1.5)$$

$$dz = z(f_{\text{maskee}}) - z(f_{\text{masker}}). \quad (9.1.6)$$

EXAMPLE 9.3

With Equations 9.1.5 and 9.1.6, a simple MATLAB program, example9_3.m, can be written to demonstrate frequency masking. As shown in Figure 9.4, the masker is a 1-kHz tone. This tone generates a new masking curve, which makes frequency components that are below this curve (two tones at 650 and 1,500 Hz) inaudible because their levels are below the masking curve. The masking threshold can be used to eliminate redundant frequency components and quantization noise to achieve higher coding efficiency.

In addition to the steady-state frequency masking illustrated in Figure 9.4, the temporal masking also occurs after the masker is not present. There are three temporal regions: (1) premasking before the presentation of the masker, (2) simultaneous masking at the start of the masker, and (3) postmasking after the stop of the masker. The effective duration of premasking is relatively short at about 20 ms, whereas the postmasking can last for about 100 to 200 ms. Simultaneous masking is dependent on the relationship between frequencies and their relative volumes. For example, if a stronger tonal component is played before a quiet sound, the quiet sound may not be heard if it appears within the order of 50 to 200 ms. These three temporal masking effects are considered in modern perceptual audio coding such as Ogg Vorbis.

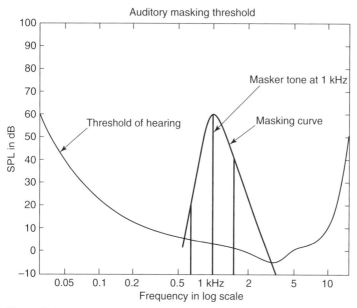

Figure 9.4 An example of frequency masking

Several new audio coding techniques have been developed to improve MP3 by using a better psychoacoustic model, temporal noise shaping, etc. These second-generation techniques include MPEG-4 AAC [56], which is used by Apple's iTunes music store and iPod. MP3 and AAC depend on similar psychoacoustic models, and key patents are held by many companies. Ogg Vorbis from the Xiph.org Foundation is a free software and is patent free. This second-generation coding technique has performance similar to MPEG-4 AAC and higher quality than MP3; however, the computational cost of Ogg Vorbis is higher than MP3. Implementation issues are discussed in [59] and [61]. In this chapter, we implement Ogg Vorbis decoder with the BF537 processor.

9.2 MP3/OGG VORBIS AUDIO ENCODING

The basic idea of perceptual audio coding techniques is to split the time-domain signal into its frequency components and to code its frequency component parameters with the number of bits determined by the psychoacoustic model. Figure 9.5 illustrates the hybrid filterbank & transform block shown in Figure 9.2 for MP3 coding, which consists of cascade of polyphase filterbank and MDCT. First, the audio signal is split into 32 subbands with a polyphase filterbank. Each subband is critically sampled at a 1/32 sampling rate to minimize the bit rate, and 36 samples are buffered into the block. To reduce block effects, the windowing technique introduced in Section 3.3.4 plays an important role. Again, note that Ogg Vorbis only uses large-size MDCT to function as a filterbank for splitting time-domain signals. The MDCT uses the concept of time-domain aliasing cancellation, whereas the

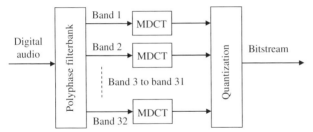

Figure 9.5 Block diagram of filterbank and transform in MP3

quadrature mirror filterbank uses the concept of frequency-domain aliasing cancellation. MDCT also cancels frequency-domain aliasing; thus MDCT achieves perfect reconstruction.

The polyphase filterbank with N subbands and critically sampled at f_s/N can be designed with pseudo quadrature mirror filters [57], which exactly cancel the aliasing from adjacent bands. The prototype low-pass filter with cutoff frequency at π/N and impulse response $h(n)$ of length L can be modulated to form N bandpass filters as [57]

$$h_k(n) = h(n)\cos\left[\frac{(k+1/2)(n-(L-1)/2)}{N} + \varphi_k\right], \quad (9.2.1)$$

for $k = 0, 1, \ldots, N-1$ and $n = 0, 1, \ldots, L-1$. The phase value φ_k is selected to cancel aliasing between adjacent bands.

EXAMPLE 9.4

The MATLAB *Filter Design Toolbox* supports the design of the polyphase filterbank by providing 11 functions to create multirate filter objects. For example, the following function

```
hm = mfilt.firdecim(N);
```

returns a direct-form FIR polyphase decimator object hm with a decimation factor of N. A low-pass Nyquist filter of gain 1 and cutoff frequency of π/N is designed by default. The magnitude response of the low-pass filter designed by example9_4.m with $L = 512$ and $N = 32$ is shown in Figure 9.6.

As shown in Figure 9.5, the MP3 coder compensates for filterbank problems by processing the subband signals with MDCT. Similar to the DFT defined in Equation 3.3.9, the discrete cosine transform (DCT) of signal $x(n)$, $n = 0, 1, \ldots, N-1$, can be expressed as

$$X(k) = a(k)\sum_{n=0}^{N-1} x(n)\cos\left[\frac{\pi(2n+1)k}{2N}\right], k = 0, 1, \ldots, N-1, \quad (9.2.2)$$

where $X(k)$ is the kth MDCT coefficient, $a(k) = 1/\sqrt{N}$ for $k = 0$, and $a(k) = \sqrt{2/N}$ for $k = 1, 2, \ldots N-1$. It is important to note that the DCT coefficient, $X(k)$, is a real number. MATLAB provides the following function for DCT:

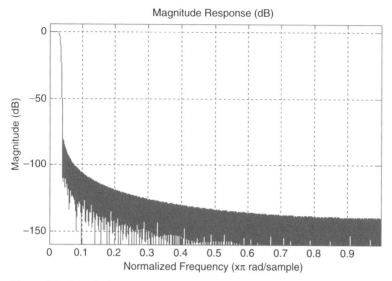

Figure 9.6 Magnitude response of prototype low-pass filter for 32-band filterbank

```
y = dct(x);
```

MDCT is widely used for audio coding techniques. In addition to an energy compaction capability similar to DCT, MDCT can simultaneously achieve critical sampling and reduction of block effects. The MDCT of $x(n)$, $n = 0, 1, \ldots, N-1$, is expressed as

$$X(k) = \sum_{n=0}^{N-1} x(n) \cos\left[\left(n + \frac{N+2}{4}\right)\left(k + \frac{1}{2}\right)\frac{2\pi}{N}\right], k = 0, 1, \ldots, N/2\text{-}1. \quad (9.2.3)$$

The inverse MDCT is defined as

$$x(n) = \frac{2}{N} \sum_{k=0}^{N/2-1} X(k) \cos\left[\left(n + \frac{N+2}{4}\right)\left(k + \frac{1}{2}\right)\frac{2\pi}{N}\right], n = 0, 1, \ldots, N-1. \quad (9.2.4)$$

For real-valued signals, MDCT coefficients in Equation 9.2.3 can be calculated with the FFT method. Similarly, IMDCT can be computed with FFT. Refer to Hands-On Experiment 9.4 for the fast implementation of IMDCT.

A block of data is segmented with an appropriate window function. For implementation of time-domain aliasing cancellation, the window needs to satisfy the following conditions to have perfect reconstruction: (1) The analysis and synthesis windows must be equal, and the length N must be an even number. (2) The window coefficients must be symmetric. There is a 50% overlap between two successive windows for transform. MP3 uses a sine window expressed as [56]

$$w(n) = \sin\left[\frac{\pi(n+0.5)}{2N}\right]. \quad (9.2.5)$$

MP3 specifies two different MDCT block lengths: a long block of 18 samples and a short block of 6 samples. There is 50% overlap between successive windows, so the window sizes are 36 and 12. The long block length allows better frequency resolution for audio signals with stationary characteristics, and the short block length provides better time resolution for transients.

AAC uses a Kaiser–Bessel derived window, and Ogg/Vorbis uses the following Vorbis window [61]:

$$w(n) = \sin\left[\frac{\pi}{2}\sin^2\left(\frac{\pi(n+0.5)}{2N}\right)\right]. \quad (9.2.6)$$

Ogg/Vorbis uses two window lengths between 64 and 8,192 and commonly uses 256 for the short window and 2,048 for the long window. Note that the windows applied to MDCT are different from the windows introduced in Chapter 3 that are used for other types of signal analysis. One of the reasons is that MDCT windows are applied twice for both MDCT and the inverse MDCT.

EXAMPLE 9.5

Both the sine window given in Equation 9.2.5 and the Vorbis window for $N = 32$ are calculated in `example9_5.m`, which uses the GUI WVTool (window visualization tool) for analyzing windows. These two windows and their magnitude responses are shown in Figure 9.7. The figure shows that the Vorbis window has better stopband attenuation.

EXERCISE 9.2

1. Evaluate the sine and Vorbis windows with different length L.
2. Use WINTool introduced in Chapter 3 to further analyze the sine and Vorbis windows.

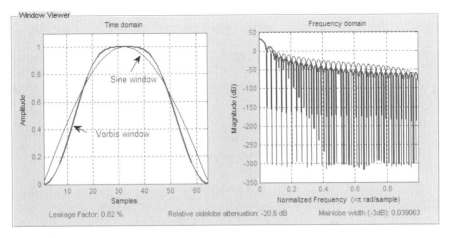

Figure 9.7 Sine and Vorbis windows

Window function and length will affect the frequency response of the filterbank. In general, a long window improves coding efficiency for audio with complex spectra; however, it may create problems for transient signals. The Ogg Vorbis encoder allows smooth change of window length and the use of a transition window (long start, short stop; or short start, long stop) to better adapt to input audio characteristics.

As shown in Figure 9.2, the psychoacoustic model analyzes the input audio signal to determine the best quantization levels for encoding each frequency band. The MDCT outputs are fed into the quantization block, which uses the psychoacoustic model to perform a bit allocation and quantization. To further compress the bit rate, a lossless coding technique is used. Lossless audio coding uses entropy code to further remove the redundancy of the coded data without any loss in quality.

Huffman encoding is a lossless coding scheme that produces Huffman codes from input symbols. Based on the statistic contents of the input sequence, the symbols are mapped to Huffman codes. Symbols that occur more frequently are coded with shorter codes, whereas symbols that occur less frequently are coded with longer codes. On average, this will reduce the total number of bits if some combinations of input sequences are more likely to appear than others. Huffman coding is extremely fast because it utilizes a look-up table for mapping quantized coefficients to possible Huffman codes. On average, an additional 20% of compression can be achieved.

9.3 MP3/OGG VORBIS AUDIO DECODING

Audio decoding is carefully defined in the standard. Actually, the Ogg/Vorbis standard only defines the decoding process. The basic structure of the MP3 decoder illustrated in Figure 9.8 includes three important functions: (1) decoder, (2) inverse quantization, and (3) synthesis filterbank. The synthesis filterbank can be obtained from the analysis filterbank $h_k(n)$ defined in Equation 9.2.1 as follows:

$$g_k(n) = h_k(L - n - 1). \tag{9.3.1}$$

In the synthesis filterbank, the signal is upsampled by a factor N and filtered by the set of filters, $g_k(n)$.

The encoded bit stream format for the compressed audio signal includes (1) header containing the format of the frame; (2) optional cyclic redundancy checksum (CRC) for error correcting; (3) side information containing information for decoding and processing the main data; (4) main data consisting of coded spectral coefficients; and (5) ancillary data holding user-defined information such as song title. The MP3 file has a standard format, which is a frame consisting of 384, 576, or 1,152 samples, and all the frames have associated header information (32 bits) and

Figure 9.8 Basic structure of MP3 decoder

Figure 9.9 Simplified Ogg Vorbis decoder

side information (9, 17, or 32 bytes). The header and side information help the decoder to decode the associated Huffman encoded data correctly.

The Ogg Vorbis bit stream consists of identification, comment, setup, and audio packets. The audio packet contains audio data, which can be decoded with the simplified diagram given in Figure 9.9. The decoding block consists of header decoding, floor decoding, residue decoding, and channel decoupling. The floor is a smooth curve to represent a spectral envelope, which is equivalent to the frequency response of the linear predictive coding filter. The decoded residue values are added with the decoded floor in the frequency-domain to complete the spectral reconstruction. The resulting frequency-domain coefficients are transformed back to the time domain with the inverse MDCT defined in Equation 9.2.4.

The Ogg Vorbis decoder is implemented with TMS320C55x in [61], which shows that the inverse MDCT needs more than 50% of processing time, the most time-consuming block of the decoder. Several implementations listed in [59] show that the optimization of inverse MDCT can achieve an efficient Ogg Vorbis decoder. Efficient implementations include using an optimized assembly routine for inverse MDCT, exploiting the inverse MDCT with FFT, or implementing it with dedicated hardware.

The inverse MDCT and windowing are performed to obtain time-domain audio samples, where the window function is defined in Equation 9.2.6. The current MDCT window is 50% overlapped with adjacent windows. As mentioned above, Ogg Vorbis uses short and long windows with length (a power-of-two number) from 32 to 8,192.

9.4 IMPLEMENTATION OF OGG VORBIS DECODER WITH BF537 EZ-KIT

This section implements the Ogg Vorbis decoder in real time with the Blackfin processor. This Ogg Vorbis decoder supports encoded files with different bit rates. The Ogg Vorbis file is transferred from the computer to the Blackfin memory. This encoded bit stream is then decoded with the Ogg Vorbis decoding algorithm explained in the previous sections and played back in real time with the EZ-KIT.

HANDS-ON EXPERIMENT 9.1

This experiment implements an Ogg Vorbis decoder on the BF537 EZ-KIT. Users who use the BF533 EZ-KIT should refer to Application Project 1 at the end of this chapter for instruction. To run the Ogg Vorbis decoder with the BF537 EZ-KIT, turn all SW7 switches to **ON**, and turn SW8 switches 1 and 2 to **ON**. Several Ogg Vorbis files (with .ogg extension) have been created and stored in directory c:\adsp\chap9\chap9_1_537\Debug logg files for experiment. In addition, users can convert a .wav file to an Ogg Vorbis file with encoding software like OggdropXpd. This software may be downloaded from websites such as http://www.rarewares.org/ogg.html.

The Ogg Vorbis decoder copies an Ogg Vorbis file from the host computer into the Blackfin memory and decodes and plays the audio output with the EZ-KIT. In directory c:\adsp\chap9\chap9_1_537, there are two VisualDSP++ project files, Vorbis_Player_BF537.dpj and libtremor537_lowmem.dpj. The first file is the main project file that sets up the audio driver and runs the Tremor library, and the second file builds the Tremor library. The Tremor library code consists of the source code of the fixed-point Ogg Vorbis decoder, which is based on the open source code available on the website http://xiph.org/. Only Vorbis_Player_BF537.dpj needs to be loaded in VisualDSP++, because this project file is linked to the Tremor library. Build and run the project. When prompted for the Ogg file to be loaded, type in liutm.ogg (or other Ogg Vorbis files). This Ogg Vorbis file is encoded with a bit rate of 112 kbit/s. Do you hear any music coming out from the headphones or loudspeakers? Reload the project and decode other Ogg Vorbis files.

The default Ogg Vorbis decoder was built with optimization turned on. We can also examine the difference when the optimizer is turned off. Load the project libtremor537_lowmem.dpj into VisualDSP++ and rebuild with optimization turned off. Make sure that the project file is **Set as Active Project** before building it. Next, build the project Vorbis_Player_BF537.dpj and run the decoder again. In the decoder.c program, we have set up a benchmark routine to measure the cycle counts (or MIPS) needed by the Ogg Vorbis decoder to complete one second of decoding. The MIPS score is stored at the memory location testsample.0, and this information can be viewed in the **Blackfin Memory** window. Load the Ogg Vorbis file liutm.ogg again and fill in the MIPS benchmark results with and without optimization as shown in Table 9.1.

We can also examine the MIPS required to decode different Ogg Vorbis files encoded at different bit rates. A set of Ogg Vorbis files shown in the first column of Table 9.2 are

Table 9.1 MIPS Benchmark of Ogg Vorbis Decoder with and without Optimization

	MIPS
Decoder (without optimization)	
Decoder (with optimization)	

Table 9.2 Bit Rate, MIPS, and Duration of the Decoded Ogg Vorbis Files

Filename	Bit Rate (kbit/s)	MIPS	Duration (100,000 bytes)
liutm64.ogg	64		
liutm128.ogg	128		
liutm196.ogg	196		
liutm256.ogg	256		

encoded at different (average) bit rates of 64, 128, 192, and 256 kbit/s. Perform the MIPS benchmark for these files and fill in the third column of Table 9.2. In addition, note that these decoded music files vary in length. Why? Examine the main file `decode.c` again and note that this program will transfer 100,000 bytes (determined by the constant NUM_BYTE_TO_DECODE) of data from the file to the Blackfin memory. Because these files are encoded at different bit rates, the lengths of the decoded files are also different. Determine the duration of the decoded music clips, and fill in the results in the last column of Table 9.2. Reload the project for decoding a new Ogg Vorbis file. Verify the duration of the music playback. Users can also increase the number of bytes to decode and verify the new duration of the decoded music clip.

Examine the file size of the Ogg files shown in Table 9.2 and determine the compression ratio of these files with reference to the original file that is sampled at 48 kHz with 16 bits per sample. Also, perform a benchmark on the memory (data and instruction) used in the Ogg Vorbis decoder when decoding the files listed in Table 9.2.

9.5 AUDIO EFFECTS

This section introduces three-dimensional (3D) audio and reverberation effects with different combinations of FIR and IIR filters. We begin with an explanation of these audio effects using block diagrams, move on to the MATLAB floating-point simulations, and finally implement these audio effects in real time with the Blackfin processor.

9.5.1 3D Audio Effects

Unlike stereo sound, which only allows lateral listening, 3D sound enables sound to be positioned around the listener's head at different directions with only a pair of loudspeakers or a headphone. This effect is achieved by using a set of FIR filters, commonly called head-related impulse responses (HRIRs) in time domain or head-related transfer functions (HRTFs) in frequency domain [58]. For example, Figure 9.10 shows that a mono sound source can be perceived as coming from 30° right with respect to the front of the listener by convolving (or filtering) the original signal with the impulse responses of two HRIR filters (left and right) that correspond to the 3D position of 30° right and 0° elevation. This simple filtering reconstructs the acoustic pressures at the listener's ears that would occur with a free-field sound source at the desired location. This system is commonly called the binaural spatializer.

The HRIRs are usually measured with microphones inside the ears of a dummy head and loudspeakers that are placed at different positions that emit white noise [54]. A popular HRIR data set was measured by MIT Media Lab [58], which consists of 710 different positions at elevations from −40° (i.e., 40° below the horizontal plane) to +90° (directly overhead). At each elevation, 360° of azimuth was sampled

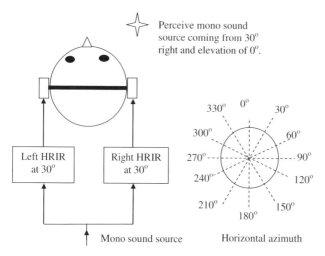

Figure 9.10 3D audio filtering for a mono sound source

in equal increments, and the definition of the angle of horizontal azimuth is shown in Figure 9.10.

EXAMPLE 9.6

This example uses the MATLAB program example9_6.m to convolve a mono sound source with a set of HRIR filters at an azimuth of 30° and 0° elevation. Plot the HRIR and note the differences between the two HRIRs at an azimuth of 30° as shown in Figure 9.11. Note that the right HRIR (bottom diagram) has a higher amplitude and responds earlier compared with the left HRIR (top diagram). The HRIR at the azimuth of 30° is transformed to frequency domain to obtain the HRTF plots as shown in Figure 9.12. Note that the right HRTF (solid line) has a higher magnitude (>10 dB) than the left HRTF (dashed line), and the peaks and notches of the HRTFs occur at different frequencies. These different spectral envelopes contain important cues to convey the spatial information of the sound source. Listen to the original and filtered signals with a headphone. Continue the experiment with different azimuths of 0°, 60°, 90°, 120°, 150°, and 180°.

EXAMPLE 9.7

The previous example shows a single-source binaural spatializer. This system can be extended to a multiple source spatializer that can localize different sound sources in 3D space. The system simply convolves each source signal with a pair of HRIRs that corresponds to the direction of the source. All the processed (left and right) signals are summed together to form the left and right input of the final binaural signals. For example, a two-source binaural spatializer is shown in Figure 9.13. The gain settings at the output of the HRIR filters provide further control of the perceived sound image. A MATLAB file, example9_7.m, is written to perform this two-source binaural spatializer.

9.5 Audio Effects **395**

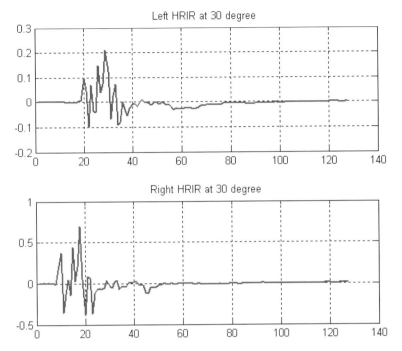

Figure 9.11 Two HRIRs at azimuth of 30°

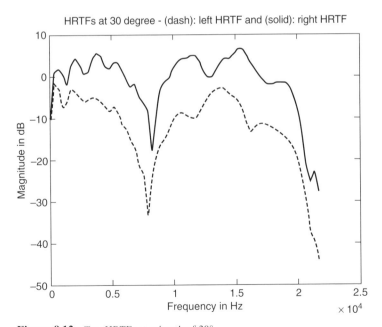

Figure 9.12 Two HRTFs at azimuth of 30°

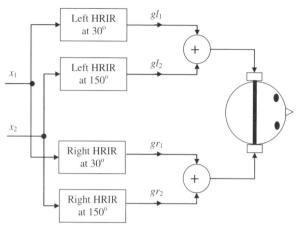

Figure 9.13 Two-source binaural spatializer

EXERCISE 9.3

1. In the previous two-source binaural spatializer, place the source x_1 at 210° instead of 30°. How do we derive the HRIR for azimuth at 210° if only the HRIRs for the right hemisphere (0° to 180°) are measured?

2. Modify the two-source binaural spatializer to include distance information. Assuming that sources x_1 and x_2 are located at one and two meters away from the listener, respectively. Note that the signal strength of the sound source is halved for every doubling of distance from the listener.

9.5.2 Implementation of 3D Audio Effects with BF533/BF537 EZ-KIT

The Blackfin EZ-KIT allows users to select and play back signals with different 3D effects in real time. The audio signal can be continuously sampled at the audio input channels of the Blackfin processor, and 3D audio processing is carried out. The 3D audio effects can be changed on the fly by pressing different switches on the EZ-KIT. This real-time processing platform supports a useful functionality test for interactive applications.

HANDS-ON EXPERIMENT 9.2

This experiment implements a real-time binaural spatializer with the BF533 (or BF537) EZ-KIT. A stereo or mono sound source is connected to the audio input channels of the EZ-KIT, and the output of the EZ-KIT is connected to a headphone. The project files are located in directory c:\adsp\chap9\exp9_2_533 for the BF533 EZ-KIT (or c:\adsp\chap9\exp9_2_537 for the BF537). Load the project file and build and run the project. The default

Table 9.3 Definition of LEDs and Switches for Different Operating Modes

Modes	BF533 EZ-KIT	BF537 EZ-KIT
Static (default when start)	LED4 off LED5–8 light up to indicate the 12 azimuth angles SW5: Every press increases the angle by 30° clockwise SW6: Every press increases the angle by 30° counterclockwise	LED1 off LED2–5 light up to indicate the 12 azimuth angles SW11: Every press increases the angle by 30° clockwise SW12: Every press increases the angle by 30° counterclockwise
Mixed	SW4 to switch to this mode LED4 blinks LED5–8 light up to indicate the 12 azimuth angles SW5: Every press increases the angle by 30° clockwise SW6: Every press increases the angle by 30° counterclockwise	SW10 to switch to this mode LED1 blinks LED2–5 light up to indicate the 12 azimuth angles SW11: Every press increases the angle by 30° clockwise SW12: Every press increases the angle by 30° counterclockwise
Pass-through	SW4 to switch to this mode	SW10 to switch to this mode
Dynamic	SW4 to switch to this mode LED4 off LED5–8 rotate to indicate dynamic movement of sound source	SW10 to switch to this mode LED1 blinks LED2–5 rotate to indicate dynamic movement of sound source

starting mode of the spatializer is the "Static" mode, where sound can be moved from one position to another with every press of a button as described in the second row of Table 9.3. The signal source is considered as a mono sound source.

The second mode is the "Mixed" mode. To switch to this mode, press the switch **SW4** for the BF533 EZ-KIT (or **SW10** for the BF537). The "Mixed" mode implements the two-source binaural spatializer as shown in Figure 9.13. A stereo wave file, mixed_48k_stereo.wav, is available in directory c:\adsp\audio_files\ to test the 3D audio effects in this mode. This binaural spatializer places two sound sources in symmetric locations, and these sound sources can be perceived as moving clockwise or counterclockwise in tandem. Press the switches indicated in Table 9.3 to perceive the movement of the two sound sources.

The third mode is the "Pass-through" mode, which passes the left and right input channels to the output channels without processing. This mode is used to compare the differences between the 3D audio effects and the "Pass-through" without processing. The final "Dynamic" mode in this project moves a single sound source around the head. The sound position is automatically changed by 30° clockwise in every second. A cross-fading method is used in this mode to avoid any sudden discontinuation that creates clicking distortion when moving from one position to another. If the switch **SW4** (for the BF533) or **SW10** (for the BF537) is pressed again, it will rotate back to the "Static" mode.

9.5.3 Generating Reverberation Effects

Room reverberation [64] can be simulated with a set of IIR filters. The reverberation algorithm simulates the sound reflections in an enclosed environment such as rooms, concert halls, etc. These audio effects enhance the listening experience of audio playback in an environment with little or poor reverberation. For example, when listening music with headphones, there is no reverberation from the surroundings that is being added to the music. We can artificially add these reverberation effects to the music and make it sound richer.

Reverberation consists of three components: direct sound, early reflections, and late reflections (or reverberation). Direct sound takes a direct (shortest) path from the sound source to the receiver (or listener). Early reflections, which arrive within 10 to 100 msec after the direct sound, are caused by sound waves reflected once before reaching the listener. The late reflection is produced after 100 msec and forms the reverberation. The reverberation time t_{60} is defined as the time needed for the impulse response to decay by 60 dB. Reverberation time is often used to characterize the reverberation of an enclosed space. A large room such as a concert hall has a long reverberation time between 1.5 and 2 s, whereas a small meeting room has a reverberation of few hundredths of a millisecond.

A digital reverberation algorithm was proposed by Schroeder [64] to model the reverberation effects in a room. As shown in Figure 9.14, this algorithm simulates room reverberation with four comb filters $C_1(z)$, $C_2(z)$, $C_3(z)$, and $C_4(z)$ connected in parallel, followed by two cascaded allpass filters $A_1(z)$ and $A_1(z)$.

The transfer function of the comb filter is expressed as

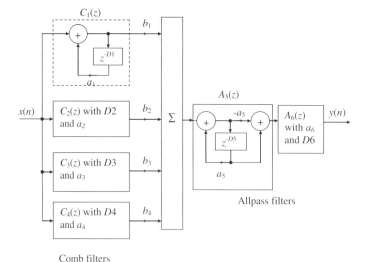

Figure 9.14 Block diagram of a digital reverberation algorithm

$$C_i(z) = \frac{1}{1 - a_i z^{-Di}}, \quad i = 1, 2, 3, 4, \tag{9.5.1}$$

which is the IIR filter with Di poles equally spaced ($2\pi/Di$) on the unit circle. The coefficient a_i determines the decay rate of the impulse response for each comb filter. The comb filter increases the echo density and gives the impression of the acoustics environment and room size. However, it also causes distinct coloration of the incoming signal. Allpass filters prevent such coloration and emulate more natural sound characteristics in a real room. The transfer function of the allpass filter is expressed as

$$A_i(z) = \frac{-a_i + z^{-Di}}{1 - a_i z^{-Di}}, \quad i = 5, 6. \tag{9.5.2}$$

By cascading the allpass filters with comb filters, the impulse response of the overall system becomes more diffuse.

EXAMPLE 9.8

The digital reverberation algorithm shown in Figure 9.14 can be simulated with MATLAB. Open the MATLAB file `example9_8.m`, which specifies the time delays of 40, 44, 48, and 52 msec for the four comb filters. The allpass filter has a delay of 7 msec. The reverberation algorithm is running at 48 kHz. Determine the length of the comb and allpass filters. Set all feedback coefficients, a_i, $i = 1, \ldots, 4$, of the comb filters as 0.75. Run the simulation with the input wave file `liutm_48k_mono.wav`, which is located in directory `c:\adsp\audio_files`, and play back the processed signal `output.wav` in directory `c:\adsp\chap9\MATLAB_ex9`. Do you perceive any reverberation effect? Use different time delays and coefficients to create different effects.

EXERCISE 9.4

1. Examine the impulse and magnitude responses of the comb filters and allpass filters separately. Compare the differences of using a longer delay and a larger coefficient.

2. Use an impulse signal as input to the reverberation algorithm shown in Figure 9.14 and save the impulse response. Examine the impulse and magnitude responses of the cascade structure. Compute the reverberation time, t_{60}, based on the parameters given in Example 9.8.

9.5.4 Implementation of Reverberation with BF533/BF537 EZ-KIT

This section implements the reverberation algorithm in real time with the Blackfin EZ-KIT. The performance of the previous floating-point MATLAB implementation must be reevaluated with the fixed-point (1.15) format before porting to the Blackfin processors. Refer to the fixed-point design and implementation of the IIR filters in Chapter 6 for details.

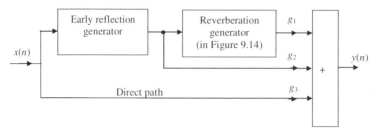

Figure 9.15 Moorer's digital reverberation structure

HANDS-ON EXPERIMENT 9.3

This experiment implements the reverberation algorithm [47] with the Blackfin processor. The project file is located in directory c:\adsp\chap9\exp9_3_533 for the BF533 (or c:\adsp\chap9\exp9_3_537 for the BF5337). In addition to the generation of reverberation shown in Figure 9.14, this project also implements the early reflection portion and direct path of the impulse response as shown in Figure 9.15. This reverberation structure, proposed by Moorer [62], produces a more realistic reverberation sound than that proposed by Schroeder (Fig. 9.14). Build and run the project. Connect stereo music to the audio input port of the EZ-KIT. Press **SW5** (for the BF533 EZ-KIT) or **SW11** (for the BF537) to listen to the reverberation version of the music and then press **SW4** (for BF533) or **SW10** (for BF537) to switch back to the pass-through mode. Three gains (g_1, g_2, and g_3) can be individually controlled to emphasize the significance of the early reflection generator, reverberation, and direct path signals. We can also turn off the contribution of different paths by setting the corresponding gains to zero.

9.6 IMPLEMENTATION OF MDCT WITH LABVIEW EMBEDDED MODULE FOR BLACKFIN PROCESSORS

MDCT divides the audio signal into overlapping frequency bands and later recombines them using inverse MDCT (IMDCT) to reproduce the signal with minimal distortion. In this experiment, we focus on IMDCT, which is used in the Ogg Vorbis decoder. First, we analyze two methods of implementing the IMDCT in LabVIEW: (1) direct implementation of IMDCT and (2) faster implementation using FFT. Next, the algorithms are benchmarked for both LabVIEW for Windows and the Blackfin EZ-KIT. Benchmarking is done by isolating IMDCT and placing it in a simple structure that can be reused to study the performance of any function within LabVIEW or the LabVIEW Embedded Module for Blackfin Processors.

9.6 Implementation of MDCT with LabVIEW Embedded Module 401

HANDS-ON EXPERIMENT 9.4

IMDCT is an overlapping algorithm that generates an output data set that is twice the size of the input signal. The resulting signal is then overlapped and combined with the last 50% of the previous packet and the first 50% of the next packet to reconstruct the signal.

Launch LabVIEW and open the file Test_IMDCT_by_Eqn.llb located in directory c:\adsp\chap9\exp9_4 to begin exploring IMDCT. Test_MDCT_Eqn.vi, inside the .llb library, allows users to test and verify that the implementation of the IMDCT is correct. The input signal is a sine wave as shown in Figure 9.16. This test signal is unique in that the past, present, and next data packets are identical.

This input signal is passed through a **Window** function to remove the beginning and ending discontinuities and then transformed with the MDCT defined in Equation 9.2.3. Next, the IMDCT defined in Equation 9.2.4 is used to transform the DCT coefficients back to the time domain. The windowing function is applied again to combine the past, present, and next signal packets to complete the reconstruction of the original input signal.

Run the VI and verify that the input and output signals are identical. Open **IMDCT subVI** shown in Figure 9.17 and analyze the graphical code used to implement the algorithm.

As shown in Figure 9.17, the outer **For Loop** iterates $2N$ times, where N is the size of the input signal, to generate each point of the output signal. Formula Node inside the inner **For Loop** calculates and accumulates all of the values according to the IMDCT. Note that the output of the algorithm has twice the number of points as the input signal.

FFT can be used to improve the speed of execution for computing IMDCT. Ideally, when packets are created with a length that is a power of two, the radix-2 FFT algorithm

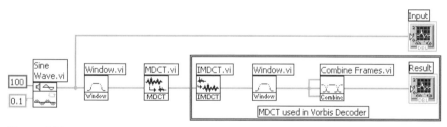

Figure 9.16 Direct implementation of MDCT and IMDCT, MDCT by Eqn.vi

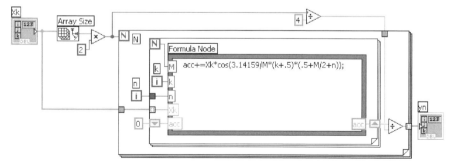

Figure 9.17 Direct implementation of IMDCT, IMDCT.vi

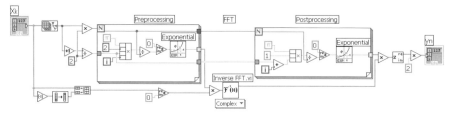

Figure 9.18 FFT implementation of IMDCT, IMDCT_FFT.vi

can be used. The calculation of IMDCT with inverse FFT is derived in the following three equations. The detailed explanation can be found in [56]. Some additional processing is necessary, as shown in the LabVIEW implementation in Figure 9.18.

$$X'(k) = X(k)e^{j\frac{2\pi}{N}kn_0}, \qquad (9.6.1)$$

$$x(n) = \text{IFFT}[X'(k)], \qquad (9.6.2)$$

$$y(n) = x(n)e^{j\frac{2\pi}{N}(n+n_0)}, \qquad (9.6.3)$$

where $n_0 = \left(\dfrac{N}{4} + \dfrac{1}{2}\right)$, N represents the number of samples to be recovered, k is the frequency index, and n is the time index.

Modify Test_MDCT_Eqn.vi by replacing both MDCT.vi and IMDCT.vi with MDCT_FFT.vi and IMDCT_FFT.vi, respectively. To do this, right-lick on MDCT.vi, select **Replace → Select A VI...**, and navigate to MDCT_FFT.vi in directory c:\adsp\chap9\exp9_4\Test_IMDCT_by_FFT.llb. Do the same for IMDCT_FFT.vi. Run the exercise again to verify that this FFT implementation of IMDCT generates the same results as the direct implementation of IMDCT.

HANDS-ON EXPERIMENT 9.5

In this experiment, we benchmark the performance of the IMDCT with a common architecture that can be used to benchmark any function or algorithm in LabVIEW or the LabVIEW Embedded Module for Blackfin Processors. DSP algorithms typically execute very fast, with submillisecond execution times. For this reason, we derive the average execution time by running the algorithm several times while measuring the total execution time. When operating on the embedded target, the final execution time is output to the standard output port with **One Button Dialog VI**.

Open IMDCT_Benchmark.vi located in Bench_IMDCT.llb in directory c:\adsp\chap9\exp9_5. This VI provides a common benchmarking architecture for testing any application. Figure 9.19 shows that within the benchmarking VI, the data set **Xk** simulates a 128-point packet of input data. The IMDCT has been placed in **For Loop** so that it can execute several iterations in order to calculate an average execution time per iteration.

9.6 Implementation of MDCT with LabVIEW Embedded Module

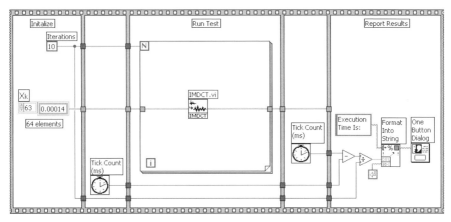

Figure 9.19 A benchmarking structure, IMDCT_Benchmark.vi

Table 9.4 Execution Time for Different Platforms

	Execution Time in Milliseconds		
	Windows	Blackfin Default	Blackfin Optimized
IMDCT.vi			
IMDCT_FFT.vi			

Use Table 9.4 to record the execution results. Fill up the benchmark table by running the VI shown in Figure 9.19 targeted to Windows. Record the resulting execution time in the top left cell of Table 9.4. Then right-click on IMDCT.vi and replace it with IMDCT_FFT.vi. Run the VI again and record the result.

After you have completed testing in Windows, change the execution target to the Blackfin EZ-KIT. Open the project IMDCT_Benchmark_Default_BF5xx.lep located in directory c:\adsp\chap9\exp9_5. Execute the VI on the embedded target. The resulting time should be sent back to the computer through the standard output port and can be seen on the **Output** tab of the **Processor Status** window. Once you have tested IMDCT.vi, again replace it with IMDCT_FFT.vi.

Finally, close the current project and open the new project IMDCT_Benchmark_Optimized_BF5xx.lep located in directory c:\adsp\chap9\exp9_5. Execute the VI on the embedded target and record the execution results. Again replace IMDCT.vi with IMDCT_FFT.vi to test the execution time of both IMDCT computation methods. Examine the configured Blackfin **Build Options** from the **Embedded Project Manager** to see which optimizations were enabled.

Once the table is complete, study the results. Which compiling optimizations effectively reduce the execution time? What methods could be used to further increase the execution speed of IMDCT? Explain. Also, consider data types and the analysis library used.

9.7 MORE APPLICATION PROJECTS

1. The BF533 EZ-KIT can be used to perform the same Ogg Vorbis decoder defined in Hands-On Experiment 9.1. However, it takes a longer time to load an Ogg Vorbis file into the BF533 EZ-KIT. Load and build the project file `Vorbis_Player_BF533.dpj` located in directory `c:\adsp\chap9\project9_1`. Perform the same operations including benchmark measurements described in Hands-On Experiment 9.1.

2. In Hands-On Experiment 9.1, benchmarking is carried out on the entire decoding process. It is useful to profile the computational load for different modules in the decoding process and identify the hot spot of the program. Activate the statistical profiler and examine the percentage of time spent in different modules. Which portion of the module accounts for the majority of the cycle count?

3. For using loudspeakers to play back 3D sound, we need an additional processing block called the cross talk canceller that cancels cross talk from the loudspeakers to the ears. Refer to reference [54] for details on the cross talk canceller. Extend Hands-On Experiment 9.2 to include the loudspeaker playback mode. Comment on the perception differences between 3D audio listening with a headphone and loudspeakers.

4. Extend the MATLAB simulation of the reverberation algorithm given in Example 9.8, using longer delays of 52, 56, 60, and 80 msec for the comb filters and delays of 10 msec for the allpass filters. Determine the lengths of the delay buffer for implementing these delays when sampling rate is 44.1 kHz.

5. The reverberation algorithm implemented in Hands-On Experiment 9.3 can be further fine-tuned by programming a room selection option. Different sets of reverberation parameters can be tabulated in the program for users to select the reverberation effects that correspond to different room sizes and reverberation times. Modify the project in Hands-On Experiment 9.3, which allows users to perform this selection on the fly with a switch on the EZ-KIT.

6. The reverberation algorithm can be integrated into the binaural spatializer to further enhance the realism of spatial reproduction. A simplified approach is to sum all the input sources and use this combined signal as input to the reverberation algorithm described in Example 9.8. The reverberation output can then be added to the output of the left and right channels of the binaural spatializer. Implement this new structure by combining and modifying the codes given in Hands-On Experiments 9.2 and 9.3.

7. Integrate the 3D audio and reverberation effects into the Ogg Vorbis decoder with the BF537 EZ-KIT. Program the switch on the EZ-KIT to select different combinations of audio effects.

8. The ability to select different audio enhancement modes is a common feature found in most portable audio players. For example, the graphic equalizer introduced in Chapter 5 can be used to enhance the audio content in a portable audio player. Integrate the FIR filter-based eight-band graphic equalizer into the Ogg Vorbis decoder project in Hands-On Experiment 9.1.

9. In general, the bass perception from low-end loudspeakers and headphones is poor. This is due to the poor frequency response of transducers at frequency below 200 Hz. A simple way to overcome this is to boost up the low-frequency bands with a graphic equalizer. However, this approach suffers from overdriving the loudspeakers and wasting of energy. A better approach is called the virtual bass system [60], which creates the low-frequency harmonics of the musical source and mixes these harmonics into the

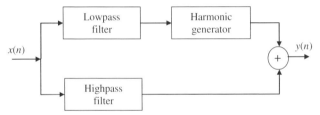

Figure 9.20 A simple virtual bass system

original music. These harmonics will give the same pitch perception as the low-frequency content, which can be reproduced accurately in most low-end loudspeakers with moderate midfrequency response. For example, if the musical signal consists of a prominent low-frequency component at 50 Hz, the virtual bass system generates frequency harmonics at 100, 150, and 200 Hz. The interfrequency difference of these harmonics will create a 100-Hz pitch that is perceived by the human ear based on psychoacoustic effects.

A simple virtual bass system can be developed as shown in Figure 9.20. It consists of two low-pass and high-pass filters of cutoff frequency 150 Hz. The sampling frequency is 48 kHz. The low-pass filter output is passed to a harmonic generator, which creates the harmonics with a simple nonlinear function such as a power function, $y(n) = x^2(n) + x^3(n)$. These harmonics are then mixed with the high-frequency components to form the enhanced bass system. Implement the virtual bass system the BF533/BF537 EZ-KIT and evaluate its effects.

10. A set of audio files located in directory `c:\adsp\audio_files\noisy_audio\` have been severely distorted. Based on the techniques learned in preceding chapters, devise a suitable technique to clean up these audio files. In some cases, advanced techniques that are not described in this book must be explored. Users can refer to the textbooks listed in the References or search the website for solutions. Implement the developed algorithms on the BF533/BF537 EZ-KIT and test its real-time performance. Obtain the signal-to-noise ratio measurements of the audio signal before and after enhancement and comment on its performance.

Chapter 10

Practical DSP Applications: Digital Image Processing

Image processing is an important application of two-dimensional (2D) signal processing. Because of the developments of fast, low-cost, and power-efficient embedded signal processors for real-time processing, digital image processing is widely used in portable consumer electronics such as digital cameras and picture phones. This chapter introduces basic concepts of digital image processing and some simple applications.

10.1 OVERVIEW OF IMAGE REPRESENTATION

A digital image (or picture) is visual information received, represented, processed, stored, or transmitted by digital systems. With advancements in microelectronics, digital image processing is widely used for applications such as enhancement, compression, and recognition. Digital image processing has many characteristics in common with one-dimensional (1D) signal processing introduced in previous chapters.

As discussed in Chapter 2, 1D digital signals are denoted by $x(n)$, which represents the value (amplitude) of signal x at discrete time n. Digital images are 2D signals that consist of picture elements called pixels. Each pixel can be represented as $x(m,n)$, where m is the row (height) index from 0 to $M-1$ (top to bottom), n is the column (width) index from 0 to $N-1$ (left to right), and $x(m,n)$ is the value of 2D space function x at the pixel location (m,n). Therefore, each digital image can be represented by $M \times N$ pixels. Similar to the 1D sampling theorem stated in Chapter 2, the 2D sampling theorem indicates that a sequence of pixels $x(m,n)$ can represent the analog picture if the sampling is sufficiently dense.

The digital image pixel coordinates are illustrated in Figure 10.1. For pixel coordinate (m,n), the first component m (row) increases downward, while the second

---

Embedded Signal Processing with the Micro Signal Architecture. By Woon-Seng Gan and Sen M. Kuo
Copyright © 2007 John Wiley & Sons, Inc.

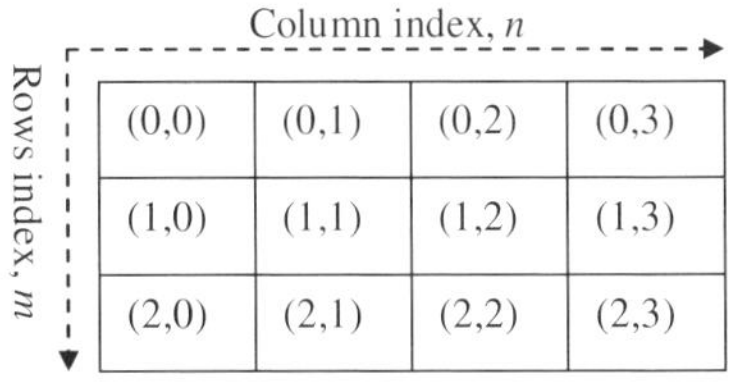

10.1 Overview of Image Representation

	Column index, n			
Rows index, m	(0,0)	(0,1)	(0,2)	(0,3)
	(1,0)	(1,1)	(1,2)	(1,3)
	(2,0)	(2,1)	(2,2)	(2,3)

Figure 10.1 An example of 3×4 digital image pixel coordinates

component n (column) increases to the right. For example, the data for the pixel in the second row, third column is stored in the matrix element (1,2). Note that the pixel coordinates used by MATLAB are integer values and range between 1 and M or N. Therefore, the pixel in the second row, third column has coordinate (2,3) in MATLAB.

Pixel values can be represented by integer or fixed-point number. For a black-and-white (B&W, or grayscale intensity) image, each pixel is represented by one number that indicates the grayscale level of that pixel. The number of levels used to represent a B&W image depends on the number of bits. For example, 8 bits ($B = 8$) are commonly used for 256 (2^B) grayscale levels to display the intensity of B&W images from 0 to 255, where "0" represents a black pixel and "255" corresponds to a white pixel. Similar to the quantization noise of digitizing 1D analog signals, encoding continuous grayscale with finite number of bits also results in quantization errors. The resolution of digital images depends on the total number of pixels ($M \times N$) and grayscale levels 2^B.

The sensation of color is determined by the light frequency spectrum, from red to violet. Each pixel of color image can be represented by three components: red (R), green (G), and blue (B) and thus is called RGB data. Proper mixing of these three basic colors can generate all other colors. Similar to B&W images, each color can be represented with 8 bits, thus forming the commonly used 24-bit RGB data (8 bits for R, 8 bits for G, and 8 bits for B). Color images are widely used in cameras, televisions, and computer displays. For example, the traditional North American television standard is 480×720 pixels. Most digital images are in the size of $764 \times 1,024$ pixels or larger for computers and high-definition televisions and 480×720 or smaller for real-time transmission via networks.

EXAMPLE 10.1

In most speech processing for telecommunications, speech is sampled at 8 kHz with 16-bit resolution. Therefore, we have 8,000 signal samples to process in a second, and we need 16,000 bytes to store 1 second of digital speech. In video processing, we have 30 frames (images) per second. Assuming that each B&W image has $764 \times 1,024$ pixels, we have more than 23 million samples to process and store in a second. These requirements increase dramatically for color images. Therefore, image or video processing requires higher complexity in terms of computational power and memory requirements.

MATLAB stores and represents B&W images as matrices, in which each element of the matrix corresponds to a single pixel in the image. For example, an image composed of M rows and N columns of different dots would be stored in MATLAB as an $M \times N$ matrix. Color images require a three-dimensional array as an $M \times N \times 3$ matrix, where the first plane in the three-dimensional array represents red, the second plane represents green, and the third plane represents blue. MATLAB file formats store true-color images as 24-bit RGB images, where the red, green, and blue components are 8 bits each. The three color components for each pixel are stored along the third dimension of the data array. For example, the red, green, and blue color components of the pixel (10,5) are stored in RGB(10,5,1), RGB(10,5,2), and RGB(10,5,3), respectively.

MATLAB provides the function `imread` to read an image. This function can read most image formats such as `bmp` (windows bitmap), `gif` (graphics interchange format), `jpg` (Joint Photographic Experts Group, JPEG), `tif` (tagged image file format, TIFF), etc. Most image file formats use 8 bits to store pixel values. When these are read into memory, MATLAB stores them as class `uint8`. For file formats that support 16-bit data, MATLAB stores the images as class `uint16`.

The *Image Processing Toolbox* [51] provides two image display functions, `imshow` and `imtool`. The function `imshow` is the fundamental image display function, and the function `imtool` starts the Image Tool, which is an integrated environment for displaying images and performing some common image processing tasks.

EXAMPLE 10.2

The following MATLAB command

```
I = imread(filename,fmt);
```

reads a grayscale or color image from the file specified by the string `filename`, and the string `fmt` specifies the format of the file. In example10_2.m, we read a JPEG file, Disney.jpg, and display it with `imtool` as shown in Figure 10.2. The `imtool` function integrates the image display tool with the pixel region tool, the image information tool, and the adjust contrast tool. It can help explore large images, such as scroll bars, the overview tool, pan tool, and zoom buttons. For example, the image tool displays Pixel infor: (639,441)[143 111 90] at the bottom left corner, where (639,441) indicates pixel location pointed at by the arrow (cursor), and [143 111 90] indicates the corresponding R, G, B values. We can move the cursor to different locations for examining its pixel values. We can also enter the whos command to see memory storage of image I as follows:

```
>> whos
   Name        Size            Bytes       Class
   I           480x640x3       921600      uint8 array
Grand total is 921600 elements using 921600 bytes
```

Intensity and true-color images can be `uint8` (unsigned 8-bit integers), `uint16` (unsigned 16-bit integers), `int16` (16-bit signed integers), `single` (single precision), or `double` (double precision). MATLAB supports conversion between the various image types.

Figure 10.2 An example of digital color image displayed by MATLAB

10.2 IMAGE PROCESSING WITH BF533/BF537 EZ-KIT

VisualDSP++ IDE supports image files in different formats such as bmp, gif, jpg, and tif to be loaded into the Blackfin memory for processing. In the subsequent experiments, we import an image and process this image with the Blackfin processor.

HANDS-ON EXPERIMENT 10.1

This experiment loads an image from the computer to the Blackfin memory with the VisualDSP++ IDE. The BF533 (or BF537) EZ-KIT is used as a platform for loading images. Click on **View → Debug Windows → Image Viewer . . .**; an **Image Configuration** window is opened as shown in Figure 10.3. Set **Source location** to **File** and type in the file name (for example, c:\adsp\image_files\Disney.jpg) for the image to be loaded into the Blackfin memory. We set **Start address** to 0x0 with **Memory stride** of 1. This starting address is the SDRAM memory of the Blackfin processor. Click on **OK** to display the image in the **Image Viewer** window. It takes a few minutes for transferring the image from the computer to the Blackfin via the USB port and displaying in the **Image Viewer** window.

Note that the image information is displayed in the **Image Configuration** window. The memory locations needed to store an image are determined by the size of the image (width × height). In addition, an RGB image takes three times the memory of a B&W image. Load different (B&W or color) images to the Blackfin processor's memory at different start addresses and display them. Some sample images can be found in directory c:\adsp\image_files\.

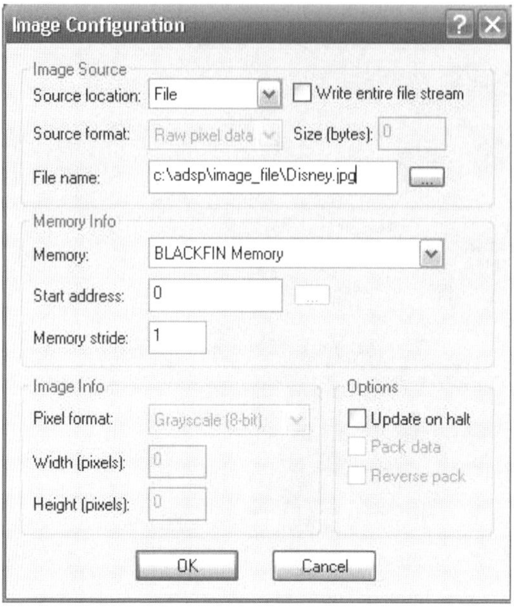

Figure 10.3 **Image Configuration** window for loading image into Blackfin memory

10.3 COLOR CONVERSION

Color is determined by the spectral distribution of the energy of illumination source and the visual sensation of the human viewer. A color space represents colors with digital pixels. A B&W image uses one number for each pixel, whereas a color image needs multiple numbers per pixel. There are several different color spaces to represent color images. The *Image Processing Toolbox* represents colors as RGB values. Because the image processing functions assume all color data as RGB, we can process an image that uses a different color space by first converting it to RGB data, processing it, and then converting the processed image back to the original color space.

The National Television System Committee (NTSC) color space is used in televisions in the USA. In the NTSC format, color images consist of three components: luminance (Y), hue (I), and saturation (Q); thus the NTSC format is also called the YIQ format. The first component, luminance, represents grayscale information, and the last two components make up chrominance (color information). The conversions between the RGB and YIQ color spaces are defined as [72]

$$\begin{bmatrix} Y \\ I \\ Q \end{bmatrix} = \begin{bmatrix} 0.299 & 0.587 & 0.114 \\ 0.596 & -0.274 & -0.322 \\ 0.211 & -0.523 & 0.312 \end{bmatrix} \begin{bmatrix} R \\ G \\ B \end{bmatrix}. \quad (10.3.1)$$

and

10.3 Color Conversion

$$\begin{bmatrix} R \\ G \\ B \end{bmatrix} = \begin{bmatrix} 1 & 0.956 & 0.621 \\ 1 & -0.273 & -0.647 \\ 1 & -1.104 & 1.701 \end{bmatrix} \begin{bmatrix} Y \\ I \\ Q \end{bmatrix}. \qquad (10.3.2)$$

The *Image Processing Toolbox* provides functions for conversion between color spaces. The function `rgb2ntsc` converts RGB images to the NTSC color space, and the function `ntsc2rgb` performs the reverse operation. In addition, the function `rgb2gray` extracts the grayscale information from a color image.

EXAMPLE 10.3

In MATLAB code `example10_3.m`, we first convert the color image shown in Figure 10.2 to a B&W image with the function `rgb2gray`. We then convert the color image from the RGB data to the NTSC format. Because luminance is one of the components of the NTSC format, we can isolate the grayscale level information in an image. The obtained B&W image is shown in Figure 10.4. Finally, we store the B&W image back to disk file `Disney_BW.jpg` using the function `imwrite`.

Human vision is more sensitive to brightness than color changes; thus humans perceive a similar image even if the color varies slightly. This fact leads to the YCbCr color space, which is widely used for digital video and computer images such as JPEG and MPEG standards. In this format, luminance information is stored as a single component Y, and chrominance information is stored as two color-difference components Cb and Cr. The value Cb represents the difference between the blue component and a reference value, and Cr represents the difference between the red component and a reference value. The relation between the RGB color space and the YCbCr color space can be expressed as

$$\begin{bmatrix} Y \\ Cb \\ Cr \end{bmatrix} = \begin{bmatrix} 0.257 & 0.504 & 0.098 \\ -0.148 & -0.291 & 0.439 \\ 0.439 & -0.368 & -0.071 \end{bmatrix} \begin{bmatrix} R \\ G \\ B \end{bmatrix} + \begin{bmatrix} 16 \\ 128 \\ 128 \end{bmatrix}. \qquad (10.3.3)$$

Figure 10.4 The B&W image obtained from the color image shown in Figure 10.2

For `uint8` images, the data range for Y is [16, 235], and the range for Cb and Cr is [16, 240]. The MATLAB function `rgb2ycbcr` converts RGB images to the YCbCr color space, and the function `ycbcr2rgb` performs the reverse operation.

10.4 COLOR CONVERSION WITH BF533/BF537 EZ-KIT

We can perform color conversion with the BF533/BF537 EZ-KIT. In the following experiments, we examine the C code that performs various color conversions.

HANDS-ON EXPERIMENT 10.2

In this experiment, we load a color image (e.g., `Disney.jpg`) into the Blackfin memory at the start address of 0x0 as shown in Hands-On Experiment 10.1. Three project files with the same name, `colorconversion.dpj`, are located in separate directories `c:\adsp\chap10\exp10_2\rgb2xxxx` (where xxxx represents `ntsc`, `cbcr`, and `gray`), and they are used to perform different color conversion experiments (e.g., RGB image to NTSC, RGB image to YCbCr, and RGB image to gray) on the BF533 (or BF537) EZ-KIT. The color conversions are based on Equations 10.3.1 and 10.3.3. Assembly programs are written to perform the color conversion, and the gray level is extracted by the C routine.

Build and run each project. From the C main function, note the location and size of the original color image and the gray image in the Blackfin memory and display them with the VisualDSP++ image viewer as shown in Figure 10.5. Compare the color conversion results with those obtained with MATLAB.

Figure 10.5 Image Viewer windows for both color and gray images

10.5 TWO-DIMENSIONAL DISCRETE COSINE TRANSFORM

The discrete cosine transform (DCT) consists of a set of basis vectors that are cosine functions. The DCT may also be computed with a fast algorithm such as FFT. Comparing the DCT with the DFT introduced in Chapter 3, the DCT has advantage that the transform is real valued. It is widely used in audio compression algorithms, as introduced in Chapter 9. In this section, we expand it to 2D DCT, which represents an image as a sum of sinusoids of varying magnitudes and frequencies. 2D DCT and 2D inverse DCT (IDCT) [70] are widely used in many image and video compression techniques such as JPEG and MPEG standards.

The 2D DCT of a rectangular $M \times N$ image is defined as

$$X(i,k) = C(i)C(k) \sum_{m=0}^{M-1} \sum_{n=0}^{N-1} x(m,n) \cos\left[\frac{(2m+1)i\pi}{2M}\right] \cos\left[\frac{(2n+1)k\pi}{2N}\right], \quad (10.5.1)$$

for row $i = 0, 1, \ldots M - 1$ and column $k = 0, 1, \ldots N - 1$, and $X(i,k)$ is the corresponding DCT coefficient for the pixel value $X(m,n)$. In Equation 10.5.1, the normalization factors are defined as

$$C(i) = \begin{cases} 1/\sqrt{M}, & i = 0 \\ \sqrt{2/M}, & \text{otherwise} \end{cases} \quad (10.5.2a)$$

$$C(k) = \begin{cases} 1/\sqrt{N}, & k = 0 \\ \sqrt{2/N}, & \text{otherwise} \end{cases} \quad (10.5.2b)$$

The 2D DCT can be applied to the rows and then the columns or started from columns to rows. For a typical image, most of the significant information is concentrated in a few lower (smaller i and k) DCT coefficients. The 2D IDCT is defined as

$$x(m,n) = \sum_{i=0}^{M-1} \sum_{k=0}^{N-1} C(i)C(k) X(i,k) \cos\left[\frac{(2m+1)i\pi}{2M}\right] \cos\left[\frac{(2n+1)k\pi}{2N}\right]. \quad (10.5.3)$$

Most image compression algorithms use a square image with $M = N = 8$. For example, JPEG image coding standard partitions the original image into 8×8 subimages and processes these small blocks one by one. In addition, the original pixel values of B&W images are subtracted by 128 to shift the grayscale levels from the original range [0 to 255] to a symmetric range [−128 to 127] before performing DCT.

The `dct2` function in the *Image Processing Toolbox* computes the 2D DCT of an image. Again, note that matrix indices in MATLAB always start at 1 rather than 0; therefore, the MATLAB matrix elements $X(1,1)$ correspond to the mathematical quantity $X(0,0)$.

414 Chapter 10 Practical DSP Applications: Digital Image Processing

EXAMPLE 10.4

The MATLAB command

    X = dct2(x);

computes the 2D DCT of image matrix x to obtain the same size DCT coefficient matrix X. In addition, the command x = idct2(X) performs the 2D IDCT of coefficient matrix X to obtain the image matrix x. In example10_4a.m, we read the B&W image Disney_BW.jpg created in Example 10.3 and perform 2D DCT of the image and IDCT of the DCT coefficients. When executing the program, we are surprised to find that we obtain a blank image. Why?

In JPEG compression, the original pixel values of B&W images are subtracted by 128 to shift the grayscale levels from the original [0 to 255] to [−128 to 127] before performing DCT. The modified program is given in example10_4b.m. Run the program, and the recovered image is shown in Figure 10.6, which is different than the original B&W image shown in Figure 10.4. Why?

In the imtool window, move the cursor around to examine pixel values at different locations. We find that the image shown in Figure 10.6 only consists of a few different values, and this is caused by finite-precision effects. Therefore, we convert the image from 8-bit integer to 32-bit floating-point format with the function im2single, perform DCT and IDCT with floating-point arithmetic, and convert the recovered image back to integer format with the function im2uint8. We run the program example10_4c.m with floating-point conversion, and we obtain a recovered image that is almost the same as the original one shown in Figure 10.4.

EXAMPLE 10.5

As mentioned above, in many image compression standards the image is divided into 8 × 8 blocks and the 2D DCT is computed for each block. The resulting 64 DCT coefficients are split into one DC (0,0) and 63 AC coefficients and then quantized, coded, and transmitted

Figure 10.6 A recovered B&W image after DCT and IDCT operations

10.5 Two-Dimensional Discrete Cosine Transform

or stored. The receiver decodes the quantized DCT coefficients, computes the 2D IDCT of each block, and then reconstructs a single image. For typical images, many of the DCT coefficients with large indexes (i,k) have values close to zero and thus can be discarded without seriously affecting the quality of the reconstructed image.

The code `example10_5.m` (adapted from **Help** menu) computes the 2D DCT of 8 × 8 blocks in the input image, saves only 10 DCT coefficients with small indexes (i,k) in each block, and then reconstructs the image with the 2D IDCT of each block. In the code, we use `T = dctmtx(8)` to obtain an 8 × 8 DCT transform matrix. If **A** is a square matrix, the 2D DCT of **A** can be computed as **T*A*T'**. This computation is faster than using the `dct2` function if we are computing a large number of small subimages. In the code, we also use the function `blkproc` to implement distinct block processing for image. We use a mask to keep only 10 DCT coefficients with the smallest indexes and clear other coefficients to zero so they do not need to be transmitted or stored. This is the basic principle of image compression with DCT. The reconstructed image is displayed in Figure 10.7, which shows some loss of quality compared with the original B&W image given in Figure 10.4; however, we only code 10 DCT coefficients instead of 64. Note that we can use MATLAB function `subimage` to display multiple images in a single figure. In this case, we can compare the original image with the reconstructed image side by side.

The 1D N-point DFT defined in Chapter 4 can be extended to 2D (M,N)-point DFT as follows:

$$X(i,k) = \sum_{m=0}^{M-1}\sum_{n=0}^{N-1} x(m,n)e^{-j\frac{2\pi}{M}im}e^{-j\frac{2\pi}{N}kn}, \qquad (10.5.4)$$

where X(i,k) is a complex-valued number. Similarly, the 2D inverse DFT can be expressed as

$$x(m,n) = \frac{1}{MN}\sum_{i=0}^{M-1}\sum_{k=0}^{N-1} X(i,k)e^{j\frac{2\pi}{M}im}e^{j\frac{2\pi}{N}kn}. \qquad (10.5.5)$$

Figure 10.7 The reconstructed image using 2D DCT and block processing

MATLAB provides the functions `fft2` and `ifft2` to support the 2D DFT and the inverse 2D DFT, respectively.

10.6 TWO-DIMENSIONAL DCT/IDCT WITH BF533/BF537 EZ-KIT

We have used an example to illustrate how to transform an image using 2D DCT, retain only 10 DCT coefficients, and use IDCT to reconstruct the image. In this section, we use a luminance quantization table recommended by the JPEG standard [71] to remove insignificant coefficients based on the percentage of retained coefficients.

A luminance quantization table for retaining approximately 50% of the DCT coefficients is provided in Table 10.1. The quantized DCT coefficients can be derived by dividing the DCT coefficients with the respective values in the quantized table, and the results are truncated to the integer values.

A different quantization table can be derived from Table 10.1 by finding a scaling factor that is proportional to the percentage of the retained DCT coefficients. This scaling factor is then multiplied with the quantized values of Table 10.1 to retain a different percentage of DCT coefficients. Depending on the percentage of DCT coefficients retained, the number of zeros in the quantized DCT coefficients will vary.

For example, to retain 1% to 50% of the DCT coefficients, the scaling factor is computed as

$$\text{Scaling factor} = \frac{50}{\text{Percentage of DCT coefficients retained}}. \qquad (10.6.1)$$

To retain 51% to 99% of the DCT coefficients, the scaling factor is computed as

$$\text{Scaling factor} = \frac{100 - \text{Percentage of DCT coefficients retained}}{50}. \qquad (10.6.2)$$

In a simple image compression scheme, we only transmit or store the nonzero DCT coefficients. We will show how to transform, quantize, and inverse transform an image with the Blackfin processor.

Table 10.1 Luminance Quantization Table

16	11	10	16	24	40	51	61
12	12	14	19	26	58	60	55
14	13	16	24	40	57	69	56
14	17	22	29	51	87	80	62
18	22	37	56	68	109	103	77
24	35	55	64	81	104	113	92
49	64	78	87	103	121	120	101
72	92	95	98	112	100	103	99

From [71].

Table 10.2 Switch Settings for Controlling Compression of B&W Image

Percentage of DCT Coefficients Retained	BF533 EZ-KIT	BF537 EZ-KIT
100%	SW4	SW10
80%	SW5	SW11
50%	SW6	SW12
30%	SW7	SW13

HANDS-ON EXPERIMENT 10.3

This experiment implements the 2D DCT and the IDCT with the BF533/BF537 EZ-KIT. In addition, we introduce a quantization process between the DCT and IDCT operations to compress the image to different levels. The quantization table used in this experiment is derived from the JPEG compression standard as shown in Table 10.1. Load a B&W image, disney2.bmp, into the Blackfin memory at address 0x0. Load the project file in directory c:\adsp\chap10\exp10_3_BF533\exp10_3_BF533.dpj for the BF533 EZ-KIT (or c:\adsp\chap10\exp10_3_BF537\exp10_3_BF537.dpj for the BF537) into VisualDSP++. Note the size of the image by clicking on **Configure** ... in the **Image Viewer** and edit the picture height and width in dct.h. Build and run the project.

The settings for the switches on the BF533/BF537 EZ-KITs are shown in Table 10.2 for selecting different compression ratios of the image. Press the switch of the EZ-KIT to select the level of compression for the original image. One of the six LEDs will light up to indicate the completion of the compress routine. Halt the program in order to view the reconstructed image. The reconstructed image is located at address 0x80000. Use the **Image Viewer** in VisualDSP++ to observe the reconstructed image. Instead of loading the image repeatedly in the image viewer, simply click on the refresh icon in the **Image Viewer** window to update the image, because all the reconstructed images are located at the same address. Observe and comment on the quality of the reconstructed image.

This experiment can be extended to the compression of a color image by performing the same compression of the chrominance components with the chrominance quantization table listed in [71].

10.7 TWO-DIMENSIONAL FILTERING

The theory of 1D systems presented in previous chapters can be extended to 2D systems. For example, digital convolution (or FIR filtering) introduced in Chapters 2 and 3 for 1D signal (such as speech) can be expanded to 2D convolution of images with 2D impulse response (or convolution kernel) of filter. 2D filtering can reduce noise and emphasize desired features of given images. In this section, we focus on 2D FIR filters.

10.7.1 2D Filtering

Similar to the 1D FIR filter, a 2D FIR filter has a finite length of impulse responses that are also called filter coefficients. The impulse response of a 2D $L_1 \times L_2$ FIR filter can be represented by the following matrix:

$$h_{l1,l2} = \begin{bmatrix} h_{0,0} & h_{0,1} & \cdots & h_{0,L_2-1} \\ h_{1,0} & h_{1,1} & & \vdots \\ \vdots & & \ddots & \\ h_{L_1-1,0} & \cdots & & h_{L_1-1,L_1-1} \end{bmatrix}. \quad (10.7.1)$$

For example, the moving-average filter introduced in Equation 2.3.2 with impulse response {1/L, 1/L, ..., 1/L} can be extended to the 2D moving-average (mean) filter. A 5×5 mean filter can be expressed as

$$h_{l1,l2} = \frac{1}{25} \begin{bmatrix} 1 & 1 & 1 & 1 & 1 \\ 1 & 1 & 1 & 1 & 1 \\ 1 & 1 & 1 & 1 & 1 \\ 1 & 1 & 1 & 1 & 1 \\ 1 & 1 & 1 & 1 & 1 \end{bmatrix}. \quad (10.7.2)$$

MATLAB provides the function `freqz2` to compute 2D frequency response. The following command

```
[H,Fx,Fy] = freqz2(h,Nx,Ny);
```

returns H, the frequency response of 2D filter h, and the frequency vectors Fx (of length Nx) and Fy (of length Ny). Fx and Fy are normalized frequencies in the range −1.0 to 1.0, where 1.0 corresponds to π radians. When used with no output arguments, `freqz2` produces a mesh plot of the 2D frequency response.

Similar to the 1D FIR filtering defined in Equation 2.3.4, the I/O equation of an $L_1 \times L_2$ FIR filter can be expressed as

$$y(m,n) = \sum_{l1=0}^{L_1-1} \sum_{l2=0}^{L_2-1} h_{l1,l2} x(m-l1, n-l2). \quad (10.7.3)$$

The concept of linear convolution of 2D filter coefficients with the image $x(m,n)$ is similar to a 1D filter shown in Example 2.5. In 2D filtering, the image $x(m,n)$ is rotated by 180° about its center element, and this is equivalent to flipping the image over the vertical axis and then flipping it again over the horizontal axis to obtain $x(-m,-n)$. The filter coefficient matrix $h_{l1,l2}$ is shifted over the rotated image, multiplying the filter coefficients with the overlapped image pixels and summing the products to obtain the filtered output at each point. A problem will occur close to the edges of the image where the filter coefficients overlap with areas outside the image. The simplest solution is to compute the filter output only when the filter coefficients completely overlay with image pixels. However, this results in a smaller size of the output image. *The Image Processing Toolbox* provides functions `imfilter` for performing 2D filtering and `conv2` for 2D convolution.

EXAMPLE 10.6

This example filters the B&W image shown in Figure 10.4 by the 5 × 5 moving-average filter defined in Equation 10.7.2. This 2D filtering example is given in `example10_6.m`. The 5 × 5 mean filter can be constructed in MATLAB with the following command:

```
h = ones(5,5)/25;
```

The frequency response of the 2D filter is computed with the function `freqz2`, which is illustrated in Figure 10.8. This figure clearly shows that the 5 × 5 moving-average filter is a low-pass filter, which will attenuate high-frequency components.

The 2D filtering of input image x with filter h can be implemented as follows:

```
y = imfilter(x,h);
```

and the filtered image y is displayed in Figure 10.9. In most images, high-frequency components are concentrated at the edges (or transition regions) of the image. Therefore, low-pass filtering of images will blur the image by smearing its edges so the boundary will no longer be as sharp as the original one.

In video processing, frame-to-frame averaging produces a similar effect as low-pass filtering in time if the image is stationary. This temporal averaging technique effectively reduces noise without smearing the image.

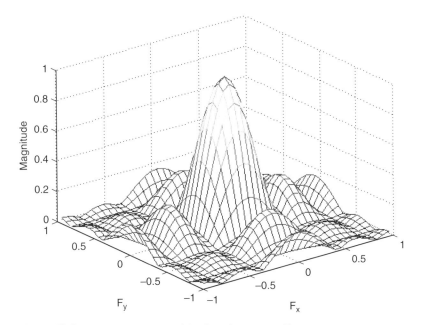

Figure 10.8 Magnitude response of 5 × 5 moving-average filter

Figure 10.9 Image filtered by an 5 × 5 moving-average (low-pass) filter

10.7.2 2D Filter Design

This section briefly introduces the design of 2D digital filters [70], with focus on designing FIR filters. Similar to 1D FIR filter design, the design of FIR filters with the windowing method involves multiplying the ideal impulse response with a window function to generate a corresponding filter, which tapers the ideal impulse response with finite length. The windowing method produces a filter whose frequency response approximates a desired frequency response. The *Image Processing Toolbox* provides two functions, fwind1 and fwind2, for window-based filter design. The function fwind1 designs a 2D filter with two 1D windows, and the function fwind2 designs a 2D filter by a specified 2D window directly. The following example uses fwind1 to create a 2D filter from the desired frequency response Hd:

```
Hd = zeros(11,11); Hd(4:8,4:8) = 1;
```

The hamming function from the *Signal Processing Toolbox* is used to create a 1D window. The fwind1 function then extends it to a 2D window as follows:

```
h = fwind1(Hd,hamming(11));
```

A more effective technique uses filter design methods introduced in Chapter 4 for designing a 1D FIR filter and then transforms it to a 2D FIR filter with the frequency transformation method, which preserves the characteristics of the 1D filter such as the transition width and ripples. The MATLAB *Image Processing Toolbox* provides a function ftrans2 for designing a 2D FIR filter with frequency transformation. For example, the following command

```
h = ftrans2(b);
```

uses the McClellan transformation. The 1D filter b must be an odd-length (type I) filter such as it can be designed by fir1, fir2, or remez in the *Signal Processing Toolbox*.

EXAMPLE 10.7

In this example, we design a 2D high-pass filter. First, we use the remez function to design a 1D filter as follows:

```
b = remez(10,[0 0.4 0.6 1],[0 0 1 1]);
```

This high-pass filter is transformed to a 2D filter, and its magnitude response is shown in Figure 10.10. The MATLAB code is given in example10_7.m.

As shown in Equation 10.7.2, the sum of the filter coefficients usually equals 1 in order to keep the same image intensity. If the sum is larger than 1, the resulting image will be brighter; otherwise, it will be darker. Because 2D filtering requires intensive computation, most commonly used 2D filters are 3×3 kernels summarized as follows [73]:

1. Moving-average filter: $h_{l1,l2} = \dfrac{1}{9}\begin{bmatrix} 1 & 1 & 1 \\ 1 & 1 & 1 \\ 1 & 1 & 1 \end{bmatrix}$

2. Sobel filter: $h_{l1,l2} = \begin{bmatrix} -1 & -2 & -1 \\ 0 & 0 & 0 \\ 1 & 2 & 1 \end{bmatrix}$ and $h_{l1,l2} = \begin{bmatrix} -1 & 0 & 1 \\ -2 & 0 & 2 \\ -1 & 0 & 1 \end{bmatrix}$

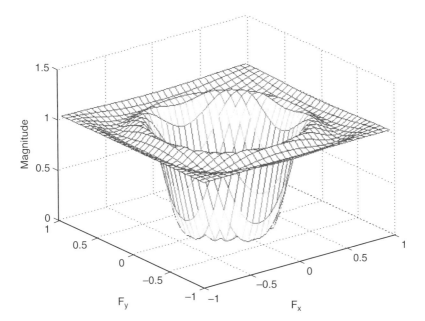

Figure 10.10 Magnitude response of the 2D high-pass filter

3. Laplacian filter: $h_{l1,l2} = \begin{bmatrix} 0 & -1 & 0 \\ -1 & 4 & -1 \\ 0 & -1 & 0 \end{bmatrix}$

Note that the Laplacian filter is a high-pass filter. When this filter moves over an image, the output of the filter is a small value (darker) when the filter overlaps with a region of similar grayscale levels. At the edges of the image, the output is large (brighter).

10.8 TWO-DIMENSIONAL FILTERING WITH BF533/BF537 EZ-KIT

This section performs 2D filtering with the BF533/BF537 EZ-KIT. In particular, we examine the effect of smoothing (low-pass filtering) and sharpening (high-pass filtering). Smoothing filters are commonly used for noise reduction in images. For example, the output of a mean filter is simply the average of the pixels in the neighborhood of the filter mask. If the filter mask is 5×5, the output is the average over the 25 pixels. Therefore, the smoothing filter results in an image that is less sharp. In contrast, sharpening an image requires a high-pass filter. This sharpening filter is used to highlight the fine detail (edges) in an image or enhance detail that has been blurred. The sharpening filter kernel has a high magnitude in its center pixel compared to its neighboring pixels. For example, the Laplacian filter is a high-pass sharpening filter.

HANDS-ON EXPERIMENT 10.4

This experiment first performs the low-pass filtering of a B&W image. Load the image `disney2.bmp` into the Blackfin memory at starting address 0x0. Specify the correct processor type and load the project file in directory `c:\adsp\chap10\exp10_4\exp10_4.dpj` into VisualDSP++. Perform 5×5 low-pass filtering on the image and observe the processed image at starting address 0x130000. From the main C file, it is noted that the image is declared as `unsigned char` (8-bit data type); however, the 2D convolution routine is implemented with the (1.15) data format. Alternately, we can scale the 8-bit data to fit the 16-bit data range of the (1.15) format by multiplying the grayscale data with 0x80. The processed image can be scaled back to [0 to 255] by dividing with 0x80 and checking for overflow or underflow. Change the filter mask to a 5×5 high-pass filter and perform image filtering. Comment on the differences between low-pass and high-pass filtered images.

10.9 IMAGE ENHANCEMENT

Digital images are usually distorted by additive noises, linear and nonlinear distortions. In this section, we introduce several image enhancement techniques to improve the perception of the given image for human viewers. We will show that different methods are required for different noises and distortions.

10.9.1 Gaussian White Noise and Linear Filtering

An image may be corrupted by noise generated by sensing, transmission, or storage process. For example, noise generating from an electronic sensor usually appears as random, uncorrelated, additive errors, which causes extreme pixel-to-pixel changes rather than the small changes normally occurring in a natural picture. A 2D random sequence is called a random field, and an identically distributed Gaussian noise field is called white noise when its mean is zero. The following MATLAB function

```
J = imnoise(I,type, . . .);
```

adds noise of a given type to the image I, where type is a string that can have one of these values:

```
'gaussian'-Gaussian white noise with constant mean and
  variance
'localvar'-Zero-mean Gaussian white noise with an
  intensity-dependent variance
'poisson'-Poisson noise
'salt & pepper'-"On and off" impulselike noise
'speckle'-Multiplicative noise
```

EXAMPLE 10.8

Suppose the B&W image shown in Figure 10.4 is corrupted by Gaussian white noise. In example10_8.m, we use the following command for adding Gaussian white noise of mean m and variance v to the original image:

```
J = imnoise(I,'gaussian',m,v);
```

An image corrupted by the Gaussian white noise with zero mean and variance 0.02 is shown in Figure 10.11.

As discussed in Example 10.5, most images have energy concentrated in the low-frequency range, whereas white noise is uniformly distributed over all frequencies. At low frequencies, the energy of white noise is less than the image, whereas the energy of noise is larger than the energy of the image at high frequencies. Therefore, we may use a low-pass filter for enhancing an image that is corrupted by Gaussian white noise.

EXAMPLE 10.9

In example10_9.m, the noisy image shown in Figure 10.11 is enhanced by the 5 × 5 moving-average filter defined in Equation 10.7.2. The filtered output image is shown in Figure 10.12. The figure clearly shows that the low-pass filter can reduce white noise at high frequencies; however, it also degrades the image because the high-frequency components presented at the edges are also smeared.

Figure 10.11 An image corrupted by Gaussian white noise

Figure 10.12 Enhanced image with a low-pass mean filter

The Wiener filter is the optimal filter for attenuating noise if the spectra of the image and noise can be estimated from the noisy image. Assume the noisy image is expressed as

$$s(m,n) = x(m,n) + v(m,n), \qquad (10.9.1)$$

where $x(m,n)$ is the noise-free image and $v(m,n)$ is the random noise field. We further assume that noise $v(m,n)$ is zero-mean with variance σ_v^2; the frequency response of general 2D Wiener filter can be defined as

$$H(\omega_i, \omega_k) = \frac{S_{xx}(\omega_i, \omega_k)}{S_{xx}(\omega_i, \omega_k) + \sigma_v^2}, \qquad (10.9.2)$$

where $S_{xx}(\omega_i,\omega_k)$ is the power spectrum density of the desired image $x(m,n)$. With the knowledge of $S_{xx}(\omega_i,\omega_k)$ and noise, we can design the optimum filter as described in Equation 10.9.2 for reducing white noise.

EXAMPLE 10.10

The MATLAB function `wiener2` designs and performs Wiener filtering of a noisy image by partitioning the image into several small windows, estimating power spectrum density and noise variance at each window, designing an adaptive Wiener filter based on statistics estimated from the corresponding local neighborhood, and filtering the noisy image as the filter slides across the image. When the local image variance is large, it performs little smoothing; otherwise, it performs more smoothing. The MATLAB code `example10_10.m` implements the Wiener filter to enhance the noisy image shown in Figure 10.11, and the filtered image is shown in Figure 10.13. Compare it with the mean filter output in Figure 10.12. It clearly shows that the Wiener filter performs better than the moving-average for removing Gaussian white noise by preserving edges and other high-frequency components of an image.

10.9.2 Impulse Noise and Median Filtering

In Example 2.8, we show that a nonlinear median filter is effective for impulse-like noises. In this section, we use a 2D median filter for removing impulse noises called "salt & pepper" in images.

Figure 10.13 Enhanced image with Wiener filters adaptive to local statistics

426 Chapter 10 Practical DSP Applications: Digital Image Processing

EXAMPLE 10.11

In example10_11.m, we use the following command to add salt & pepper noise into the B&W image I:

```
J = imnoise(I,'salt & pepper',d);
```

where d is the noise density. The salt & pepper noise consists of random pixels with extreme values close to white (salt, 255) or black (pepper, 0) for an 8-bit image, which is a good mimic of impulse noise. An image corrupted by the salt & pepper noise with density d = 0.05 is shown in Figure 10.14.

In the program, we first use the following command to design a 3×3 mean filter:

```
h = ones(3,3)/9;
```

We then use this linear smoothing filter for enhancing the corrupted image shown in Figure 10.14. The filtered output image is displayed in Figure 10.15, which shows that the "salt" and "pepper" dots reduce their intensity by spreading out to become larger dots.

We introduced 1D median filters in Chapter 2 for effectively reducing impulse noise and compared their performance with a moving-average filter. An N-point 1D median filter ranks the samples in the buffer based on their values and outputs the middle of the sequence in the buffer. Similarly, instead of smoothing a noisy image by averaging the pixels in its neighborhood with a 2D moving-average filter, each output pixel of the 2D median filter is the median of the values in the neighborhood pixels after all numbers have been ranked.

Figure 10.14 An image corrupted by salt & pepper noise with density 0.05

10.9 Image Enhancement   427

Figure 10.15 Reducing salt & pepper noise with a linear filter

Figure 10.16 Performance of median filter for reducing salt & pepper noise

EXAMPLE 10.12

In example10_12.m, we use a 3 × 3 median filter to remove the salt & pepper noise as shown in Figure 10.14. MATLAB provides the function medfilt2 for implementing a 2D median filter as follows:

```
Y = medfilt2(J_SP,[3 3]);
```

where J_SP is the image corrupted by the salt & pepper noise. The output image is shown in Figure 10.16. Compare it with the linear filter output given in Figure 10.15. We show that the 2D median filter has the ability to preserve edges and to remove impulse noises.

Figure 10.17 Corrupted image (a) with salt & pepper noise, (b) after low-pass filtering, and (c) after median filtering

HANDS-ON EXPERIMENT 10.5

This experiment performs low-pass and median filtering with the Blackfin processor to remove the salt & pepper noise shown in Figure 10.14. An image file (e.g., `disney2.bmp`) is first loaded into the Blackfin memory, and the image is corrupted with salt & pepper noise. Set the target processor in the **Project Options** and load the project file `exp10_5.dpj` in directory `c:\adsp\chap10\exp10_5` into VisualDSP++. Build and run the project. Display the corrupted image at starting address `0x130000`. Next, display the images after low-pass filtering and median filtering at starting addresses `0xA30000` and `0xB30000`, respectively. These three displays are shown in Figure 10.17. Compare these to the results obtained with the MATLAB program.

10.9.3 Contrast Adjustment

Adjustment of image intensity is very useful to enhance visual information from images that are too bright (overexposed) or too dark (underexposed). We may change the grayscale values and thus alter the contrast of the image with a linear or nonlinear transformation. Useful information for contrast adjustment is the histogram, which represents the distribution of pixel values. We can compute the histogram of a given image and modify the grayscale values of the image. Basically, the histogram counts the occurrence of gray levels in an image and plots it against the gray level bin. MATLAB provides the function `imhist` for displaying the histogram of an image. The following command

```
imhist(I);
```

10.9 Image Enhancement

displays a histogram for the image I whose number of bins are specified by the image type. If I is an 8-bit grayscale image, `imhist` uses 256 bins as a default value.

EXAMPLE 10.13

As shown in Figure 10.18, the original image of a dining room is underexposed; in particular, the tablecloth is too dark to reveal any detail. We use the MATLAB script `example10_13.m` to display this B&W image and compute its histogram as shown in Figure 10.19. This histogram shows that most pixel values are distributed in the range of 20 to 100, which indicates that the image is too dark, as shown in Figure 10.18.

Histogram equalization enhances the contrast of images by equally distributing the grayscale values so that the histogram of the output image approximately matches a specified histogram. Histogram processing is commonly used in image enhancement and image segmentation. The *Image Processing Toolbox* provides the function `histeq` to enhance contrast with histogram equalization.

Contrast adjustment can be done by mapping the original image to a new contrast range. The MATLAB function

```
J = imadjust(I,[low_in; high_in],[low_out; high_out]);
```

maps the pixel values in image I between `low_in` and `high_in` to new values in J between `low_out` and `high_out`. Values below `low_in` and above `high_in` are clipped, which eliminates extreme values. The principle of this intensity adjustment is illustrated in Figure 10.20. As shown in the figure, pixel values between [`low_in`; `high_in`] are mapped to [`low_out`; `high_out`]. If the slope is less than unity, the image contrast in this range is compressed; if the slope is greater than unity, the image contrast in this range is enhanced.

Figure 10.18 An underexposed dining room picture

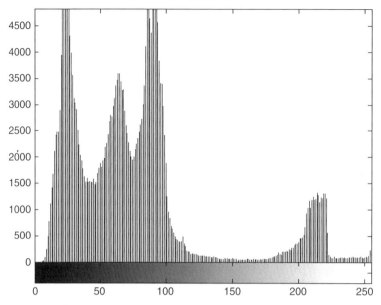

Figure 10.19 Histogram of image shown in Figure 10.18

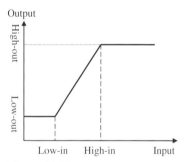

Figure 10.20 Adjustment of image intensity with clipping threshold

EXAMPLE 10.14

As shown in Figure 10.19, the bin index 100 corresponds to the high_in value of 0.4. To make the image brighter, we stretch it to the high_out value of 1. MATLAB code example10_14.m adjusts the intensity of image with the following command:

```
K = imadjust(J,[0 0.4],[0 1]);
```

This command scales pixel values of 0.4 to 1, and this brightens the image. The enhanced image is shown in Figure 10.21, which displays the details of the tablecloth after adjustment of the image intensity. A histogram of the stretched image is shown in Figure 10.22. This equalized histogram shows that the pixel values are distributed more evenly.

10.9 Image Enhancement   **431**

Figure 10.21 A corrected image with intensity adjustment

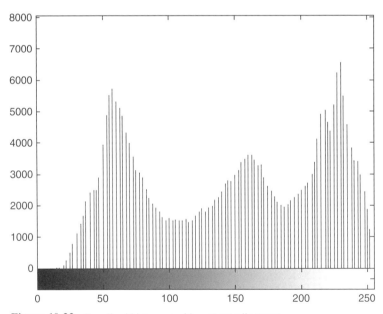

Figure 10.22 Equalized histogram with contrast adjustment

The function `imadjust` also supports a nonlinear gamma correction using the following command:

```
J = imadjust(I,[low_in;  high_in],[low_out;  high_out],
gamma);
```

This function maps the pixel values of I to new values in J using gamma to specify the shape of the curve describing the nonlinear relation between the values in I and J. If gamma is less than 1, the mapping is weighted toward higher (brighter) output values. If gamma is greater than 1, the mapping is weighted toward lower (darker) output values. In the MATLAB code, we also use a gamma value of 0.6 for brighter correction.

10.10 IMAGE ENHANCEMENT WITH BF533/BF537 EZ-KIT

This section explores histogram equalization of a given image with the Blackfin processor. The first step is to compute the histogram of an image. For a B&W image, a 256-element array is set up to record the occurrence of the particular gray level in the image.

The second step is to derive a mapping function. Instead of using the mapping function as shown in Figure 10.20, we compute a mapping function that is based on the normalized sum of the histogram. Here, we set up another array to store the sum of all the histogram values. For example, element #1 of the array contains the sum of histogram elements #0 and #1, and element #2 of the array contains the sum of histogram elements #0 to #2, and so on for the rest of the elements in this array. This array is then normalized with reference to the total number of pixels in the image.

The last step in the histogram equalization is to map the input intensity to a more uniformly spread output intensity with the mapping function derived in the previous step. In the following experiment, we show the steps used to perform histogram equalization on the Blackfin processor.

HANDS-ON EXPERIMENT 10.6

In this experiment, we load a new image `truck.bmp` into the Blackfin memory. Build and run the project file `exp10_6.dpj` in directory `c:\adsp\chap10\exp10_6` with VisualDSP++. A histogram equalization routine `hist_eq.asm` based on the normalized sum of histogram is used in this project to enhance the original image in Figure 10.23(a) and produce a clearer image in Figure 10.23(b). The enhanced image shows better contrast of the image because the terrain features are clearly visible as compared to the original image. The histograms of these images are also plotted in Figure 10.23(c) and 10.23(d) for the original image and equalized image, respectively. These plots show that the histogram of the equalized image has a more uniform distribution of gray levels, and this wider spread has led to higher contrast in the image compared with the original image.

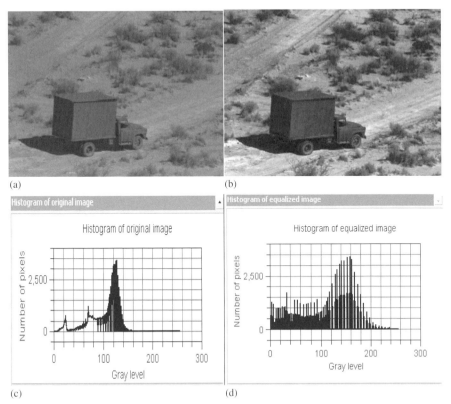

Figure 10.23 (a) Original image, (b) enhanced image with histogram processing, (c) histogram of original image, and (d) histogram of equalized image

10.11 IMAGE PROCESSING WITH LABVIEW EMBEDDED MODULE FOR BLACKFIN PROCESSORS

Images are commonly processed with the same signal processing principles we have introduced in previous chapters. Image processing can be performed in real time, for example, to compress live video to store it to disk, or as a postprocessing step such as manually touching up a digital photo before printing.

The following experiments utilize graphical system design to prototype and implement signal processing algorithms in Windows and then on the Blackfin EZ-KIT. First, we open the application and execute it with LabVIEW targeted to Windows for conceptual analysis and development. The execution target is then changed to the Blackfin processor, allowing the execution of the same code on the EZ-KIT. This ability to program and develop algorithms independent of the intended hardware target is the basis for true graphical system design and code-sharing across platforms.

434 Chapter 10 Practical DSP Applications: Digital Image Processing

HANDS-ON EXPERIMENT 10.7

The extensive libraries of analysis functions in LabVIEW can be applied to both one- and multidimensional data. In this experiment, we explore the effects of basic mathematical operations on a 2D image. The subtraction operation is used to invert the image. Division allows us to reduce the number of quantized values present in the image. Multiplication affects the contrast. We also explore thresholding, which converts the grayscale image to black and white. These types of operations are common when preparing an image for additional processing or to highlight specific details. For example, inverting an image can often make dark or faint features easier to see with the human eye.

Open `Basic Image Processing.vi` located in directory `c:\adsp\chap10\exp10_7`. Be sure that LabVIEW is targeted to Windows. The target can be changed by selecting **Switch Execution Target** from the **Operate** menu. Open the block diagram and examine the graphical code. Note that the original image, represented by an array, has 100 × 96 pixels with values ranging from 0 to 255. The value of the tab control is used to specify which processing operation in the **Case Structure** is executed. Manually change the visible case of the **Case Structure** to view the **Invert** case on the block diagram. Image inversion is calculated with the following equation and implemented in Figure 10.24.

$$\text{Inversed Image} = 255 - \text{Input Image}. \qquad (10.11.1)$$

Polymorphism allows the **Subtract** function to receive inputs of different representation and dimensionality and handle them appropriately. The polymorphic **Subtract** function in LabVIEW allows the original image to be subtracted from the scalar value by simply wiring the two values to its inputs. The result is then displayed on the **Image Result** indicator. The other processing cases contain algorithms that require some additional function blocks. The **Multiply** case, for example, needs additional code to protect against numerical overflow. The **Threshold** case uses the **Greater Than or Equal To** function to evaluate whether the pixel value is above or below the specified threshold. Each of these cases produces a resulting image that is displayed on the same **Image Result** intensity graph.

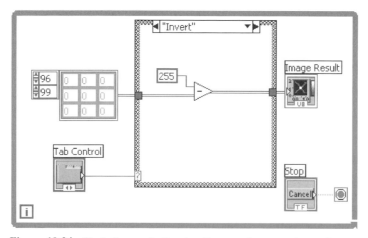

Figure 10.24 Block diagram for `Basic Image Processing.vi`

10.11 Image Processing with LabVIEW Embedded Module for Blackfin Processors 435

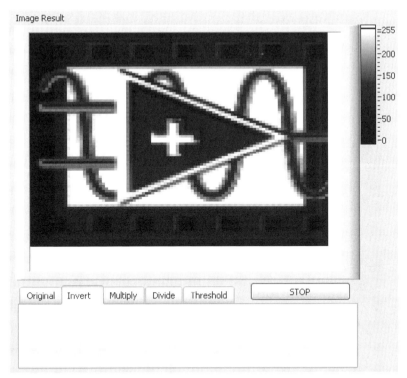

Figure 10.25 Front panel for `Basic Image Processing.vi`

Run the VI from the Front Panel (Fig. 10.25). Move through the tabs one at a time and examine the effects of each processing algorithm. Are the resulting images what you expected?

A common form of image optimization is reducing the number of colors used to represent the image. Using the **Divide** tab, what value for Y produces a four-color image? Does it correspond to the algebraic equation 255 / Y = 4, when you solve for Y?

Change the LabVIEW execution target to the BF53x and open the **Embedded Project Manager** window. Open `Basic Image Processing_BF5xx.lep` located in directory `c:\adsp\chap10\exp10_7`.

As discussed above, the input is an 8-bit grayscale image containing 100 × 96 pixels. However, the default LabVIEW Embedded Module for Blackfin Processors debugging mechanism only reports the first 128 data points for any array. This number must be increased to greater than 9,600 to properly see the results of the processed image. To change the default debugging array size, navigate to **Target → Configure Target → Debugging Options** from the **Embedded Project Manager** window. Increase **Max Array Elements** to 10000. This will allow the entire image to be retrieved during debug execution. Increasing the **Debug update period** to 1000 ms (1 s) will also improve performance by causing fewer interrupts to the processor during execution. Click **OK** and return to the **Embedded Project Manager** window.

Click on the **Debug** execution button to download, run, and link the processor to the LabVIEW front panel. Click through the tabs to verify that the same results are achieved on the Blackfin processor as in Windows. This example demonstrates how LabVIEW abstracts the hardware target from design and thus allows the user to focus on the algorithms and concepts. Click on the **Threshold** tab. Can you calculate how much more data it takes to represent a 100 × 100 pixel, 8-bit, 256-color level image in memory compared to a 1-bit, 2-color level image?

HANDS-ON EXPERIMENT 10.8

This experiment implements a 2D image filter that consists of a 3 × 3 kernel matrix. We utilize the concepts of graphical system design to implement and test the filter in LabVIEW targeted to Windows and then port the design to the Blackfin processor.

Open the 2D Conv Image Processing-BF5xx.lep in directory c:\adsp\chap10\exp10_8. As in Hands-On Experiment 10.7, this VI can be executed with the LabVIEW Embedded when targeted to Windows or to the Blackfin processor. When analyzing the block diagram, we see that the image filter kernels are the same as those designed earlier in Example 10.7.

The low-pass filter implementation is shown in Figure 10.26. **2D Convolution subVI** performs the point-by-point convolution of the image and the 3 × 3 filter kernel, and then divides by the scalar (9 in this case). The scalar is used to reduce the overall multiplied effect of the filter kernel to 1. Open **2D Convolution subVI** to see how the graphical program performs the 2D convolution.

Open the front panel and run the VI. Click on the **Custom** tab shown in Figure 10.27. The Laplacian filter is performed by default. Change the filter to the Sobel filter with the parameters found in Example 10.7. Note that there are two Sobel filter implementations,

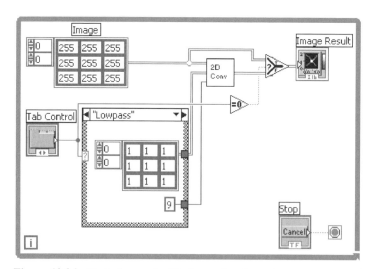

Figure 10.26 Block diagram for 2D Convolution Image Processing.vi

10.11 Image Processing with LabVIEW Embedded Module for Blackfin Processors

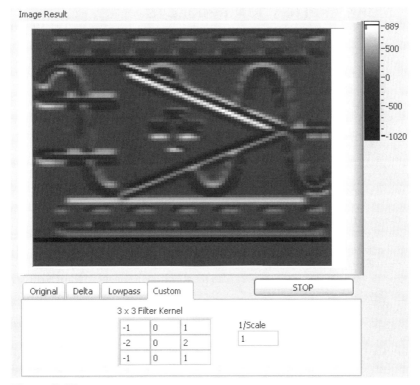

Figure 10.27 Front panel for 2D Convolution Image Processing.vi

which are not symmetric about both axes in contrast to the delta and low-pass filters. How do the resulting images differ when the orientation of the Sobel filter is changed?

To run the project on the Blackfin processor, change the execution target from Windows to the Blackfin EZ-KIT. Open the 2D Conv Image Processing-BF5xx.lep file in directory c:\adsp\chap10\exp10_8 from the **Embedded Project Manager** window.

As discussed above, the input is an 8-bit grayscale image containing 100 × 96 pixels. Again, the default LabVIEW Embedded Module for Blackfin Processors debugging parameters must be changed properly to see the results of the processed image. In the project **Debugging Options**, increase **Max Array Elements** to 10000 and **Debug update period** to 1000 ms.

Run the VI on the Blackfin by clicking on the **Debug** execution button. As in Hands-On Experiment 10.7, be sure that the algorithms on the Blackfin processor behaves the same as they did in LabVIEW targeted to Windows.

Explore the values being passed through wires on the block diagram with probes. The **Custom** tab is useful for testing common 3 × 3 image filter implementations as well as your own designs. What 3 × 3 filter values allow you to invert the image as shown in Figure 10.25?

10.12 MORE APPLICATION PROJECTS

1. In this chapter, all the images used in examples and experiments are transferred from the files in the computer to the SDRAM in the Blackfin processor. The BF533 EZ-KIT comes with a video connector that allows users to transfer images from the image/video capturing devices such as digital cameras. A sample program from Analog Devices is provided in directory c:\Program Files\Analog Devices\VisualDSP 4.0\Blackfin\ EZ-KITs\ADSP-BF533\Examples\Video Input to capture the image through the parallel port interface of the Blackfin processor. A peripheral DMA is set up as stop mode, which stops capturing after 50,000 transfers to move the image from the PPI to the SDRAM, and the PPI port is configured to ITU-656 input mode. Open the project file BF533_EZ_KIT_Video_Receive_C.dpj in VisualDSP++. Before building the project, make sure that switch #6 of SW3 in the BF533 EZ-KIT is set to **ON**. Build the project. Connect the video NTSC/PAL composite video blanking sync (CVBS) signal to the video input jack, which is located at the bottom right of the video in/out connector (AVIN1) on the BF533 EZ-KIT. The CVBS combines the color, luminance, and synchronization information into one signal. Click on the **Run** icon to start the capture of the image from the imaging device. Click on the **Halt** icon to stop the processor. Open the image in the image viewer with the following parameters: **Start address** = 0x1; **Stride** = 2; **Horizontal pixels** = 720; **Vertical Pixels** = 525; **Pixel format** = Gray Scale (8-bit). Click **OK** to display the captured image from the camera. Capture images under different lighting conditions and perform contrast adjustment to these images.

2. The PPI of the Blackfin processor can also be used to display an image from the SDRAM to an external display monitor. In this project, we look into how to display a color bar pattern stored in the SDRAM memory to the PPI, which in turn connects to the monitor. Configure the peripheral DMA as 2D DMA, autobuffer, and 16-bit transmit DMA; perform the data transfer from the SDRAM to the PPI. Load the project video_out. dpj, which is located in directory c:\Program Files\Analog Devices\VisualDSP 4.0\Blackfin\EZ-KITs\ADSP-BF533\Examples\Video Output. Make sure that switch #6 of SW3 on the BF533 EZ-KIT is set to **ON**. Connect a video display monitor to the video output jack (found on the top row of the video in/out connect) on the BF533 EZ-KIT [29]. Run the project and see the color bar pattern displayed on the video monitor. Overwrite the SDRAM with another image or sequence of images and display them.

3. The gamma correction is commonly used in compensating the nonlinearity of display devices. To display the image in linear output, the gamma correction prewraps the RGB color image with a gamma value. For example, a gamma value of 2.2 is used in most computer monitors. For an 8-bit RGB sample, a 256-value table is used to map the image. The equations for gamma correction are given as follows:

$$R_g = gR^{1/\gamma}, \quad G_g = gG^{1/\gamma}, \quad B_g = gB^{1/\gamma}, \qquad (10.12.1)$$

where g is the conversion gain factor, γ is the gamma value, and RGB are the color components. Implement the gamma correction on any RGB image and display the image on the computer monitor by modifying the program in Project 2.

4. Modify the code in Hands-On Experiment 10.3 to compress a captured image from a digital still camera in the BF533 processor. The compression ratio can be selected from the switches in the BF533 EZ-KIT as indicated in Table 10.2. Use the image viewer in

VisualDSP++ to view the reconstructed image. In addition, perform low-pass filtering on the reconstructed image to smooth out any blocking edges.

5. To estimate the quality of a reconstructed image compared with the original image, a peak signal-to-noise ratio (PSNR) measurement is commonly used in image compression. PSNR is computed from the mean square error (MSE) of the reconstructed image as follows

$$\text{MSE} = \frac{1}{NM} \sum_{i=0}^{N-1} \sum_{j=0}^{M-1} [x(i,j) - R(i,j)]^2, \qquad (10.12.2)$$

$$\text{PSNR} = 20\log_{10}\left(\frac{255}{\sqrt{\text{MSE}}}\right), \qquad (10.12.3)$$

where $x(i,j)$ is the original $N \times M$ image and $R(i,j)$ is the reconstructed image obtained by decoding the encoded version of $x(i,j)$. Note that the error metrics are computed on the luminance signal only and the peak image level is white (255). A detailed description of PSNR and its application in image coding can be found in [68]. Compute and display the PSNR for the reconstructed images in Hands-On Experiment 10.3 using the EZ-KIT.

6. Extend Project 5 by displaying the pixel-by-pixel error image between the original image and the reconstructed image. Use the image viewer of VisualDSP++ to display the error image. Can you observe any difference as the compression ratio increases? Can you devise a different error computation to increase the visible difference?

7. A color image is converted from RGB color space to YCbCr space as shown in Hands-On Experiment 10.2. A 2D low-pass filter is then applied only to the luminance (Y) in the YCbCr color space. After filtering, convert the YCbCr back to RGB color space and display in the image viewer of VisualDSP++.

8. A set of noisy image files are located in directory c:\adsp\image_files\noisy_image\. Based on the image processing techniques learned in this chapter, devise a suitable technique to clean up each of these image files. In some cases, advanced techniques that are not described in this book must also be explored. Users can refer to the textbooks listed in the References or search the website for solutions. Implement the algorithm with the BF533/BF537 EZ-KIT and test its performance. Compare the performance with MATLAB simulations.

Appendix A

An Introduction to Graphical Programming with LabVIEW

Contributed by National Instruments

This appendix provides an overview of National Instruments LabVIEW and the LabVIEW Embedded Module for Blackfin Processors software.

A.1 WHAT IS LABVIEW?

LabVIEW is a full-featured graphical programming language and development environment for embedded system design. LabVIEW provides a single graphical design tool for algorithm development, embedded system design, prototyping, and interfacing with real-world hardware. Additional modules have been designed to expand the core functionalities of LabVIEW to real-time operating systems, DSP, and FPGA programming, making LabVIEW an ideal platform for signal processing algorithm design and implementation.

This section briefly introduces the LabVIEW development environment from installation and basic programming to system development. This material provides a high-level overview of the concepts necessary to be successful with LabVIEW and the LabVIEW Embedded Module for Blackfin Processors. Additional references to supplemental manuals and resources are provided for more in-depth information.

A.1.1 A Picture Is Worth a Thousand Lines of Code

Graphical system design and prototyping opens the door to the masses in embedded development. The industry is confirming that higher levels of abstraction in

---

Embedded Signal Processing with the Micro Signal Architecture. By Woon-Seng Gan and Sen M. Kuo
Copyright © 2007 John Wiley & Sons, Inc.

embedded tools are needed, suggesting electronic system-level (ESL) design is an answer and citing concepts such as hardware/software co-design, predictive analysis, transaction-level modeling, and others. Does this sound like simplifying the problem or just making it more complex?

Fundamentally, the problem is that the domain experts—scientists, researchers, and engineers developing new algorithms—are at the mercy of the relatively few embedded system developers who are experienced in dealing with today's complex tools. The industry requires a fundamental change to empower the thousands of domain experts to experiment and prototype algorithms for embedded systems to prove success early before they are overtaken with the complexity of implementation.

The test and measurement industry saw this same phenomenon. Using a graphical modeling and programming approach, engineers can more quickly represent algorithms and solve their problems through block diagrams. This is especially true for embedded systems development. With a graphical programming language, engineers succinctly embody the code that is actually running on the hardware. For example, they can represent parallel executing loops simply by placing two graphical loop structures next to each other on the diagram. How many lines of code would it take to represent that in today's tools? How many domain experts could look at text code and quickly interpret exactly what is happening? Only a truly innovative user experience—like Windows for PCs or the spreadsheet for financial analysis—can truly empower the masses to move into a new area.

Domain experts also need a simple and integrated platform for testing algorithms with real-world I/O. The reason these embedded systems fail so often is because there are so many unknowns during development that are not discovered until the end of the development cycle.

After seeing this shift in the test and measurement industry from traditional text-based programming approaches like BASIC and C to LabVIEW, it seems more obvious than ever. Graphical system design is essential to open embedded systems to more people.

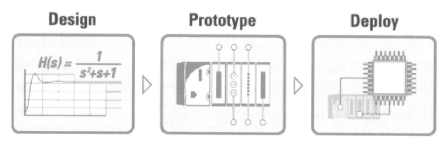

Figure A.1 Graphical system design in LabVIEW

A.1.2 Getting Started with LabVIEW

The LabVIEW development system is the core software with which other modules, toolkits, and drivers can be installed to expand functionality and capability. For the purposes of this book, the LabVIEW Embedded Module for Blackfin Processors should also be installed, adding the ability to generate Blackfin-compatible C code that can be targeted to the Blackfin EZ-KIT using the Analog Devices VisualDSP++ tool chain under the hood.

A.1.3 Install Software

The installation order is as follows:

1. Install VisualDSP++ 4.0 for LabVIEW
2. Install the LabVIEW Embedded 7.1
3. Install the LabVIEW Embedded Module for Blackfin Processors 1.0

A.1.4 Activating Software

The software will default to a 60 day evaluation version. If you have purchased a copy of the software and have received an activation code, follow these steps to activate both LabVIEW Embedded Module for Blackfin Processors and Visual DSP++ 4.0 at once.

1. Navigate to the NI License Manager. Start by selecting **Start → Programs → National Instruments → NI License Manager**. The NI License Manager will open, showing all NI software installed on the computer that can be activated.
2. Expand the LabVIEW category to expose the LabVIEW Embedded Module.
3. Right click on the **Embedded Module** and select **Activate**. Follow the steps provided in the **NI Activation Wizard** to activate the software.

A.1.5 Connecting Hardware

After the appropriate software has been installed, attach the power connector to the Blackfin EZ-KIT and then connect the EZ-KIT to the computer using the USB cable. Windows will detect the EZ-KIT and install the appropriate driver.

A.1.6 Running an Example Program

Open and run the first example program by opening **LabVIEW Embedded Edition** and changing the execution target on the introduction screen from **LabVIEW** to the appropriate Blackfin EZ-KIT, labeled **Analog Devices ADSP-BF5xx, VDK Module**.

Open the **Embedded Project Manager** by selecting **Tools → Embedded Project Manager → File → Open Project**. All examples can be found in the c:\Program Files\National Instruments\LabVIEW 7.1 Embedded\examples\ lvemb\Blackfin directory. In the **Embedded Project Manager**, select **File → Open Project** and navigate to the fundamentals directory to open the Pushbutton - polling (BF5xx).lep project. Click on the **Execute** button to compile, link, and download the application to the Blackfin EZ-KIT. Refer to the **Introduction to LabVIEW Embedded Edition 7.1** and the **LabVIEW Embedded Module for Blackfin Processors** sections of this appendix for more information.

A.2 OVERVIEW OF LABVIEW

This section introduces the key components of the LabVIEW environment and provides step-by-step experiments in developing LabVIEW programs and graphical user interfaces.

A.2.1 The LabVIEW Environment

The LabVIEW development environment allows you to develop graphical programs and graphical user interfaces easily and effectively. A program in LabVIEW is called a Virtual Instrument (VI), and acts as an individual function similar to a function defined in the C programming language. The main executable VI is called the top-level VI and VIs used as modular subroutines are called subVIs. Every VI is made of two key components, the *front panel* and *block diagram*, shown in Figure A.2. The front panel is the graphical user interface and the block diagram contains the graphical code that implements the functionality of the VI. Once you create and save a VI, it can be used as a subVI in the block diagram of another VI, much like a subroutine in a text based programming language.

A.2.2 Front Panel

The front panel contains objects known as controls and indicators. Controls are the inputs that provide data to the block diagram as will be discussed in Section A.2.3, while output data from the block diagram is viewed and reported on front panel indicators. In Figure A.3(a), two numeric array controls, **Vector A** and **Vector B**, allow the user to input two numeric vectors. The result of the dot product is displayed as a single scalar value in the **Dot Product Result** numeric indicator. There is also a Boolean control button that will stop the VI.

The **Controls** palette in Figure A.3 (b) contains all the front panel objects that you can use to build or modify the user interface. To open the **Controls** palette, right click on free space within the front panel. The **Controls** palette can remain visible while you work by clicking on the pushpin icon in the upper left corner.

A.2 Overview of LabVIEW 445

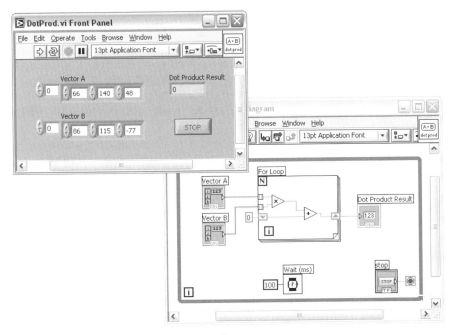

Figure A.2 LabVIEW front panel and block diagram

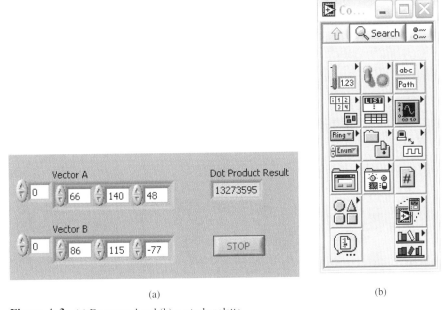

Figure A.3 (a) Front panel and (b) controls palette

A.2.3 Block Diagram

The block diagram contains the graphical code that defines how the VI will operate. Front panel controls and indicators have a corresponding terminal block on the block diagram. The corresponding terminals for array controls **Vector A**, **Vector B** and the **Boolean stop** button provide input to the block diagram, while the Dot Product Result numeric indicator terminal allows the result of the algorithm to be displayed.

Wires define how data will flow between terminals and other nodes on the block diagram. VIs execute based on the concept of data flow, meaning that a node, subVI or structure will not execute until data is available at all its inputs. Data wire color and thickness are used to show data type (floating point number, integer number, Boolean, string, etc) and dimensionality (scalar or array). For example, in Figure A.4 (a), the input vectors both enter the left side of the **For Loop** with a thick orange wire representing double precision numeric 1D *arrays*. The wire exiting the **For Loop** is a thin orange wire representing a double precision numeric *scalar*.

When creating or modifying a block diagram, the **Functions** palette provides access to all functions, structures, and subVIs available in LabVIEW. Right-click on empty space on the block diagram to see the **Functions** palette, shown in Figure A.4 (b). Again, the pushpin can be used to keep the palette from disappearing.

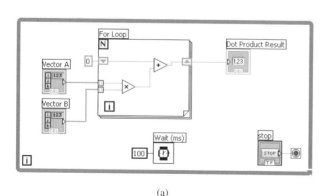

(a)   (b)

Figure A.4 **Functions** palette for block diagram objects

A.2.4 Debugging in LabVIEW (Windows)

Many debugging features in LabVIEW make it ideal for designing and testing algorithms before moving to a hardware target like the Blackfin EZ-KIT. LabVIEW continuously compiles code while it is being developed, allowing errors to be detected and displayed before the application is executed. This method of error checking provides immediate feedback by ensuring that subVIs and structures have been wired properly during development. Programming errors are shown in the form of broken or dashed wires or a broken **Run** arrow toolbar icon. When the **Run** arrow toolbar icon appears broken and grey in color, click it to see the list of errors preventing the VI from executing, shown in Figure A.5. Use the **Show Error** button or double-click on the error description to highlight the problematic code on the block diagram or the object on the front panel that is causing the error.

There are additional debugging features unique to LabVIEW that help you understand the operation of a VI during runtime. The block diagram debugging features include the use of breakpoints to pause code during execution, adding probes to see the data value present on wires, and highlight execution mode, a debugging feature that slows execution and shows the actual flow of data along wires and through structures. These tools assist in understanding and confirming program logic and program flow.

A.2.5 Help

Several help options exist for programming and function reference. These features can be accessed through the **Help** menu on both the front panel and block diagram tool bar. Context Help can be activated by selecting **Help → Show Context Help**. Context Help displays a floating help window that changes based on the current block diagram or front panel element under the cursor. Launch the comprehensive *LabVIEW Help* by selecting **Help→VI, Function, & How-To Help**. View the

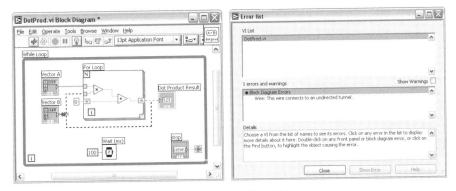

Figure A.5 Debugging broken wires and using the Error List

extensive library of LabVIEW manuals by selecting **Help→Search the LabVIEW Bookshelf**.

The National Instruments support website, www.ni.com, is also a great resource for finding answers to questions. Thousands of KnowledgeBase entries and searchable forums on the NI web site offer interactive up-to-date help resources.

HANDS-ON EXPERIMENT A.1

In this experiment, we will create the VI shown in Figure A.2 that performs the dot product of two input vectors. This simple VI will introduce the concepts of dataflow programming in LabVIEW and graphical constructs such as arrays, looping structures, and shift registers. Start by opening LabVIEW and creating a new VI. Notice that the front panel user interface has a gray workspace and when the block diagram is opened, it has a white workspace. We will begin by first creating the user interface on the front panel and then coding the dot product algorithm on the block diagram.

Begin on the front panel by right-clicking on the free space to display the **Controls** palette. Click on the **Pushpin** to give the palette its own window. First, we will create **Vector A** by placing an **Array** shell on the front panel. You will quickly become comfortable with the location of controls and indicators in their respective palettes. For now, take advantage of the palette search feature to find functions and controls if you are not sure where they are located. Find the **Array** shell by opening the **Controls** palette and clicking on **Search,** located at the top of the palette. Type "array" in the search field and note that as you type, the list box populates with your search results. Locate the **Array** control in the list and double-click on the entry. The window will change to the appropriate subpalette and the **Array** shell will be highlighted. Click on the **Array** shell icon and drag your cursor where you want to place it on the front panel. Arrays are created empty, that is, without an associated data type. Define the array data type by dropping a numeric, Boolean, or string control within the array shell. From the **Numeric** subpalette on the **Controls** palette, select a numeric control and drop it into the **Array** shell, as shown in Figure A.6. The color indicates the data type. In this case, orange indicates that the array holds double precision numbers. Double-click on the **Array** label to change its text to **Vector A**. The array can be expanded vertically or horizontally to show multiple elements at once. Expand **Vector A** horizontally to show three elements.

Create **Vector B** by creating a copy of **Vector A**. You can click on the edge **Vector A** to select it and then select **Edit → Copy** and then **Edit → Paste** to create a copy. Place the copy below the original (see Fig. A.7). Change the label for this control to **Vector B**.

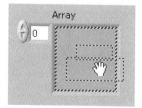

Figure A.6 Creating an Array control

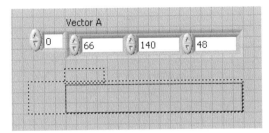

Figure A.7 Copying objects holding CTRL while dragging the object

Now that the controls, or inputs, to the VI have been created, we need an indicator to show the output of the dot product operation. On the **Controls** palette, select the **Numeric Indicator**, available by selecting **Controls → Numeric → Numeric Indicator**, and place it on the front panel. Name this indicator **Dot Product Result**. Finally, place a **Stop** button on the front panel so that the user can terminate the application. This control is found on the **Controls → Boolean** subpalette.

After the front panel graphical user interface has been completed, we turn our attention to completing the dot product functionality on the block diagram. If the block diagram is not already open, select **Window → Show Block Diagram**. Block diagram code executes based on the principle of dataflow. The rule of dataflow states that functions execute only when all of their required inputs are populated with data. This programming architecture allows both sequential and parallel execution to be easily defined graphically. When the block diagram for the dot product is complete, as shown in Figure A.4(a), data passes from the inputs on the left to the outputs on the right and follows the dataflow imposed by the wires connecting the block diagram objects. The next step is to create the block diagram by first placing the correct structures and functions and then connecting them together with the appropriate wiring.

Place a **For Loop** available on the **Structures** palette in the center of the block diagram. It may be necessary to move the controls and indicators out of the center to do so. A **For Loop** in LabVIEW behaves like a for loop in most text based programming environments, iterating an integer number of times as specified by the input terminal N, $\boxed{N}$.

After the **For Loop** has been placed and sized, the next step is to place both the **Multiply** $\boxed{\times}$ and the **Add** $\boxed{+}$ functions within the loop. Move **Vector A** and **Vector B** to the left of the loop and begin wiring **Vector A** through the **For Loop** and into the top input of the **Multiply** function. The mouse cursor automatically changes to the wiring tool when it is placed above the input or output of an icon or another wire. Use the wiring tool to connect inputs and outputs by clicking and drawing wires between them. Now, wire **Vector B** to the bottom input of the **Multiply** function, and the output of the Multiply function to the bottom input of the **Add** function. The product of each pair of multiplied values will be cumulatively added by means of a shift register every time the loop iterates. Now create a shift register on the left side of the **For Loop** by right-clicking and selecting **Add Shift Register**. Shift registers will be added to each side of the loop automatically. A shift register allows values from previous loop iterations to be used in the next iteration of the loop. Wire the output of the **add** operation to the **right shift register**. The shift register will then change to the color of the wire. Now right-click on the **left shift register** and create a constant, wire the **left shift register** to the top of the **add**, and wire the **right shift register** to the **Dot Product Result**. Notice in Figure A.8, the shift register input is initialized with a value of zero, and during each iteration of the **For Loop**, the product of the elements from **Vector A** and **Vector B** is summed and stored in the shift register . This stored value will appear at the shift register input during the next iteration of the loop. When the **For Loop** has iterated through all of the vector elements, the final result will be passed from the right shift register to the **Dot Product Result**. Can you now see how the graphical code within this **For Loop** is performing the dot product of the two input arrays?

Finish the program by adding a **While Loop**, available on the **Functions → Structures** subpalette that encloses all of the existing code on the block diagram. This structure will allow the code inside to run continuously until the stop condition is met. For our VI, wire the **stop** button to the stop condition so that it can be used to stop the loop from the front panel thus, ending execution of the entire VI. Finally, we must add functionality in the **While Loop** to keep it from running at the maximum speed of the computer's processor. Find the **Wait Until Next ms Multiple** function and wire in 100 ms of delay. This will suspend execution of the VI for 100 ms so that the processor can handle other operations and programs. Your VI should look similar to the completed block diagram shown in Figure A.8.

If you have successfully completed the Dot Product VI, the **Run** arrow on the toolbar should appear solid and white, indicating that there are no errors in the code that would prevent the program from compiling. A broken **Run** arrow means that there is an error that you can investigate and fix by clicking on this error to report what is causing the problem. Click on the **Run** arrow to run the VI and experiment by placing different values into **Vector A** and **Vector B**. Compute the dot product of two small vectors by hand and compare the results with VI you created. Do they match?

Click on the **stop** button to halt the VI execution. Click on **View → Block Diagram** to switch to the block diagram. Enable the highlight execution feature by clicking on the toolbar icon with the light bulb icon . Now press **run** again. What do you notice? How can this feature help you during development? Right-click on a wire and select **Probe**. You can now see a small dialog box that displays the data value of that wire as the VI executes.

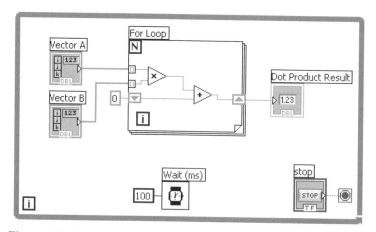

Figure A.8 Block diagram of the **vector dot product VI**

Experiment with the other debugging features of the LabVIEW environment and save the program when you are finished.

HANDS-ON EXPERIMENT A.2

In this experiment, you will create the VI as shown in Figure A.9. This VI generates a one-second sinewave of a given frequency, displays its time-domain plot, and outputs the signal to the sound card.

Open a new blank VI. Start by placing the `Sine Waveform.vi` on the block diagram. Right-click on the **frequency** input of this function and select **Create → Control**. This automatically creates a numeric control of double representation with the name **frequency**. Switch to the front panel to see the actual control. Do the same for the **amplitude** input. Finally, right-click on the **Sampling Info** input and select **Create → Constant**. This input happens to be a cluster data type that allows you to specify the sampling frequency and the number of samples for the generated sinewave. Clusters are a type of data unique to LabVIEW but are similar to structures in text based programming. They allow you to bundle together different data elements that are related in some way. Clusters can consist of the same or different data types. Search the **LabVIEW Help** for more information on clusters. Change both values within the cluster to 44,100. The first is **Fs**, the sampling rate, and the second is **#s**, the number of samples to generate. Setting both values to the same number generates 1 second of 44,100 sample/second data as shown in Figure A.9.

Now that we have all of the necessary inputs, navigate to the front panel and place a **Waveform Graph** on the front panel. Navigate back to the block diagram and wire the **signal out** terminal of the Sine Waveform VI to the **Waveform Graph**. The VI should now run. Test it once. Zoom in on the graph to see a smaller piece of the data. Consider clicking and changing the right-most value on the x axis to 0.005.

In order to play the sinewave on the sound card, we must use the `Snd Write Waveform.vi`. Right-click on the **sound format** input and select **Create → Constant**. The default values will need to be changed to mono, 44100, and 16-bit. Then, wire the output signal to the **mono 16-bit** input on the top right of the Snd Write Waveform VI. Finish the VI by

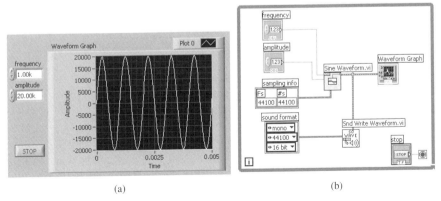

(a)   (b)

Figure A.9 Screen shots of the (a) signal generator and (b) its block diagram

enclosing the block diagram code with a **While Loop** so that it will run until the stop button is pressed. Right-click the termination terminal of the loop and select **Create → Control** so that a **stop** button appears on the front panel. Arrange the front panel objects so that they look like those shown in Figure A.9.

As in the previous experiment, fix any errors that might be causing the **Run** arrow to be broken. Plug speakers or headphones into audio output of the computer so that you can listen to the generated sinewave signal. Click on the **Run** button to start the VI. Experiment by changing values of the input controls on the front panel. Does the VI behave as expected?

A.3 INTRODUCTION TO THE LABVIEW EMBEDDED MODULE FOR BLACKFIN PROCESSORS

This section examines the state-of-the-art software that allows algorithms to be developed in the LabVIEW graphical environment and downloaded onto the Blackfin EZ-KITs.

A.3.1 What Is the LabVIEW Embedded Edition 7.1?

The LabVIEW Embedded Edition 7.1 adds the ability to cross-compile a LabVIEW Embedded project for a specific embedded target. This is performed by integrating LabVIEW Embedded and the tool chain specific to the embedded target. First, LabVIEW analyzes the graphical block diagram code and instead of compiling it directly for Windows, generates an ANSI C code representation for the top-level VI and each subVI in the VI hierarchy. This task is performed by the **LabVIEW C Code Generator**, which converts LabVIEW data wires and nodes into variables and structures in their C code equivalent. The LabVIEW C Runtime Library is used when functions cannot be simplified to basic C constructs. The **Inline C Node** generates C code identical to that entered on the LabVIEW block diagram. LabVIEW Embedded Modules, such as the LabVIEW Embedded Module for Blackfin Processors, integrate the processor specific tool chain allowing the generated C code to be compiled and run on the Analog Devices Blackfin processor.

A.3.2 What Is the LabVIEW Embedded Module for Blackfin Processors?

The LabVIEW Embedded Module for Blackfin Processors allows programs written in the LabVIEW to be executed and debugged on Blackfin BF533 and BF537 processors. The software was jointly developed by Analog Devices and National Instruments, to take advantage of the NI LabVIEW Embedded technology and the convergence of capabilities of the Blackfin EZ-KIT, for control and signal processing applications. With the LabVIEW Embedded Module for Blackfin Processors,

A.3 Introduction to the LabVIEW Embedded Module for Blackfin Processors

you can design and test algorithms in Windows and then compile and run them on the Blackfin Processor. Advanced debugging features such as viewing of live front panel updates via JTAG, serial, or TCP/IP, as well as single-stepping through code with VisualDSP++ allow complex embedded systems to be rapidly designed, prototyped, and tested.

A.3.3 Targeting the Blackfin Processor

This section describes the process for changing the execution target and using the LabVIEW Embedded Project Manager. Begin by launching LabVIEW 7.1 Embedded by navigating to **Start → All Programs → National Instruments LabVIEW 7.1 Embedded Edition**. Once open, you will see the menu shown in Figure A.10. Change the **Execution Target** option to the appropriate Blackfin processor.

Use the **New** button to create a new blank LabVIEW VI for the selected target; however, when targeting embedded devices such as the BF533 or BF537, the **Embedded Project Manager** should be opened first. The **Embedded Project Manager** is used to manage your LabVIEW application. Open the **Embedded Project Manager** by selecting **Tools → Embedded Project Manager**. After the project manager window opens, select **File → Open Project** and navigate to the project file DotProd - BF5xx.lep located in the directory c:\adsp\chap1\exp1_4. Note that LabVIEW embedded project files have ".lep" as the file extension. You should see the embedded project as shown in Figure A.11.

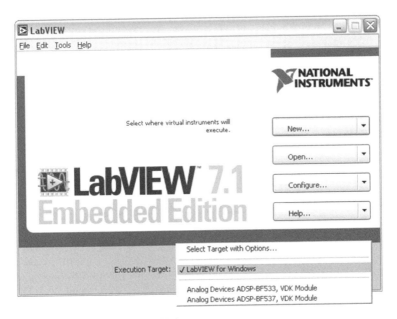

Figure A.10 LabVIEW Embedded startup screen

454 Appendix A An Introduction to Graphical Programming with LabVIEW

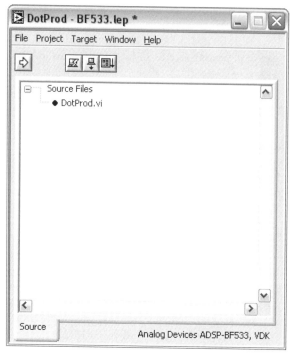

Figure A.11 LabVIEW Embedded project manager

The Embedded Project Manager provides a way to manage the target settings and all of the files included in the application. First, check the **Target** settings to be sure the **Blackfin Target** is set up appropriately. The two most important target settings are **Target → Build Options** and **Target → Configure Target**. Settings changed in the **Configure Target** dialog box are unique to the development machine, while **Build Options** are unique to the individual LabVIEW Embedded project file. Any time you switch between BF533 and BF537 targets, you will need to verify target configuration.

A.3.4 Build Options

The **Build Options** dialog box, shown in Figure A.12, is separated into three configurations: **General**, **Advanced**, and **Processor**. The **General** tab allows the **Debugging** mode to be configured and provides some options for code generation optimizations. The **Advanced** tab provides higher level debugging and optimization options to be set, as well as compiler and linker flag options for more experienced users. The **Processor** tab contains additional processor specific options for advanced users, including one of the most critical options for proper code compilation, the **Silicon revision**. The silicon revison chosen in the project and that of your Blackfin

A.3 Introduction to the LabVIEW Embedded Module for Blackfin Processors 455

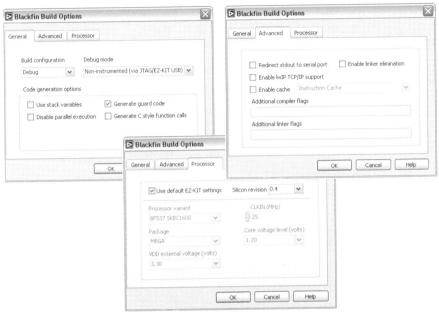

Figure A.12 LabVIEW Embedded target build options

processor must match. The processor model and revision can be found on the silk-screen of the Blackfin chip at the center of the EZ-KIT circuit board.

A.3.5 Target Configuration

The **Target Configuration** dialog box is separated into two configuration tabs, **Target Settings** and **Debug Options,** as shown in Figure A.13. The **Target Settings** must be verified each time the target is changed from one processor type to another. The **Debug Options** pane provides configuration settings for update rate, communications port, and advanced debug options. Remember, these settings are unique to the development machine, so they are not saved in the project file and not changed when a new project file is loaded.

A.3.6 Running an Embedded Application

LabVIEW embedded applications can be executed by clicking on the **Run** button in the **Embedded Project Manager**. Clicking the **Run** button on an individual VI will bring the **Project Manager** window to the front or ask you to create a project file if there is no project associated with the VI.

LabVIEW Embedded applications are first translated from graphical LabVIEW to C code, compiled, and then downloaded to the embedded controller for execution.

456 Appendix A   An Introduction to Graphical Programming with LabVIEW

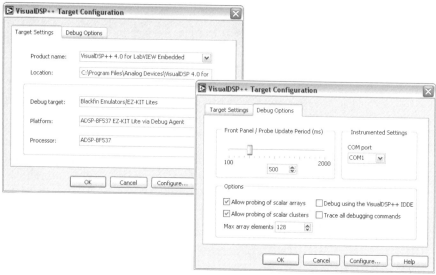

Figure A.13 LabVIEW Embedded target configuration options

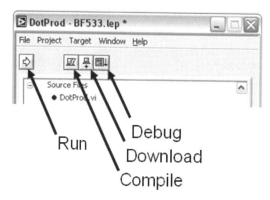

Figure A.14 LabVIEW Embedded target execution options

Using the buttons in the **Embedded Project Manager** window, you can specify which steps in the embedded tool chain to complete. Refer to Figure A.14 for a description of the **Compile** and **Run** button options. Each option will complete all of the necessary tasks before it in the compilation chain. For example, the **Compile** button will generate C and compile the files in the project, while the **Debug** button will generate C, compile, download, run, and begin debugging.

For additional reference material and more specific help using LabVIEW Embedded for Blackfin Processors, refer to the *Getting Started with the LabVIEW Embedded Module for Analog Devices Blackfin Processors* manual [52] and the *LabVIEW Embedded Module for Analog Devices Blackfin Processors Release*

Notes [53]. These manuals are also installed in the National Instruments folder on your computer and can be found at `\National Instruments\LabVIEW 7.1 Embedded\manuals\Blackfin_Getting_Started.pdf` and `National Instruments\manuals\Blackfin_Release_Notes.pdf`. The manuals can also be downloaded from `http://ni.com/manuals`.

A.3.7 Debugging in the LabVIEW Embedded Module for Blackfin Processors

Different debugging options are available when using the LabVIEW Embedded Module for Blackfin Processors to develop a Blackfin application. Debugging modes available in the module include the standard non-instrumented debugging that employs the JTAG/EZ-KIT USB port, as well as a unique instrumented debugging mode that uses the serial or TCP port. The debugging method can be selected using the **Target → Build Options → Debug Mode** menu shown previously in Figure A.12. Debugging provides a window into the processes and programs running on the DSP during execution. The LabVIEW Embedded Module for Blackfin Processors allows you to interact with and view the values of front panel controls and indicators, as well as probed data wires on the block diagram. This type of debugging control allows the programmer to visualize and understand every step of an algorithm graphically, which can be more intuitive than conventional register-based debugging. Data is transferred from the target to the Windows environment based on the polling rate and vector size specified in the **Target Configuration** tab, shown in Figure A.13.

A.3.7.1 Noninstrumented Debugging

Debug via JTAG/EZ-KIT USB is the same type of debugging available in the VisualDSP++ environment, but with the added benefit of LabVIEW's interactive debugging features. Keep in mind that JTAG mode interrupts the processor to perform debug data transfers, which can disrupt real-time processes. The faster the update rate and the more data transferred in each vector, the more JTAG debugging could interfere with the execution of your process. For this reason, Instrumented Debugging was developed for faster communication between the host and embedded processor. Refer to the release notes for additional noninstrumented debugging troubleshooting information.

A.3.7.2 Instrumented Debugging

Instrumented debugging allows the user to fully interact with a running embedded application without disrupting the processor during execution. This feature, unique to the LabVIEW 7.1 Embedded Edition, is achieved by adding additional code, as much as a 40% increase in overall code size, to the application before compilation.

This added code captures and transfers values over an interface other than USB, such as serial or TCP ports. Occasionally, this form of debugging may slow down a time critical process, but it is extremely useful for debugging applications at higher data rates with the greatest ease and flexibility. When using Instrumented Debugging, the USB cable must remain attached for execution control and error checking during debugging.

HANDS-ON EXPERIMENT A.3

In this experiment, you will create the VI shown in Figure A.15, which calculates the dot product of two vectors and displays the result on the front panel. This VI will be compiled, downloaded, and executed on the Blackfin processor with front panel interaction available through the debugging capabilities of the LabVIEW Embedded Module for Blackfin Processors. The dot product result will also be passed back to the host computer through the standard output port using the **Inline C Node**.

Open the **Embedded Project Manager** window by clicking on **Tools → Embedded Project Manager** and create a new LabVIEW embedded project through the **File → New Project** menu. Name the project dotproduct.lep and click **OK**. The LabVIEW execution target should set to execute on the Blackfin processor. Verify this by selecting **Target → Switch Execution Target**, and make sure that the appropriate Blackfin EZ-KIT is selected. Next, add a new VI to the project by choosing **File → New** . . . , selecting the Blank VI option and naming the file as DotProd.vi. Click **OK**. Notice that the file has been added to your project and its front panel has been opened.

The next step is to recreate parts of the Hands-On Experiment A.1 following those instructions, or you may start with a saved version of that VI. The goal is to modify the VI to that found in Figure A.15. For the purposes of this exercise, we will replace the **For Loop** implementation of the dot product with the dot product LabVIEW analysis function. The controls and indicators will be identical to those used in the previous exercise and the outer **While Loop** will be reused.

Once you have recreated the new DotProd.vi complete with controls, indicators, and while loop, place the Dot Product.vi, located on the **Analyze → Mathematics → Linear**

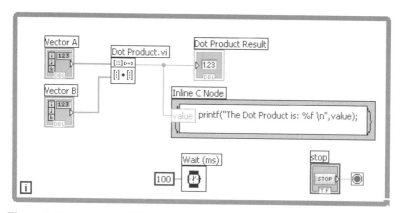

Figure A.15 Dot product VI block diagram written in LabVIEW Embedded

A.3 Introduction to the LabVIEW Embedded Module for Blackfin Processors   459

Algebra palette, in the center of the VI. Click to select and then click to drop the `Dot Product.vi` on the block diagram. Configure the function inputs and outputs by wiring **Vector A** to the **X array** input, **Vector B** to the **Y array** input, and **Dot Product Result** to the **X*Y** output of the Dot Product VI. Click and drag each of these icons to move and arrange them as seen in Figure A.15.

Next, we will add console output capability to our application using an **Inline C Node**, which will allow us to enter text-based C code directly into the LabVIEW application. Navigate to the **Structures** palette, select the **Inline C Node**, and drag the mouse across the block diagram to create a rectangular region roughly the size shown in Figure A.15. Click inside the **Inline C Node** and enter `printf("The Dot Product is: %f \n",value);`, which will output to the debugging window the value of the dot product during every iteration of the **While Loop**. To pass the value of the dot product to this C code, right-click on the left border of the **Inline C Node** and select **Add Input**. This creates a new input terminal for the C node, which is named as value to match the name referenced in the C code.

Verify that you have enclosed the dot product controls and indicators with a **While Loop** structure to allow the code to execute continuously. Finally, add time delay to the loop if you have not already done so. Add a **Wait (ms)** function, located in the **Time, Dialog & Error** palette and create a constant input of 100 milliseconds to provide a short pause between the processing of each iteration of the loop.

After the VI has been wired correctly, clicking on the **Run** arrow will open the **Embedded Project Manager** window. Select **Build Option** from the **Target** menu to open the **Blackfin Build Options** dialog box. You can choose what level of debugging support to use, as well as various compilation options for the project. For **Debug mode**, choose **Non-instrumented (via JTAG/EZ-KIT USB)** from the pull-down menu and make sure that the **Build configuration** is set to **Debug**. This option allows for debugging our application without the need for any additional cables or hardware. You will also need to uncheck **Redirect stdout to serial port** on the **Advanced** tab. Before running the application open the **Processor Status** dialog box from the **Target** menu to view the console output transferred over the standard output port from the application.

After the build options have been set for the project, click on the **Debug** button to compile, build, download, run, and debug the application. Switch to the front panel and observe how the calculated value for the dot product changes as you enter different values for the input vectors. Note also the console output in the **Processor Status** dialog box. Do the values calculated by this VI match those you calculated in Hands-on Experiment A.1?

HANDS-ON EXPERIMENT A.4

In this final experiment, we will modify an audio pass-through example to generate and output a custom sinewave to the audio out port of the Blackfin EZ-KIT using the LabVIEW code shown in Figure A.16. To begin, launch the **Embedded Project Manager** and open the `Audio Talkthrough - BF5XX.lep` project file. Select **File → Save Project As** and save the project as `Sine Generator.lep`. From the **Embedded Project Manager** window, open `Audio Talkthrough - BF53X.vi` and select **File → Save As** to save the VI as `Sine Generator.vi` in the same directory as your embedded project file. We are now ready to add the sinewave generator functionality to the code.

460 Appendix A   An Introduction to Graphical Programming with LabVIEW

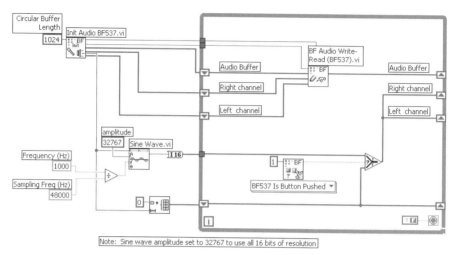

Figure A.16   Completed `Sine Generator.vi` block diagram

Start by first adding the `Sine Wave.vi`, located on the **Analyze → Signal Processing → Signal Generation** palette, to the block diagram. Place this VI outside of the **While Loop**, below the **Init Audio** function. At this point you should expand the size of the **While Loop** to make more room for additional code. This is easily done by placing the mouse over the **While Loop** and dragging down the lower handle until the loop is approximately the same size as shown in Figure A.16. Move the mouse over the sinewave output of `Sine Wave.vi`, and notice that the terminal has an orange color, denoting a floating-point data type. Because the Blackfin processor is a fixed-point processor, we need to convert the floating-point array of samples into a fixed-point format, as explained in Chapter 6. To accomplish this, place a **To Word Integer** VI, located on the **Numeric → Conversion** palette, next to the **sine wave** output terminal and wire them together. Next, a wire between the output of the conversion function to the border of the **While Loop**. What data type is this wire?

You must now define the signal parameters for the sinewave you want to generate. To ensure that the sinewave generated is the appropriate number of samples, connect a wire between the **samples** input of the `Sine Wave.vi` and the **Half buffer size** output of the `Init Audio.vi`. Refer to Figure A.16 if you need help locating the terminals. Next, define the amplitude of the sinewave by right-clicking the **amplitude** terminal and selecting **Create → Constant** from the shortcut menu. The amplitude has a default value of 1. Finally, we need to define the frequency of sinewave. From the **Numeric** palette, place two numeric constants and a **Divide** function and wire the constants to the inputs of the **Divide** function. Wire the output of the **Divide** function to the **Frequency** input terminal of the `Sine Wave.vi`. The units that this VI uses for frequency are cycles/sample, which we will calculate by dividing the desired sinewave frequency in Hz by the sampling frequency of our audio system. Use 1000 for the desired frequency and 48000 for the sampling frequency. At this point, the block diagram should look like Figure A.17.

Now that the sinewave generation is implemented, we need to add in the ability to play the signal using the audio capabilities built into the Blackfin EZ-KIT when the user holds down a push button. Place an **Initialize Array** function from the **Array** palette below the **Sine Wave** function and wire the **dimension size** terminal to the **half buffer size** output of

A.3 Introduction to the LabVIEW Embedded Module for Blackfin Processors 461

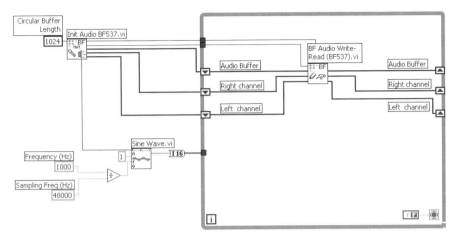

Figure A.17 Sinewave generator—intermediate stage

the Init Audio.vi. You can connect to the same wire that connects the Init Audio and Sine Wave VIs. Create a constant for the **element** input and change it to 0 of I16 data type by right-clicking on the constant and selecting **Representation → I16**. This creates an empty array of the same size as the generated sinewave that we can use to output a silent signal when the button is not pressed. Connect the output of the **Initialize Array** function to the border of the **While Loop**.

Right-click the tunnel created on the **While Loop** border and select **Replace with Shift Register**, then wire the left shift register to the right shift register. Next, we need to determine if a button has been pushed. Place a BF is Button Pushed VI, located on the **Blackfin → EZ-KIT → Button** palette, inside the **While Loop** and select the instance appropriate for the hardware from the pull-down menu. This VI will return a Boolean output telling us if a particular button is currently being pressed. To choose which button to monitor, create a constant on the **button number** input. Valid ranges are 1–4 for the BF537 EZ-KIT and 4–7 for the BF533 EZ-KIT, corresponding to the PB and SW numbers of the buttons, respectively. To choose which signal to output, add a **Select** function located on the **Comparison** palette and wire the **button pushed** output to the **s** input of the **Select** function. Now, wire the **f** input of the **Select** function to the wire connecting the two shift registers we just created, and wire the **t** input of the **Select** function to the converted sine array. Delete the connections between the right and left audio channel outputs of the BF Audio Write-Read.vi and the right shift registers. Finally, connect the output of the **Select** function to the right shift registers for both the right channel and the left channel. The block diagram is now complete, and should match Figure A.15.

Compile, download, and run the application on the Blackfin processor and test the VI by pressing the button you chose to monitor and listen for the generated sinewave to be output on the audio out line of the Blackfin EZ-KIT. Experiment with generating different sinewave frequencies. Can you hear the difference? Why do most combinations parameters distort the audio tone? Consider the combination of buffer size, sampling rate, and the frequency of the sinewave. How would you modify the VI to remove this phenomenon?

Appendix B

Useful Websites

This Appendix provides many useful websites that are related to the contents of this book. A companion website, www.ntu.edu.sg\home\ewsgan\esp_book.htm, has been set up by the authors to provide more information and latest updates on software tools, installations of software, more hands-on exercises, solutions to selected problems, and many useful teaching and learning aids for instructors and students.

Websites	Details
www.analog.com	Analog Devices Inc. (ADI) main website
www.intel.com/design/msa/index.htm	Intel's MSA website
www.analog.com/processors/index.html	Tools and resources website for ADI processors
www.analog.com/processors/blackfin/index.html	Blackfin processor main website
www.analog.com/processors/universityProgram/includedproducts.html	ADI university program
www.analog.com/processors/blackfin/technicalLibrary/manuals/index.html	ADI Blackfin processor manuals
www.analog.com/processors/blackfin/technicalLibrary/index.html	ADI Blackfin technical library
www.analog.com/processors/blackfin/evaluationDevelopment/blackfinProcessorTestDrive.html	VisualDSP++ development software trial download for Blackfin processors
www.analog.com/ee-notes	Application notes
http://forms.analog.com/Form_Pages/support/dsp/dspSupport.asp	DSP technical support
www.analog.com/salesdir/continent.asp	DSP sales and distributors

---

Embedded Signal Processing with the Micro Signal Architecture. By Woon-Seng Gan and Sen M. Kuo
Copyright © 2007 John Wiley & Sons, Inc.

Appendix B Useful Websites

Websites	Details
www.analog.com/processors/technicalSupport/ICanomalies.html	ADI hardware anormalies
www.analog.com/BOLD	Blackfin online learning and development (BOLD)
www.analog.com/visualaudio	VisualAduio algorithm development tool
Processor.support@analog.com	Technical support email
www.ni.com	National Instruments (NI) main website
www.ni.com/labview/blackfin/	NI LabVIEW Embedded website
www.mathworks.com	The Mathworks main website
www.bdti.com/products/reports_msa.html	BDTI MSA report

Appendix C

List of Files Used in Hands-On Experiments and Exercises

This appendix lists all the experiment and exercise files used in this book. The book is written in such a way that the main text of each chapter is supplemented with examples, quizzes, exercises, hands-on experiments, and exercise problems. These files are located in the directories:

    c:\adsp\chap{x}\exp{x}_{no.}_<option>   for Blackfin and LabVIEW experiments

    c:\adsp\chap{x}\exercise{x}_{no.}   for Blackfin and LabVIEW exercises

    c:\adsp\chap{x}\MATLAB_ex{x}   for MATLAB examples and exercises

where {x} indicates the chapter number and {no.} indicates the experiment number; <option> states the BF533 or BF537 EZ-KIT. To reference these supplementary files, we tabulate them as shown below.

Embedded Signal Processing with the Micro Signal Architecture. By Woon-Seng Gan and Sen M. Kuo
Copyright © 2007 John Wiley & Sons, Inc.

Appendix C List of Files Used in Hands-On Experiments and Exercises

Chapter	Experiments/Exercises	Brief Description	Platform Needed
1	Experiment 1.1	— Explore VisualDSP++ environment	— Blackfin simulator
	Experiment 1.2	— Vector addition	— Blackfin simulator
	Experiment 1.3	— Graphic display	— Blackfin simulator
	Experiment 1.4	— Dot product in LabVIEW Embedded	— LabVIEW Embedded
2	Experiment 2.1	— Explore SPTool	— MATLAB (SPTool)
	Experiment 2.2	— Moving-average filter in MATLAB	— MATLAB (SPTool)
	Experiment 2.3	— Moving-average filter in Blackfin simulator	— Blackfin simulator
	Experiment 2.4	— Moving-average filter in EZ-KIT	— EZ-KIT
	Experiment 2.5	— Moving-average filter in LabVIEW	— LabVIEW
	Experiment 2.6	— Moving-average filter in LabVIEW Embedded	— LabVIEW Embedded +EZ-KIT
	Exercise 2.1	— Generate sine wave and compute SNR	— MATLAB
	Exercise 2.2	— Import and display signals	— MATLAB (SPTool)
	Exercise 2.3	— Moving-average filter exercises	— MATLAB (SPTool)
	Exercise 2.4	— Filter design exercises	— MATLAB (SPTool)
	Exercise 2.5	— Draw signal-flow graph	
	Exercise 2.6	— Compare moving-average and median filters	— MATLAB (SPTool)
	Exercise 2.7	— Filters to remove hum	— MATLAB (SPTool)
	Exercise 2.8	— Filters to remove noisy speech signal	— MATLAB (SPTool)
	Exercise 2.9	— Filters to remove noisy square wave	— MATLAB (SPTool)
	Exercise 2.10	— Moving-average filter in Blackfin simulator	— Blackfin simulator
	Exercise 2.11	— Moving-average filter in EZ-KIT	— EZ-KIT
3	Experiment 3.1	— Explore filter visualization tool	— MATLAB (FVTool)
	Experiment 3.2	— Frequency analysis in SPTool	— MATLAB (SPTool)
	Experiment 3.3	— Explore window visualization tool	— MATLAB (WVTool)

Appendix C List of Files Used in Hands-On Experiments and Exercises

Chapter	Experiments/ Exercises	Brief Description	Platform Needed
	Experiment 3.4	— Filtering and analysis in SPTool	— MATLAB (SPTool)
	Experiment 3.5	— Notch filter to attenuate undesired tone	— MATLAB (SPTool)
	Experiment 3.6	— Frequency analysis in Blackfin simulator	— Blackfin simulator
	Experiment 3.7	— Real-time frequency analyzer	— EZ-KIT
	Experiment 3.8	— Windowing on frequency analysis	— LabVIEW
	Experiment 3.9	— Windowing on frequency analysis using LabVIEW Embedded	— LabVIEW Embedded + EZ-KIT
	Exercise 3.1	— Exercises on z-transform	
	Exercise 3.2	— Exercises on z-transform of filters	
	Exercise 3.3	— Exercises on transfer function	
	Exercise 3.4	— Direct-form I and II IIR filters	
	Exercise 3.5	— Stability of IIR filters	
	Exercise 3.6	— Relate IIR filter to moving-average filter	
	Exercise 3.7	— Transfer function and frequency response	— MATLAB
	Exercise 3.8	— Compute gain and magnitude response	— MATLAB
	Exercise 3.9	— DFT exercises	
	Exercise 3.10	— Magnitude response	— MATLAB
	Exercise 3.11	— Plotting magnitude spectrum	— MATLAB
	Exercise 3.12	— Spectrum viewer in SPTool	— MATLAB (SPTool)
	Exercise 3.13	— Window analysis in WINTool	— MATLAB (WINTool)
	Exercise 3.14	— Exercises on notch filter	— MATLAB
	Exercise 3.15	— More exercises on notch filter	— MATLAB
	Exercise 3.16	— Exercises on IIR filter	— MATLAB
	Exercise 3.17	— Tone generation using IIR filter	— MATLAB
	Exercise 3.18	— Spectrogram in Blackfin simulator	— Blackfin simulator

Appendix C   List of Files Used in Hands-On Experiments and Exercises   **467**

Chapter	Experiments/ Exercises	Brief Description	Platform Needed
	Exercise 3.19	— Real-time frequency analysis in EZ-KIT	— EZ-KIT
4	Experiment 4.1	— Filter design using FDATool	— MATLAB (FDATool)
	Experiment 4.2	— Design bandpass FIR filter and export to file	— MATLAB (FDATool)
	Experiment 4.3	— Design bandstop IIR filter	— MATLAB (FDATool)
	Experiment 4.4	— Adaptive line enhancer using Blackfin simulator	— Blackfin simulator
	Experiment 4.5	— Adaptive line enhancer using EZ-KIT	— EZ-KIT
	Experiment 4.6	— Adaptive line enhancer using LabVIEW	— LabVIEW
	Experiment 4.7	— Adaptive line enhancer using LabVIEW Embedded	— LabVIEW Embedded + EZ-KIT
	Exercise 4.1	— Determine filter types and cutoff frequencies	
	Exercise 4.2	— Design filters using FDATool	— MATLAB (FDATool)
	Exercise 4.3	— Signal-flow graph exercise	
	Exercise 4.4	— Design FIR filters	— MATLAB
	Exercise 4.5	— Design bandpass FIR filters	— MATLAB (FDATool)
	Exercise 4.6	— Filter to enhance speech signal	— MATLAB (FDATool)
	Exercise 4.7	— Frequency response of Chebyshev IIR filter	— MATLAB
	Exercise 4.8	— Design other IIR filters using FDATool	— MATLAB (FDATool)
	Exercise 4.9	— System identification	— MATLAB
	Exercise 4.10	— Adaptive line enhancer	— MATLAB
	Exercise 4.11	— Effect of step size on adaptive filter	— MATLAB
	Exercise 4.12	— Adaptive noise cancellation	— MATLAB
	Exercise 4.13	— Adaptive line enhancer in Blackfin simulator	— Blackfin simulator

468 Appendix C List of Files Used in Hands-On Experiments and Exercises

Chapter	Experiments/ Exercises	Brief Description	Platform Needed
	Exercise 4.14	— Adaptive line enhancer in EZ-KIT	— EZ-KIT
5	Experiment 5.1	— Assembly programming and debugging in Blackfin	— Blackfin simulator
	Experiment 5.2	— Circular buffering in assembly	— Blackfin simulator
	Experiment 5.3	— Arithmetic and logical shifting in assembly	— Blackfin simulator
	Experiment 5.4	— Create linker description file using Expert Linker	— Blackfin simulator
	Experiment 5.5	— Design 8-band graphic equalizer (FIR) in floating point	— MATLAB (FDATool)
	Experiment 5.6	— Converting floating point coefficients to fixed point	— MATLAB (FDATool)
	Experiment 5.7	— Design graphic equalizer using Blackfin simulator	— Blackfin simulator
	Experiment 5.8	— Implement graphic equalizer using EZ-KIT	— EZ-KIT
	Experiment 5.9	— Creating a stand-alone graphic equalizer	— EZ-KIT
	Experiment 5.10	— Implement graphic equalizer using LabVIEW.	— LabVIEW
	Experiment 5.11	— Implement graphic equalizer using LabVIEW Embedded	— LabVIEW Embedded + EZ-KIT
	Exercise 5.1	— Addition operations in simulator	— Blackfin simulator
	Exercise 5.2	— Multiply operations in simulator	— Blackfin simulator
	Exercise 5.3	— More exercises on graphic equalizer	— Blackfin simulator
	Exercise 5.4	— Predefined gain control in graphic equalizer	— Blackfin simulator
6	Experiment 6.1	— Add/Subtract and saturation in Blackfin	— Blackfin simulator

Appendix C List of Files Used in Hands-On Experiments and Exercises **469**

Chapter	Experiments/ Exercises	Brief Description	Platform Needed
	Experiment 6.2	— Multiplication in Blackfin	— Blackfin simulator
	Experiment 6.3	— Using the guard bits of the accumulator	— Blackfin simulator
	Experiment 6.4	— FIR filtering using multiply-add instruction	— Blackfin simulator
	Experiment 6.5	— Truncation and rounding modes	— Blackfin simulator
	Experiment 6.6	— 32-Bit multiplication	— Blackfin simulator
	Experiment 6.7	— Quantization of signal	— MATLAB
	Experiment 6.8	— Quantization of FIR filter's coefficients	— MATLAB (FDATool)
	Experiment 6.9	— Quantization of IIR filter	— MATLAB (FDATool)
	Experiment 6.10	— Limit cycle in IIR filter	— MATLAB
	Experiment 6.11	— Benchmarking using BTC	— EZ-KIT
	Experiment 6.12	— Statistical profiling	— EZ-KIT
	Experiment 6.13	— Code and data size	— EZ-KIT
	Experiment 6.14	— Designing 8-band graphic equalizer (IIR) in floating point	— MATLAB (FDATool)
	Experiment 6.15	— Converting floating-point to fixed-point coefficients	— MATLAB (FDATool)
	Experiment 6.16	— Implementing fixed-point 8-band graphic equalizer	— Blackfin simulator
	Experiment 6.17	— Real-time graphic equalizer	— EZ-KIT
	Experiment 6.18	— IIR graphic equalizer in LabVIEW	— LabVIEW
	Experiment 6.19	— IIR graphic equalizer in LabVIEW Embedded	— LabVIEW Embedded + EZ-KIT
	Exercise 6.1	— Exercises on graphic equalizer	— Blackfin simulator
	Exercise 6.2	— Benchmarking real-time performance of graphic equalizer	— EZ-KIT
7	Experiment 7.1	— Configuring CODEC	— BF533 EZ-KIT
	Experiment 7.2	— TDM and I^{2}S modes	— BF533 EZ-KIT

470 Appendix C List of Files Used in Hands-On Experiments and Exercises

Chapter	Experiments/Exercises	Brief Description	Platform Needed
	Experiment 7.3	— 1D DMA setup (sample mode)	— BF533 EZ-KIT
	Experiment 7.4	— 2D DMA setup (block mode)	— BF533 EZ-KIT
	Experiment 7.5	— Memory DMA transfer	— Blackfin simulator
	Experiment 7.6	— Memory fill using DMA	— Blackfin simulator
	Experiment 7.7	— Locking of way in instruction cache	— Blackfin simulator
	Experiment 7.8	— Cache modes	— EZ-KIT
	Experiment 7.9	— Signal generation using look-up table	— Blackfin simulator
	Experiment 7.10	— Real-time signal generation	— EZ-KIT
	Experiment 7.11	— DTMF tone generation using LabVIEW	— LabVIEW
	Experiment 7.12	— DTMF tone generation using LabVIEW Embedded	— LabVIEW Embedded + EZ-KIT
	Exercise 7.1	— Ping-pong buffering	— EZ-KIT
	Exercise 7.2	— Exercises on signal generation	— Blackfin simulator
	Exercise 7.3	— Exercises on realtime signal generation	— EZ-KIT
8	Experiment 8.1	— Index versus point style	— EZ-KIT
	Experiment 8.2	— Benchmarking memory	— EZ-KIT
	Experiment 8.3	— Using pragmas	— EZ-KIT
	Experiment 8.4	— Using intrinsic functions	— EZ-KIT
	Experiment 8.5	— Circular buffering	— EZ-KIT
	Experiment 8.6	— Using DSP library functions	— EZ-KIT
	Experiment 8.7	— Profile-guided optimization	— EZ-KIT
	Experiment 8.8	— Linear assembly code	— EZ-KIT
	Experiment 8.9	— Using hardware loop in assembly	— EZ-KIT
	Experiment 8.10	— With dual MACs	— EZ-KIT
	Experiment 8.11	— With multi-issue instructions	— EZ-KIT

Appendix C List of Files Used in Hands-On Experiments and Exercises **471**

Chapter	Experiments/Exercises	Brief Description	Platform Needed
	Experiment 8.12	— With separate data sections	— EZ-KIT
	Experiment 8.13	— With software pipelining	— EZ-KIT
	Experiment 8.14	— Voltage-frequency scaling	— EZ-KIT
	Experiment 8.15	— Decimation and interpolation	— Blackfin simulator
	Experiment 8.16	— Real-time decimator and interpolator	— EZ-KIT
	Experiment 8.17	— Real-time signal mixing	— EZ-KIT
	Experiment 8.18	— Sample rate conversion in LabVIEW	— LabVIEW
	Experiment 8.19	— Sample rate conversion in LabVIEW Embedded	— LabVIEW Embedded + EZ-KIT
	Exercise 8.1	— Enhancing software pipelining	— EZ-KIT
	Exercise 8.2	— Examining the DSP library FIR function	— EZ-KIT
	Exercise 8.3	— Benchmarking block size and filter length	— EZ-KIT
	Exercise 8.4	— Exercise on sample rate conversion	— Blackfin simulator
	Exercise 8.5	— Exercise on real-time signal mixing	— EZ-KIT
9	Experiment 9.1	— Ogg Vorbis decoding	— BF537 EZ-KIT
	Experiment 9.2	— 3D audio effects	— EZ-KIT
	Experiment 9.3	— Reverberation effects	— EZ-KIT
	Experiment 9.4	— MDCT/IMDCT in LabVIEW	— LabVIEW
	Experiment 9.5	— Benchmark IMDCT	— LabVIEW Embedded + EZ-KIT
	Exercise 9.1	— Bit rate computation	
	Exercise 9.2	— Window type and length for audio coding	— MATLAB (WINTool)
	Exercise 9.3	— Binaural spatializer	— MATLAB
	Exercise 9.4	— Reverberation effects	— MATLAB
10	Experiment 10.1	— Display image in VisualDSP++	— EZ-KIT
	Experiment 10.2	— Color conversion	— EZ-KIT

Appendix C List of Files Used in Hands-On Experiments and Exercises

Chapter	Experiments/Exercises	Brief Description	Platform Needed
	Experiment 10.3	— 2D DCT, quantization and 2D IDCT	— EZ-KIT
	Experiment 10.4	— 2D image filtering	— EZ-KIT
	Experiment 10.5	— Image enhancement	— EZ-KIT
	Experiment 10.6	— Contrast adjustment	— EZ-KIT
	Experiment 10.7	— Mathematical operations for image in LabVIEW	— LabVIEW
	Experiment 10.8	— 2D convolution in LabVIEW Embedded	— LabVIEW Embedded + EZ-KIT

Appendix D

Updates of Experiments Using VisualDSP++ V4.5

With the recent release of VisualDSP++ V4.5, there are some minor changes to the experiments listed in this book. This appendix lists the differences and necessary changes required in the Blackfin programming code from VisualDSP ++ V4.0 (described in this book) to VisualDSP++ V4.5.

- The **Create LDF Wizard** option is no longer available for the VisualDSP++ V4.5 for Blackfin processor. In order to add `.ldf` file in an existing project, open the existing project in VisualDSP++ V4.5. Click on **Project → Project Options…** and navigate down in the left window to enable **Add Startup Code/LDF**. Select **Add an LDF and startup code** and click **OK**. A new `.ldf` file with the same filename as the project is generated. Double click on the `.ldf` file to assign the memory section and settings as described in V4.0 (described in this book).

- For those programs that define the DMA settings in the `init.c` file. Type mismatch error is reported when building these projects in V4.5. The error message is described as "A value of type "volatile unsigned short *" can't be assigned to an entity of type "void *"." Simply add (void *) before the variable. For example: change "pDMA1_START_ADDR = sDataBufferRx" to "pDMA1_START_ADDR = (void*)sDataBufferRx".

- When defining the size of the BTC channel in V4.5, it is necessary to add in 8 more words, for example, `fract16 BTC_CHAN0[VECTOR_SIZE+8]`. In V4.0, 4 more words are used instead.

- In V4.5, a new BTC library, `libbtc532.dlb` is located in directory c:\Program Files\Analog Devices\VisualDSP 4.5\Blackfin\lib. Please use this new library when building project that uses BTC under V4.5.

Embedded Signal Processing with the Micro Signal Architecture. By Woon-Seng Gan and Sen M. Kuo
Copyright © 2007 John Wiley & Sons, Inc.

Appendix D Updates of Experiments Using VisualDSP++ V4.5

- In V4.5, the default .ldf files is found in directory c:\Program Files\ Analog Devices\VisualDSP 4.5\Blackfin\ldf.
- In V4.5, the flash drivers (mentioned in Chapter 5) for BF533 and BF537 are located at directories c:\Program Files\Analog Devices\ VisualDSP 4.5\Blackfin\Examples \ADSP-BF533 EZ-Kit Lite\ Flash Programmer \BF533EzFlashDriver.dxe and c:\Program Files\Analog Devices\VisualDSP 4.5\Blackfin\Examples\ ADSP-BF537 EZ-Kit Lite\Flash Programmer\BF537EzFlash-Driver.dxe.
- In V4.5, the DSP library function (mentioned in Chapter 8) is now located at directory c:\Program Files\Analog Devices\VisualDSP 4.5\ zBlackfin\lib\src\libdsp.
- In V4.5, the sample program on capturing and displaying image (mentioned in Chapter 10) through the parallel port are listed in directories c:\Program Files\Analog Devices\VisualDSP 4.5\Blackfin\Examples\ ADSP-BF533 EZ-Kit Lite\Video Input (C) and c:\Program Files\Analog Devices\VisualDSP 4.5\Blackfin\Examples\ ADSP-BF533 EZ-Kit Lite\Video Output (ASM), respectively.

References

References on Digital Signal Processing and DSP Processors

[1] J. G. ACKENHUSEN, *Real-Time Signal Processing: Design and Implementation of Signal Processing Systems*, Upper Saddle River, NJ: Prentice Hall, 1999.

[2] N. AHMED and T. NATARAJAN, *Discrete-Time Signals and Systems*, Englewood Cliffs, NJ: Prentice Hall, 1983.

[3] A BATEMAN and W. YATES, *Digital Signal Processing Design*, New York, NY: Computer Science Press, 1989.

[4] D. GROVER and J. R. DELLER, *Digital Signal Processing and the Microcontroller*, Upper Saddle River, NJ: Prentice Hall, 1999.

[5] D. HARLIONO and W. S. GAN, Implementation of dynamic voltage and frequency scaling on blackfin processors, in *Proc. Int. Symp. Intelligent Signal Processing and Communication Systems*, pp. 193–196, 2005.

[6] S. HAYKIN, *Adaptive Filter Theory*, 4th Ed., Upper Saddle River, NJ: Prentice Hall, 2002.

[7] IEEE, IEEE standard for binary floating-point arithmetic, *IEEE Standard 754-1985*, pp. 1–17, 1985.

[8] N. KEHTARNAVAZ and N. J. KIM, *Digital Signal Processing System-Level Design Using LabVIEW*, Newness Publisher, 2005.

[9] S. M. KUO and D. R. MORGAN, *Active Noise Control Systems—Algorithms and DSP Implementations*, New York, NY: Wiley, 1996.

[10] S. M. KUO, B. H. LEE, and W. S. TIAN, *Real-Time Digital Signal Processing*, 2nd Edition, Chichester: Wiley 2006.

[11] S. M. KUO and W. S. GAN, *Digital Signal Processors—Architectures, Implementations, and Applications*, Upper Saddle River, NJ: Prentice Hall, 2005.

[12] S. M. KUO and C. CHEN, Implementation of adaptive filters with the TMS320C25 or the TMS320C30, in *Digital Signal Processing Applications with the TMS320 Family*, Vol. 3, P. Papamichalis, ed., Englewood Cliffs, NJ: Prentice Hall, 1990. pp. 191–271, Chap. 7.

[13] P. LAPSLEY, J. BIER, and A. SHOHAM, *DSP Processor Fundamentals: Architectures and Features*, New York NY: IEEE Press, 1999.

[14] J. H. MCCLELLAN, R. W. SCHAFER, and M. A. YODER, *DSP First: A Multimedia Approach*, 2nd Ed., Englewood Cliffs, NJ: Prentice Hall, 1998.

[15] S. K. MITRA, *Digital Signal Processing: A Computer-Based Approach*, 2nd Ed., New York, NY: McGraw-Hill, 2001.

Embedded Signal Processing with the Micro Signal Architecture. By Woon-Seng Gan and Sen M. Kuo
Copyright © 2007 John Wiley & Sons, Inc.

[16] A. V. OPPENHEIM, R. W. SCHAFER, and J. R BUCK, *Discrete-Time Signal Processing*, 2nd Ed., Upper Saddle River, NJ: Prentice Hall, 1999.

[17] S. J. ORFANIDIS, *Introduction to Signal Processing*, Englewood Cliffs, NJ: Prentice Hall, 1996.

[18] P. PEEBLES, *Probability, Random Variables, and Random Signal Principles*, New York, NY: McGraw-Hill, 1980.

[19] J. G. PROAKIS and D. G. MANOLAKIS, *Digital Signal Processing—Principles, Algorithms, and Applications*, 3rd Ed., Englewood Cliffs, NJ: Prentice Hall, 1996.

[20] J. R. TREICHLER, C. R. JOHNSON, JR., and M. G. LARIMORE, *Theory and Design of Adaptive Filters*, New York, NY: Wiley, 1987.

[21] B. WIDROW, J. R. GLOVER, J. M. MCCOOL, J. KAUNITZ, C. S. WILLIAMS, R. H. HERN, J. R. ZEIDLER, E. DONG, and R. C. GOODLIN, Adaptive noise canceling: principles and applications, *Proc. IEEE*, vol. 63, pp. 1692–1716, Dec. 1975.

[22] B. WIDROW and S. D. STEARNS, *Adaptive Signal Processing*, Englewood Cliffs, NJ: Prentice Hall, 1985.

References from Analog Devices

[23] Analog Devices Inc., *ADSP-BF533 Blackfin Processor Hardware Reference Manual*, Rev 3.1, May 2005.

[24] Analog Devices Inc., *ADSP-BF537 Blackfin Processor Hardware Reference*, Rev 2.0, Dec. 2005.

[25] Analog Devices Inc., *ADSP-BF53x/BF561 Blackfin Processor Programming Reference*, Rev 1.1, Feb. 2006.

[26] Analog Devices Inc., *ADSP-BF533: Blackfin® Embedded Processor Data Sheet*, Rev. C, May 2006.

[27] Analog Devices Inc., *ADSP-BF537: Blackfin® Embedded Processor Data Sheet*, Rev. 0, April 2006.

[28] Analog Devices Inc., *Blackfin Processor Instruction Set Reference*, Rev 3.0, June 2004.

[29] Analog Devices Inc., *ADSP-BF533 EZ-KIT Lite Evaluation System Manual*, Rev 2.0, Jan. 2005.

[30] Analog Devices Inc., *ADSP-BF537 EZ-KIT Lite Evaluation System Manual*, Rev 1.1, Aug. 2005.

[31] Analog Devices Inc., *Getting Started with Blackfin Processors*, Rev 2.0, Sept. 2005.

[32] Analog Devices Inc., *Getting Started with ADSP-BF537 EZ-KIT Lite*, Rev 1.0, Jan. 2005.

[33] Analog Devices Inc., *VisualDSP++ 4.0 User's Guide*, Rev 1.0, Jan. 2005.

[34] Analog Devices Inc., *VisualDSP++ 4.0 Getting Started Guide*, Rev 1.0, Jan. 2005.

[35] Analog Devices Inc., *VisualDSP++ 4.0 C/C++ Compiler and Library Manual for Blackfin Processors*, Rev 3.0, Jan. 2005.

[36] Analog Devices Inc., *VisualDSP++ 4.0 Linker and Utilities Manual*, Rev 1.0, Jan. 2005.

[37] Analog Devices Inc, *VisualDSP++ 4.0 Assembler and Preprocessor Manual*, Rev.1.0, Jan. 2005.

[38] Analog Devices Inc., *Workshop Manual for the Blackfin Processor: Innovative Architecture for Embedded Media Processing*, June 2004.

[39] Analog Devices Inc., Guide to Blackfin® Processor LDF

Files, *Application Notes* EE-237, 2004.

[40] Analog Devices Inc, Fast floating-point arithmetic emulation on the blackfin processor platform, *Engineer To Engineer Note*, EE-185, 2003.

[41] Analog Devices Inc, *AD1836A: Multichannel 96 kHz CODEC*, Rev. 0, 2003.

[42] Analog Devices Inc, Turning C source code for the Blackfin processor compiler, *Application Note*, EE-149, 2003.

[43] D. KATZ and R. GENTILE, *Embedded Media Processing*, Newnes Publisher, 2005.

[44] W. KESTER, *Mixed-Signal and DSP Design Techniques*, Newnes Publisher, 2004.

[45] D. KATZ, T. LUKASIAK and R. GENTILE, Enhance processor performance in open-source applications, *Analog Dialogue*, vol. 39, Feb. 2005.

[46] K. SINGH, Using cache memory on Blackfin processors, *Analog Devices Application Note*, EE-271, 2005.

[47] J. TOMARAKOS, and D. LEDGER, Using the low-cost, high performance, ADSP-21065 digital signal processor for digital audio applications, *Analog Devices Application Note*, 12/4/98.

References from the MathWorks

[48] MathWorks Inc., *Getting Started with MATLAB*, Version 7, 2004.

[49] MathWorks, Inc., *Signal Processing Toolbox User's Guide*, Version 6, 2004.

[50] MathWorks, Inc., *Filter Design Toolbox User's Guide*, Version 3, 2004.

[51] MathWorks Inc., *Image Processing Toolbox User's Guide*, Version 5, 2006.

References from National Instruments

[52] National Instruments, *Getting Started with the LabVIEW Embedded Module for Analog Devices Blackfin Processors*, Part Number: 371655A-01, Version 1.0, Feb. 2006.

[53] National Instruments, *LabVIEW Embedded Module for Analog Devices Blackfin Processors*, Part Number:371656A-01, 324214A-01, Version 1.0, Feb. 2006.

References on Audio Signal Processing

[54] D. R. BEGAULT, *3-D Sound for Virtual Reality and Multimedia*, Boston, MA: Academic Press, 1994.

[55] J. BOLEY, A perceptual audio coder (and decoder), article available from http://www.perceptualentropy.com/coder.html.

[56] M. BOSI, *Introduction to Digital Audio Coding and Standard*, New York, NY: Springer, 2002.

[57] M. BOSI, Audio coding: basic principles and recent developments, in *Proc. 6th HC Int. Conf.*, 2003, pp. 1–17.

[58] W. G. GARDNER, and K. D. MARTIN, HRTF measurements of a KEMAR, *Journal of the Acoustical Society of America*, vol. 97, pp. 3907–3908, 1995. HRTF measurements and other data files can be downloaded from MIT Media Lab, http://sound.media.mit.edu/KEMAR.html.

[59] A. KOSAKA, H. OKUHATA, T. ONOYE, and I. SHIRAKAWA, Design of Ogg Vorbis decoder system for embedded platform, *IEICE Trans. Fundamentals*, vol. E88, pp. 2124–2130, Aug. 2005.

[60] E. LARSEN and E. M. AARTS, *Audio Bandwidth Extension*, New York, NY: Wiley, 2004.

[61] E. MONTNEMERY and J. SANDVALL, *Ogg/Vorbis in Embedded Systems*, MS Thesis, Lunds Universitet, 2004.

[62] J. A. MOORER, About the reverberation business, *Computer Music Journal*, vol. 3, pp. 13–28, 1979.

[63] D. PAN, A tutorial on MPEG/audio compression, *IEEE Multimedia*, vol. 2, pp. 60–74, 1995.

[64] M. R. SCHROEDER, Digital simulation of sound transmission in reverberant spaces, *Journal of the Acoustical Society of America*, vol. 47, pp. 424, 1970.

[65] B. WESÉN, A DSP-based decompressor unit for high-fidelity MPEG-audio over TCP/IP networks, 1997; available on http://www.sparta.lu.se/~bjorn/whitney/compress.htm#Psychoacoustics

[66] D. T. YANG, C. KYRIAKAKIS, and C. J. KUO, *High-Fidelity Multichannel Audio Coding*, New York, NY: Hindawi Publishing, 2004.

[67] E. ZWICKER and H. FASTL, *Psychoacoustics-Facts and Models*. Springer-Verlag, 1999.

References on Image Processing

[68] V. BHASKARAN and K. KONSTANTINIDES, *Image and Video Compression Standards—Algorithm and Architectures*, Norwell. MA: Kluwer Academic Publisher, 1997.

[69] T. BOSE, *Digital Signal and Image Processing*, Hoboken, NJ: Wiley, 2004.

[70] R. C. GONZALEZ and R. E. WOODS, *Digital Image Processing*, 2nd Ed. Upper Saddler River, NJ, Prentice Hall, 2002.

[71] ITU-T.81, *Information Technology—Digital Compression and Coding of Continuous-Tone Still Images—Requirements and Guidelines*, 1993.

[72] J. S. LIM, *Two-dimensional Signal and Image Processing*, Englewood Cliffs, NJ: Prentice Hall, 1990.

[73] J. W. WOODS, *Multidimensional Signal, Image, and Video Processing and Coding*, Burlington, MA: Academic Press, 2006.

Index

2's complement, 194, 218
2D
   DCT, 413, 416
   DFT, 415
   DMA, 293, 296, 298, 301
   filter, 418, 420, 422
   inverse DCT, 413
   inverse DFT, 415
   signal, 406
3D audio, 393, 396

A

Accumulator (ACC0, ACC1), 165, 170–174, 194–195, 224, 225, 244, 249
A/D converter (conversion), see analog-to-digital converter
Adaptive
   algorithm, 139
   channel equalization, 148
   filter, 139, 142
   linear prediction (predictor), 146
   line enhancer (ALE), 146, 151, 152, 155
   noise cancellation, 147, 150
   system identification, 143
Adders, 34, 41, 63, 133, 166
Address pointer register (P0–P5), 174
Addressing modes, 163, 175, 176, 199
Advanced audio coding (AAC), 382
Algebraic syntax, 167
Aliasing, 27, 366, 371, 387
Allpass filter, 398
Amplitude spectrum, see magnitude spectrum
Analog filter, 129
Analog-to-digital converter (ADC), 28, 52, 57, 151, 207, 236–239, 249, 251–252, 274–282, 321
Ancillary data, 390

Antialiasing filter, 274
Anti-symmetric, 43, 121, 122
Application specific integrated circuits (ASIC), 4
Arithmetic and logic unit (ALU), 4, 166, 169–170, 224, 354
Arithmetic
   data type, 338
   precision, 7
   shift, 167, 193–194, 341
ASCII (text) format, 30, 47, 127
Assembly
   code optimization, 331, 357
   instruction, 339
   programming (code), 330–331, 349–357
Audio codec, 381
Autocorrelation function, 28

B

Background telemetry channel (BTC), 102–103, 190, 208, 256–257, 334, 365
Band-limited, 200
Bandpass filter, 39, 40, 48, 113, 118, 124, 126–127, 138, 147, 150, 200, 268, 387
Bandstop (band-reject) filter, 39, 48, 113, 114, 118, 124, 130, 137
Bandwidth, 6, 27, 71, 84, 87, 91, 96, 114, 275, 371
Bark, 384
Barrel shifter, 166, 167
Benchmark, 331, 334, 335, 338, 340, 342, 344–346, 351, 352, 358, 365, 370
Biased rounding, see rounding
Bilinear transform, 130
Binary
   file, 30
   point, see radix point

Embedded Signal Processing with the Micro Signal Architecture. By Woon-Seng Gan and Sen M. Kuo
Copyright © 2007 John Wiley & Sons, Inc.

Biquad, 66, 260–261, 345
Bit rate, 381, 382, 384, 386, 391–393
Blackfin
  EZ-KIT, 12, 52, 102, 152, 206, 211, 294–296, 312, 319, 331, 333–334, 338, 340, 342, 346, 351–356, 362, 365, 369–374, 391, 396, 399, 409, 412, 416, 422, 432
  LabVIEW Embedded Module, 18, 54, 105, 155, 211, 266, 321, 371, 400, 433
  Simulator, 8, 14, 50, 54, 98, 151, 189, 195, 202, 261, 317, 365
Blackman window, 84
Block
  FIR filter, 256
  floating-point, 235
  processing, 250, 252–254, 256, 258, 294–296, 344–346, 364
Boot load, 210
Branch, 179, 181
Breakpoint, 11, 192, 225, 226, 319, 332
Buses, 165–166, 183, 188, 288, 303, 361
Butterworth filter, 130, 132, 138

C
Cache
  hit, 305, 306, 309, 312
  line, 305–312
  memory, 2, 5, 164, 182, 185, 186, 303–317
  miss, 305, 306–307, 309–310, 311–312
Cacheability and protection look-aside buffer(CPLB), 313–315
Canonical form, 133, 135
Cascade form, 47, 65, 67, 133, 134, 260, 262, 264, 268, 368, 386, 398, 399
Causal, 42, 60, 66
C compiler, 189, 204, 231, 233, 330–333, 336–339, 346–347
CD player, 26, 382
Center frequency, 114, 124, 147
Channel bandwidth, 382
Channel equalization, 139, 148
Chebyshev filter, 132
Chirp signal, 138
Circular
  buffer, 174, 177, 178, 192, 339, 341–342
  pointer (index), 341–342

Coefficient quantization, 241–242, 249, 260
Coder/decoder (CODEC), 13, 52, 103, 188, 274–276, 281–282, 291–297, 369, 371
Color conversion, 407, 410
Comb filter, 71, 75, 398–399
Compiler optimization, 331–332
Complex-conjugate, 77, 78, 91, 133
Complex instruction set computer (CISC), 255
Contrast adjustment, 428
Convergence,
  factor, see step size
  speed, 140, 141
Code optimization, 330–332
Core clock (CCLK), 183, 255, 257–259, 293, 295, 303, 361–364
Critical band, 383
Crosstalk, 148
Cutoff frequency, 67, 113, 124, 132, 268, 366, 387
Cycle
  accurate, 8
  count, 12, 14, 54, 198, 208, 256–257, 295, 312, 334, 340, 345, 351, 353, 354, 357, 365, 370
  registers (CYCLES,CYCLES2), 208, 226, 255, 257
Cyclic redundancy checksum (CRC), 390

D
D/A converter (conversion), see digital-to-analog converter
Data
  address generator (DAG), 174, 177, 186, 283
  alignment, 336–338
  register (R0 to R7), 166
  type, 338, 340–341, 344
Debugger (debugging), 189–190, 195, 198, 209, 213
Decimation, filter, 366–369, 371–374
Decoding, 3, 382, 390–393
Delay, 146, 252, 254
  unit, 34, 41, 61
  (line) buffer, 41, 64, 264, 331, 334, 342, 344, 350, 355
Descriptor mode, 289–290

Index **481**

Digital
  filter, 31, 34, 36, 37, 54, 64, 75, 112, 115, 124, 129–130, 136, 139, 146, 203, 241
  full scale (dBFS), 276, 280
  frequency, 26, 28
  image, 406
  signal, 26–28, 33, 47, 60, 70, 76, 236, 238, 240, 275, 317, 382, 406
  signal processors, 4, 227, 256,
  signal processing (DSP), 1, 6, 25, 31, 33, 41, 52, 61, 76, 113, 133, 163, 166, 189, 225, 251, 255, 336, 343, 402
Digital-to-analog converter (DAC), 28, 52, 57, 207, 239, 249, 251–253, 274–282, 321
Direct form, 65–66, 118, 133–135, 137, 260, 387
Direct memory access (DMA), 164, 186, 187, 287–303, 304–305, 315–317, 338, 362–363, 364,
  access bus (DAB), 184, 188, 288, 303
  core bus (DCB), 184, 188, 288, 303
  external bus (DEB), 184, 188, 288, 303
Discrete
  cosine transform (DCT), 387, 401
  Fourier transform (DFT), 76, 78, 83, 387
Discrete-time
  Fourier transform (DTFT), 70
  Signal (sample), 27, 237
Double (ping pong) buffer, 252–254, 258, 293
DTMF (dual-tone multi-frequency)
  frequencies, 323
  receiver (or decoder), 322–323, 325–326
  generator (or encoder), 322–324, 326
Dual MAC, 173, 353, 354–357
Dynamic range, 219–221, 231, 234–236, 238–241, 245, 249

E
Edge frequency, 115, 117
Elliptic filter, 130, 132, 137, 260
Embedded
  system, 1, 7, 112, 164, 217, 255, 330–331, 358, 360
  signal processing (processor), 1, 6, 18, 57, 189, 199, 203, 217

Encoding, 382, 390, 392, 407
Energy consumption, see power consumption
Equiripple filter, 127, 206
Erasable programmable ROM (EPROM), 184
Error surface, 140
Excess mean-square error (MSE), 142, 149
Expectation
  operation, 140
  value, see mean
Expert linker, 195–196
Expressions window, 198
External
  access bus, 188–189
  bus interface unit (EBIU), 184, 188

F
Fast Fourier transform (FFT), 78, 238, 254, 264, 319, 343, 369, 383, 388, 391, 400–402, 413
Filter
  design and analysis tool (FDATool), 117, 126, 137, 201, 203, 241, 243, 249, 260–262
  design functions, 124, 131
  specifications, 113, 115
  visualization tool (FVTool), 72
Filterbank, 386, 390
Finite impulse response, see FIR filter
Finite wordlength (precision) effects, 67, 120, 133–134, 141, 414
FIR filter, 37, 62, 64, 120, 122, 200, 202, 205, 207, 211, 213, 226, 230, 241–242, 245, 249, 256–258, 260, 266, 331, 333–336, 337–339, 340–342, 344–346, 350–357, 364, 365–367, 373, 393, 417, 420
First-in first-out (FIFO), 283–284, 288, 297, 307
Fixed-point, 5, 7, 164, 166, 202–205, 217, 222, 229, 231, 235, 240–244, 249, 260, 261, 263
Flash, 3, 13, 183, 190, 209, 210
  programmer, 210
Floating-point, 5, 7, 201–203, 217, 231–234, 240–244, 255, 261, 265
Folding frequency, see Nyquist frequency
Fourier series (window) method, 124

482 Index

Fractional
  arithmetic, 339, 340,
  number (value), 5, 173, 204, 218–220, 223, 231, 319, 339, 340–341
Frame pointer (FP), 174
Frequency
  index, 76, 78, 83, 99, 402
  masking, 385
  resolution, 76, 78, 99, 255, 389
  response, 61, 70, 106, 113, 121, 206, 208, 241, 245–246, 267, 390, 418, 424
  voltage scaling, 258, 363–365
  warping, 130

G

Gamma correction, 438
General-purpose inputs/outputs (GPIO), 188
Gradient, 141–142
Grayscale, 407, 410, 413, 422, 428, 434, 437
Graphic equalizer, 200–201, 205–207, 211–216, 260–270
Graphical
  development environment, 18
  user interface, 18, 25, 72, 84, 117
Group delay, 72, 74, 120, 369
Guard bits, 224, 225

H

Hanning filter, 38, 47, 61, 201
Hardware loop, 352–357
Head-related
  transfer function, 393
  impulse response, 393–396
Hertz, 26
High-pass filter, 113, 121, 201, 421, 422
Hilbert transformer, 121
Histogram, 428
Huffman code, 390

I

I^2S, 275, 281–282
Ideal filter, 113
IIR Filters, 64, 69, 88, 92, 96, 129, 133, 241–243, 246–248, 260–269, 331, 345, 398–399

Image
  compression, 413–416
  enhancement, 422, 432
Impulse
  function, 37
  noise, 45, 46, 425
  response, 37, 62, 64, 120, 243, 245–246, 260, 264–265 268, 387, 393, 398–399, 400, 418, 420
Impulse-invariant method, 129
Infinite impulse response, see IIR filters
Inline, 332, 342–343
Input
  clock (CLKIN), 361–362
  output (I/O), 1, 5, 7, 13, 34, 37, 52, 62, 65, 68, 112, 121, 133, 164, 343, 359, 418
  vector, 140
Instruction
  fetch, 180, 182
  load/store, 354, 356
Integrated development of debugging environment (IDDE), 7, 9, 190, 195, 200, 314
Interpolation, filter, 366–369, 371–374
Interprocedural optimization, 332
Interrupt service routine (ISR), 180, 188, 256, 287, 294–295
Intersymbol interference, 148
Intrinsics, 330, 338, 339–341, 346
Inverse
  MDCT, 388, 391, 400–403
  discrete Fourier transform (IDFT), 77
  z-transform, 60, 63

J

Joint Test Action Group (JTAG), 20, 108
JPEG, 408, 411, 413, 416

K

Kaiser window, 86–87, 389

L

L1 memory, see cache memory
L2 memory, 182, 184, 288
Laplace transform, 129,
Laplacian filter, 422, 436
Latency, 251–252
Leakage factor, 142

Index **483**

Leaky LMS algorithm, 142, 146
Learning curve, 142, 149
Least-mean-square, see LMS algorithm
Least recently used (LRU), 307–309
Least significant bit (LSB), 218, 223, 227, 236–237, 240
Limit cycle oscillation, 247–248
Line-fill buffer, 186, 307, 311
Linear
   convolution, 41–43, 62, 121, 418
   phase filter, 72, 120, 121
   profiling, see Profiling
   time-invariant, 45, 63, 112
   system, 44, 135
Linearity (superposition), 44
Linker, 189, 195–196, 332
   description file, 189, 195
Little endian, 175, 197
LMS algorithm, 141, 143, 151, 154, 155
Loader file, 209–210
Logical shift, 193–194
Look-up table, 317–318, 370, 390
Loop
   bottom (LB0, LB1), 178, 181
   counter (LC0, LC1), 178, 181
   top (LT0, LT1), 178, 181
   optimization, 333, 336, 356
Low-pass filter, 88, 113, 115, 121, 201, 241, 268, 274, 366, 387, 419, 422, 428, 436

M

MAC unit, see multiply-accumulate
Magnitude
   distortion, 115
   (amplitude) response, 70–71, 89, 91, 96, 108, 113, 115, 124, 126, 130, 132, 137, 201, 213, 241, 243, 245, 260, 262, 387, 419, 421
   spectrum, 77–78, 84, 99–100, 103, 114, 238, 369
Mapping properties, 129
Marginally stable, 67
Maskee, 384
Masker, 384–385
Masking threshold, 383, 385
Mean, 28, 29, 237, 247
Mean-square error (MSE), 140, 142, 439
Median filter, 46, 59, 426, 428

Memory
   bank, 185, 310, 333, 337, 355
   buffer, 36, 252, 254
   map, 182, 195, 196, 197, 258, 355
   mapped I/O, 184
   mapped register (MMR), 188, 290, 314
   management unit (MMU), 164, 313
Micro signal architecture (MSA), 4, 5, 163, 164
Microcontrollers, 4, 163, 165
Million instructions per second (MIPS), 255–256, 331, 345, 392
Million multiply accumulate computation (MMAC), 255, 259, 358
Modified
   DCT (MDCT), 383, 386–389, 400–403
   Harvard architecture, 182
Most significant bit (MSB), 218–219, 230, 281, 284
Moving average filter, 26, 34, 45, 47, 49, 50, 52, 54, 63, 67, 70, 88, 418, 421, 426
Moving window, 37
MPEG, 382–383, 411, 413
MP3, 1, 381–383, 386, 390
MSE surface, see error surface
MSE time constant, 142
Multipliers, 4, 41, 34, 63, 13, 165–166, 170
Multiply-accumulate (MAC), 164, 171, 226–227, 230, 256, 331, 341, 342, 350, 356

N

Negative symmetry, see anti-symmetric
Noise, 26, 200, 213, 384, 393, 423, 425
Nonlinear filters, 45
Norm, 245–246
Normalized,
   (digital) frequency, 28, 81, 124, 246
   LMS algorithm, 141, 154
   step size, 141
Notch filter, 91, 94, 114
Nyquist frequency, 27, 99, 121, 124, 318–319

O

Offline processing, 250, 251
Ogg Vorbis, 3, 381–383, 386, 391
Operand, 170, 174, 229, 231

484 Index

Oscillatory behavior, 67, 243
Overflow, 169, 222–225, 236, 244–247, 339, 422, 434
Oversampling, 275

P
Parallel
  form, 135
  instruction, 172, 349, 353–354, 357,
  peripheral interface (PPI), 164, 188, 287–288
Parametric equalizer, 96
Partial-fraction expansion, 135
Passband, 96, 113, 115–116, 132, 201
  ripple, 116, 260, 268
Peak
  filter, 96
  signal-to-noise ratio (PSNR), 439
Performance
  (or cost) function, 140
  surface, 140
Peripheral access bus (PAB), 188
Phase,
  distortion, 72, 120
  lock loop (PLL), 361–365
  response (shift), 70, 243, 260
  spectrum, 78
Pipeline, 180, 181, 268,
Pipeline viewer, 190, 215
Pixel, 406
Polar form, 61
Pole, 66, 67, 91, 96, 132, 134–135, 261
Pole-zero cancellation, 67
Positive symmetric, see symmetric
Power, 30
  consumption, 5, 190, 231, 255, 258–259, 330, 358–360, 363
  spectrum density (PSD), 82, 425
  estimator, 111, 141
  management, 2, 164, 258, 360, 362
  mode, 361–363
Pragma, 333, 336–338
Probe wire, 20
Processing time, 67, 123, 252, 294–297, 331, 339, 391
Profiler (profiling), 190, 198–199, 209, 266, 295–296, 317, 331, 347
Profile-guided optimization (PGO), 190, 346–349

Program
  counter, 178–180, 193
  execution pipeline, see pipeline
  fetch pipeline, see pipeline
  sequencer, 178, 180
Pseudo code, 331, 350, 356
Psychoacoustics, 383
  model, 383, 386, 390
  masking, 383
Pulse code modulated (PCM), 3, 282
Pulse-width modulated (PWM), 322

Q
Quadrature mirror filterbank (QMF), 387
Quantization,
  effect, 67, 241
  errors (noise), 236, 237, 240–242, 247, 249, 260–261, 275, 407
  parameter, 201, 204, 242–243, 261
  step (interval, resolution), 236–237

R
Radix-2 FFT, 79
Radix point, 218–220, 232
Random-access memory (RAM), 2, 315
Random signals, 28, 423
Read-only memory (ROM), 2, 183–184
Realization of
  FIR filter, 41, 122
  IIR filter, 65, 67, 133, 135
Real-time
  application, 164
  clock (RTC), 164, 187, 359–360, 363
  constraint, 2, 250, 252, 253, 294, 304, 316, 332
  embedded system, 6, 250, 254, 330, 364
Receive clock (RCLK), 276, 282, 286
Reconstruction filter, 274
Rectangular window, 84
Recursive algorithm, 37, 63, 67, 140
Reduce-instruction-set-computer (RISC), 5, 164, 255
Region of convergence, 60
Register, 12, 165–178, 179–182, 188, 191–195, 198, 208, 223–230, 244, 257, 276–280, 282–286, 289–293, 296–303, 309–310, 312–315, 350–351, 354, 361–365

Release
  build, 209
  mode, 213
Repeat operation, 123, 177, 179, 226–227, 350
Resonator (peaking) filter, 114
Return
  from interrupt (RTI), 180
  from subroutine (RTS), 179
Reverberation, 398–400
RGB color, 407, 411
Rounding, 167, 170, 227–229, 236, 240, 244, 247–249
Run time library, 207, 264, 343–344, 346, 349, 357, 366

S
Sample-by-sample (or sample) processing, 250–252, 256, 294–295, 344–346
Sample rate conversion, 365, 369, 371
Sampling
  frequency (rate), 26, 78, 99, 124, 200, 255–256, 275, 365–369, 371–373, 382, 386
  period, 26, 61, 99, 122, 250–252, 255, 295
  theorem, 27, 406
Saturation
  arithmetic, 167, 339
  error, 249
  flag, 167
  mode, 222, 247, 249, 341
Sawtooth wave, 317–321
Scaling factor, 77, 84, 220–221, 238, 245–246, 416
Scratchpad memory, 182, 187, 314, 317
Serial
  peripheral interface (SPI), 164, 188, 281, 287, 288–289
  port (SPORT), 164, 188, 276, 281–289, 291–294
Set-associative, 306, 310
Settling time, 113
Short-time Fourier transform, 81
Sigma-delta ADC, see analog-to-digital converter
Signal
  buffer, 64, 121–123, 133, 249, 260
  generator, 317–322

Signal-to-quantization-noise ratio (SQNR), 238, 239, 245, 275, 382
Signal-to-noise ratio, 30, 36, 89, 382
Signal processing tool (SPTool), 31, 38, 82, 89, 94, 117, 127, 383
Sign bit, 193–194, 203–204, 218, 223, 225, 230, 232, 234, 235–236, 334
Sine window, 388
Sine wave, 26, 27, 106, 317, 318–321, 322–323, 401
Single instruction, multiple data (SIMD), 164
Sinusoidal
  signal, see sine wave
  steady-state response, 74
Smoothing filter, see low-pass filter
SNR, see signal-to-noise ratio
Sober filter, 421, 436
Software
  loop, 350–351
  pipelining, 356–357
Sound pressure level (SPL), 384
Spectral leakage, 84, 87
Speech, 1, 6, 26, 28, 48, 251, 366, 407
Square wave, 49, 83, 317–321, 322
s-plane, 129–130
Stability, 61, 120, 139, 140, 156, 241
  condition, 66, 67, 136
Stack, 174, 176, 182, 187, 338
Stack pointer (SP), 174, 176, 192
Stall condition, 181
Standard deviation, 29
Static RAM (SRAM), 182, 185, 186, 257, 303–305, 307, 310
Statistical profiling, see profiling
Status registers, 229, 363
Steady-state response, 74, 246
Steepest descent, 140
Step size, 140, 141, 148, 152
Stochastic gradient algorithm, see LMS algorithm
Stopband, 113–115
  ripple (or attenuation), 116, 201, 260, 268, 389
Successive approximation, see analog-to-digital converter
Sum of products, 12, 121
Superposition, 44
Symmetry, 43, 78, 121, 122

Synchronous dynamic RAM (SDRAM), 13, 184, 363, 409
System
   Clock (SCLK), 184, 187, 258, 284–287, 293, 361–364
   gain, 70, 75, 245

T

Tag field (address), 306, 307–309, 310–311
Tapped-delay line, see delay buffer
Temporal masking, 385
Threshold of hearing, 384, 385
Throughput, 167, 168, 173 181
Time delay, 72, 120
Time division multiplex (TDM), 275, 281–282
Timer, 164, 188
TMS320C55x, 391
Transmit clock (TCLK), 276, 283, 286
Transfer function, 62, 64, 66, 70, 91, 96, 112, 133, 136, 148, 398
Transfer rate, 7, 293
Transient response, 113
Transition band, 115, 132, 265
Transversal filter, see FIR filter
Truncation, 227–229, 236, 240, 244, 247–249
Twiddle factor, 77

U

Unbiased rounding, see rounding
Uniform distribution, 29, 432
Unit circle, 61, 70, 129, 136, 139, 242, 399
Unit-impulse function, 37
Universal asynchronous receiver transmitter (UART), 164, 188, 287–288
Univeral serial bus (USB), 14, 20, 108, 207, 209, 211, 409

V

Variance, 28, 29, 237, 247, 423, 424
Victim buffer, 186, 311–312
Video instructions, 174, 339
Virtual instruments (VI), 18, 213, 325–326, 373
VisualDSP++, 7, 51, 98, 102, 109, 152, 154, 190–191, 222, 224, 229, 258, 261, 264, 266, 268, 299, 317, 331–332, 335, 339, 341–342, 344, 347, 369, 392, 409, 412, 417, 422, 428, 432
   Kernel (VDK), 190
Voltage-controlled oscillator (VCO), 361
Voltage regulator, 259, 364
Vorbis window, 389

W

Watchdog timer, 167, 187
Weight vector, 140–142
Wiener filter, 424
Window design and analysis tool (WINTool), 84
Window function, 83, 343, 388, 390, 420
Wordlength, 169, 227, 236, 238–239, 241, 249
Worst-case execution time (WCET), 364
Write
   back, 311–313
   through, 311–313

Y

YCbCr color spaces, 411, 412
YIQ color spaces, 410

Z

Zero, 66, 67, 91, 96
Zero-mean, 29, 151
Zero-overhead loop, 178, 181, 349
Z-plane, 31, 70, 94, 129, 130
Z-transform, 60, 62, 64, 70, 76, 129